The E
Frequ
d
Vibration on People

Edited by Colin H. Hansen

ISBN 0 906522 45 5

5 Wates Way, Brentwood, Essex CM15 9TB, United Kingdom

Table of Contents

CHAPTER 3. PHYSIOLOGICAL EFFECTS OF LOW FREQUENCY NOISE

Introduction

The effect of low frequency noise and vibration on people is an important issue for communities around many industrial facilities and on transportation systems, and there are a number of research groups throughout the world who have been actively researching these effects. However, up to the present time, there has been no attempt to bring together all the various papers into a unifying book. It is hoped that the current book will partly fill that gap by including and commenting on all papers on the effects of low frequency noise and vibration on people, which have been published in the Journal of Low Frequency Noise Vibration and Active Control from the beginning of the year 2000 up to and including 2005.

This book is divided into two sections containing a total of five chapters, section 1 with three chapters for low frequency noise and section 2 with two chapters for low frequency vibration. Each chapter is concerned with one particular topic and where a paper is concerned with more than one topic, it is referred to in later relevant chapters but only reproduced in the chapter where it is first referenced. At the beginning of each chapter, each relevant paper is listed and the listing is followed immediately by a short summary. Following the paper listing, the original and sometimes revised papers are reproduced in full in chronological order. Chapter 1 is concerned with perception thresholds and acceptability levels of low frequency noise, chapter 2 is concerned with annoyance and sleep disturbance as a result of low frequency noise and chapter 3 is concerned with studies on the physiological effects of low frequency noise. The final two chapters are concerned with low frequency vibration, with chapter 4 being concerned with perception thresholds, comfort levels and annoyance levels and the final chapter being concerned with physiological effects. The contents of each chapter are discussed in more detail in the following paragraphs.

As well as physiological effects, sleep disturbance effects and annoyance, it is of considerable interest to know the threshold levels of human perception of low frequency noise and how threshold levels may differ from annoyance levels for various groups of people. Chapter 1 and the four papers included therein report on investigations of threshold level determination and levels of acceptability of low frequency noise. The first paper is concerned with acceptability levels, the second with threshold levels for normal test subjects, the third with the effect of masking noise on threshold levels and the fourth with threshold levels for people identified as low frequency noise sufferers. For this latter group it was found that threshold levels were very close to acceptability levels, which is in contrast to studies of normal adult populations.

By far the most effort in the study of low frequency noise (LFN) and vibration has been concentrated in the area of annoyance and sleep disturbance and chapter 2 contains sixteen papers devoted to the topic. The first paper in chapter 2 is concerned with a comparison of the levels of unpleasantness with levels of detectability for low frequency tonal noise, with the interesting result for some test subjects that the two levels were very close together. Paper 15 reports on a similar study using different frequency tones and found that only for low frequency noise sufferers (previous complainants) were the levels of acceptability close to the levels of detectability, even though the levels of detectability were similar for sufferers of LFN and non-sufferers. Paper 4 also reports on the different levels of annoyance between sufferers and non-sufferers for a number of typical low frequency sounds including traffic noise. The authors indicate that sufferers are always more easily annoyed than non-sufferers. This conclusion is also supported by the results reported in paper 11. The smaller annoyance thresholds for LFN sufferers may be because sufferers have already been sensitized to low frequency noise as pointed out in paper 5, which also comments on the correlation between low frequency vibration and infrasound, indicating the although infrasound cannot be heard, it can sometimes be sensed as a result of vibrations it may generate.

In chapter 2, papers 2, 9, 13 and 14 are concerned with the objective measurement of low frequency noise and how this correlates with annoyance. Paper 2 is concerned with evaluating the annoyance of various low frequency sounds in classrooms and the authors conclude that the dB(C) minus dB(A) method to assess annoyance may not be valid for low level noise. A later paper (paper 15) suggests a more complex formula, based on the dB(A) and dB(C) weightings for evaluating the annoyance of music noise with a heavy beat. Papers 9 and 13 compare the effectiveness of a number of objective methods for determining the annoyance of low frequency noise and vibration and paper 14 proposes a new method for assessing annoyance of LFN.

A number of papers in chapter 2 are concerned with finding out reasons why people complain about LFN and ways of resolving the complaints. It is not unusual for LFN sufferers to continue complaining even after the noise source has been removed and for others to complain about LFN when no disturbance can be detected with instrumentation. Thus a number of studies have been devoted to trying to understand what makes people complain (papers 6, 8, 12 and 16) about LFN, what are the most common sources of LFN causing complaint (paper 7) and how best to resolve such complaints (papers 3, 8 and 16). Finally, paper 10 is concerned with the effects of low frequency noise generated by blasting at distances from 30 m to 1 km from the blast site.

The seven papers in Chapter 3 are concerned with physiological effects of low frequency noise. The first three papers are concerned with the effects of LFN on vibration levels measured on the skin at various locations on the body. Paper 4 reports on oppressive feelings experienced by test subjects when closer than 50 m to a blast site; at larger distances the physiological effects were not experienced. Paper 5 is concerned with the effect of infrasound on blood pressure, heart rate and subjective feelings, while papers 6 and 7 are concerned with the effect of low frequency noise on spatial skills and brain function.

Chapter 4 is concerned with perception and comfort thresholds of low frequency vibration for articulated vehicle drivers (paper 1) and for laboratory test subjects experiencing a range of low frequency hand vibration environments (paper 2).

Chapter 5 is concerned with the physiological effects of low frequency vibration. The first paper reports on results of a study on vibration disease caused by excessive exposure to hand arm vibration and the second and third papers are concerned with the effect of low frequency vibration on the autonomic nervous system, heart rate, respiratory rate, salivation and subjective symptoms, with fairly inconclusive results.

Section 1.
Effect of Low Frequency noise on people

Chapter 1: Perception thresholds for low frequency noise

The hearing thresholds at which test subjects first perceive low frequency sound is discussed for a number of test environments in the following papers.

The unpleasantness of tonal sound at frequencies from 10 Hz to 500 Hz was evaluated using twenty seven female and twelve male test subjects and acceptable levels of sound for various living spaces was also evaluated by the same subjects. In some cases it was found that acceptable limits were close to very low levels of unpleasantness in some situations.

Perception thresholds of tonal noise at 10, 20 and 40 Hz and band limited white noise over the frequency ranges, 2 to 10Hz, 2 to 20 Hz and 2 to 40 Hz were investigated experimentally using ten male test subjects. It was found that the G-weighted sound level was lower than the perception thresholds defined in ISO 7196 for both tonal and band limited noise.

Hearing threshold levels for 20, 31.5, 40 and 50 Hz were measured for one female and four male test subjects in the presence of masking noise consisting of ambient noise and then 60 to 100 Hz band pass noise. The authors conclude that noise at frequencies above 50 Hz is effective in masking tonal noise at 50 Hz and below.

Twelve members of the Japanese noise-sufferers society were used to assess the sensory thresholds and subjectively maximum acceptable SPLs for a living room. It was found that the thresholds of the sufferers were not necessarily any lower than other adults, but the levels of acceptability for sufferers were very close to their threshold levels, which is not the same result for studies on other adult subjects.

Unpleasantness and acceptable limits of low frequency sound

Yukio Inukai, Norio Nakamura and Hideto Taya
National Institute of Bioscience and Human-Technology, AIST, Higashi 1-1, Tsukuba, Ibaraki 305-8566, Japan

ABSTRACT

Equal unpleasantness sound pressure levels of pure tones from 10 Hz to 500Hz were obtained by the method of adjustment in which subjects adjusted sound pressure levels equating to subjective degrees of unpleasantness on a 4-steps rating scale. In addition, the maximum acceptable sound pressure levels were measured at each frequency, assuming four types of situations, that is, a living room, a bedroom, an office and a factory. It was found that the acceptable limits were equivalent to very low levels of unpleasantness in some situations. The third order polynomial models of physical variables were well fitted to the subjective response data.

INTRODUCTION

The terms "loudness" and "annoyance" are often used to represent the overall subjective evaluations of low frequency noise. Several former studies (for example, Bryan[1], and Andresen and Møller[2]) reported some difference between loudness and annoyance judgments in the case of low frequency noise. On the other hand, a recent work (Kuwano, Fastl and Namba[3]) reported as follows. "It was found that the impressions of loudness, annoyance and unpleasantness of synthesized sounds are similar to each other and mainly determined by loudness level... On the other hand, in the case of recorded road traffic noise, there was some difference between loudness and annoyance judgments." They suggested that the difference might be resulted from aesthetic and/or cognitive aspects of the sound. In our previous studies, however, it was found that not only loudness, but also oppression and vibration feelings are dominant responses of even synthetic sound in the low frequency range noise. Unpleasantness was considered as one of the most representative responses of these negative feelings (Nakamura and Inukai[4]). The purpose of this study is to obtain equal unpleasantness contours and acceptable limits for different living situations, and to compare them with equal loudness contours that appeared in other researchers' previous work (Lydolf and Møller[5]).

METHOD

Stimuli

The stimuli presented were pure tones at 1/3 octave frequencies in the range from 20Hz to 250Hz and at 500Hz. Sixteen pure tones in total were used as stimulus sounds.

Subjects

Twenty-seven female and twelve male subjects aged between 19 and 62 participated in the experiment.

Apparatus

The experiment was carried out in a newly constructed 22.75 cubic metre pressure-field chamber (3.5 x 2.5 x 2.6 m). The chamber was constructed in concrete with sixteen 46cm diameter loudspeakers mounted in the wall and driven by sixteen 150W amplifiers. The background noise level of the new chamber was less than 16dB(A). Absorption rates were 0.6/125Hz and 0.95/500Hz. There was a small chair to sit on. The stimuli were generated with a sine/noise generator (B&K 104) and presented to subjects through the l6 loudspeakers in the chamber. The levels were controlled by the subjects by manipulating a remote main volume controller of the power amplifiers. The controller can set levels of stimuli in 1dB steps. Sound pressure levels were measured by the microphone in the chamber as shown in Figure 1 and calibrated at the position of the subject's head.

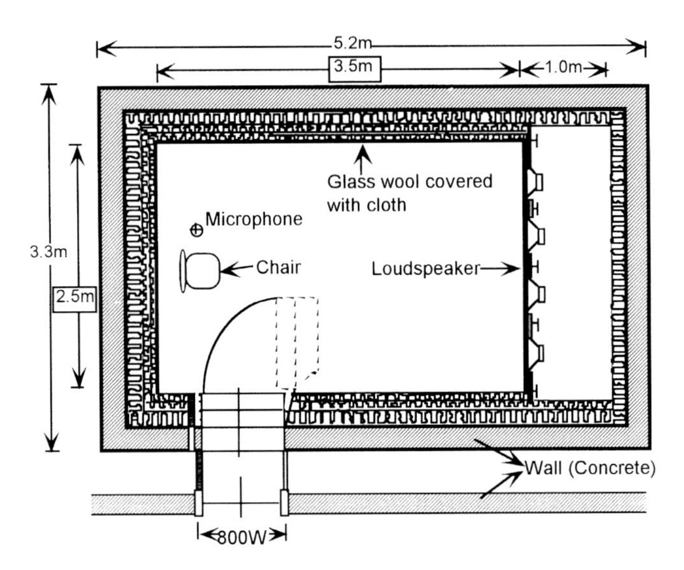

Figure 1. Plan view of the pressure-field chamber used for the experiment. The height in the chamber was 2.6m

Levels of unpleasantness

The 4 levels of unpleasantness were defined on a 5-point rating scale. The scale points were labelled as shown in Figure 2.

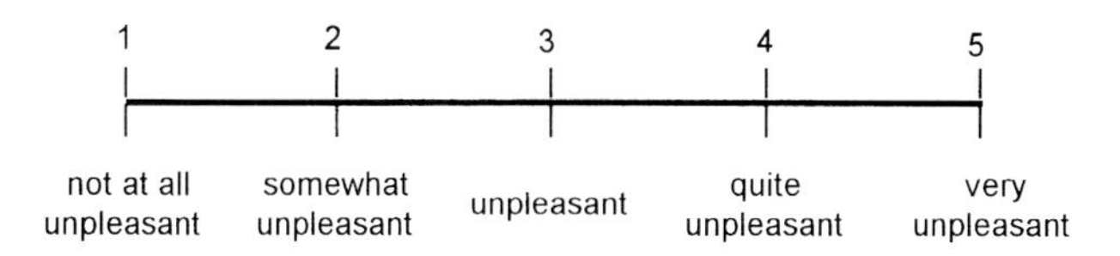

Figure 2. Unpleasantness rating scale

Assumed situations for acceptable limits of sound pressure levels

Four different situations were assumed for the measurements of acceptable limits of sound pressure levels. The instructions given to the subjects were to produce their acceptable limit levels of a given stimulus sound and to imagine the following situations.

a) A living room, reading a newspaper silently sitting on a sofa
b) A bedroom, lying down on the bed to get to sleep
c) An office, doing deskwork
d) A factory, doing physical labour

The test room better simulated an office or living room. Special settings, for instance a bed or drill, were not used for the above situations. So the bedroom and factory situations were extrapolations of the test room and were subjective in nature.

Procedure for the measurement of unpleasantness

Subjects were required to produce a level of a given stimulus, reflecting their subjective unpleasantness by manipulating the volume controller. The level produced was adjusted by the subject to represent a given category number on the rating scale in Figure 2. The category numbers used were 2 (= somewhat unpleasant), 3 (= unpleasant), 4 (= quite unpleasant) and 5 (= very unpleasant). One of the four category numbers from 2 to 5 was assigned by the experimenter to the subjects in each experiment.

Procedure for measurement of acceptable limits

Subjects were required to produce a level of a given sound stimulus, which reflected their acceptable limit of sound pressure level for the chosen situation by manipulating the volume controller. The sound pressure levels produced were measured by the computer controlled recording system. In each experiment, one of the above mentioned four situations was assigned and the instruction given by the experimenter to the subjects . Every situation was assigned once to all the subjects.

RESULTS AND DISCUSSION

Threshold, equal unpleasantness levels and acceptable maximum limits of SPL

Observed sound pressure levels were averaged over subjects for each rating grade and hearing threshold. Figure 3 shows the obtained mean values as symbol marks filled with black. Averaged sound pressure levels of acceptable maximum limits were also calculated over subjects for each living situation. Their values are shown in Figure 3, by empty marks. Their standard deviations are indicated as smaller symbols, at the bottom of the figure.

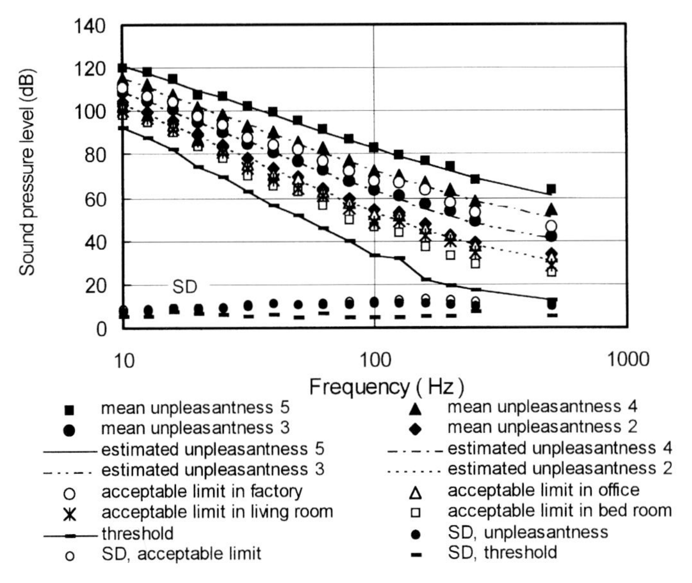

Figure 3. Equal unpleasantness contours and acceptable limits of sound pressure levels for assumed different living situations

Prediction of equal unpleasantness contours

Assuming third order polynomials of the physical variables, the parameters of the prediction formula of unpleasantness ratings, y, were obtained as follows.

$$y=(0.312-0.176x+0.0360x^2)L-43.5+37.4x-10.7x^2+0.988x^3, \quad (1)$$

Where L means sound pressure level in dB (L) at frequency h and x means the logarithm of frequency h. The goodness of fit was shown by adjusted $R^2=0.988$. Then, in order to obtain equal unpleasantness contours, equation (1) was transformed to equation (2). The estimate of sound pressure level (L) for a given unpleasantness rating grade (y) was calculated by equation (2).

$$L=(y+43.5-37.4x+10.7x^2-0.988x^3)\ /\ (0.312-0.176x+0.0360x^2), \quad (2)$$

The estimated equal unpleasantness contours were shown as four types of curves in Figure 3.

Relations between unpleasantness levels and acceptable maximum limits

The levels of unpleasantness equivalent to acceptable limits were estimated by substituting their sound pressure levels in the formula (1). The estimates of unpleasantness ratings obtained were shown in Figure 3. The figure showed that the unpleasantness ratings of acceptable limit levels were constant to a given situation over different frequencies. On the other hand, the unpleasantness ratings were different depending on situation. This result shows that unpleasantness can be used as a common scale for the evaluation of low frequency noise.

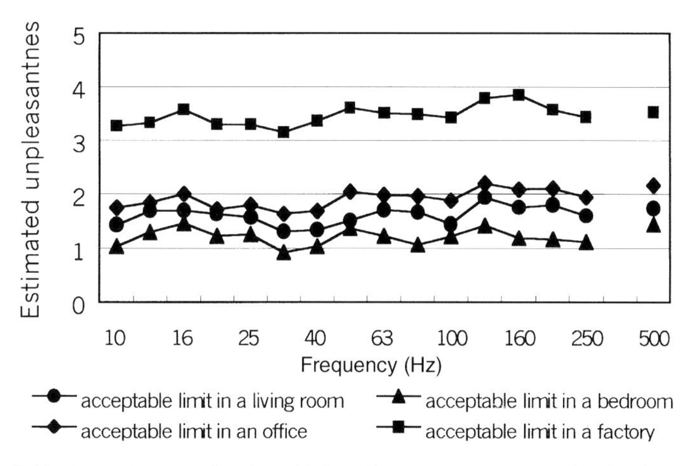

Figure 4. Unpleasantness estimates obtained from sound pressure levels of acceptable limits for different living situations

Relations between the results of this study and previously obtained results

To compare our estimated unpleasantness contours with loudness contours, our results were superimposed in Figure 5 on loudness contours cited from the paper of Lydolf and Møller[5]. The result shows that the gradients of the equal unpleasantness contours obtained were, in general, very similar to those of the equal loudness contours from 20 to 40 phon in the frequency range below 500 Hz. However, the gradients of equal unpleasantness contours were less steep or slightly gentler than the equal loudness contours for the lower frequencies below 100Hz in the pressure fields. This tendency was clearer in lower level unpleasantness and loudness contours. This means that loudness contours may not be enough to evaluate unpleasantness at low frequencies below 100 Hz. However, the difference may not necessarily result from the difference of loudness and unpleasantness, because the gradient of the threshold curve of our experiment was also milder than the results of others (Lydolf and Møller[5]; Yeowart and Evans[6]). One of the reasons might be the difference in subjects. Our subjects included many housewives who might be more sensitive to low frequency noise. For a more precise discussion, further experiments will be necessary concerning sex and age differences in sensitivity to low frequency noise.

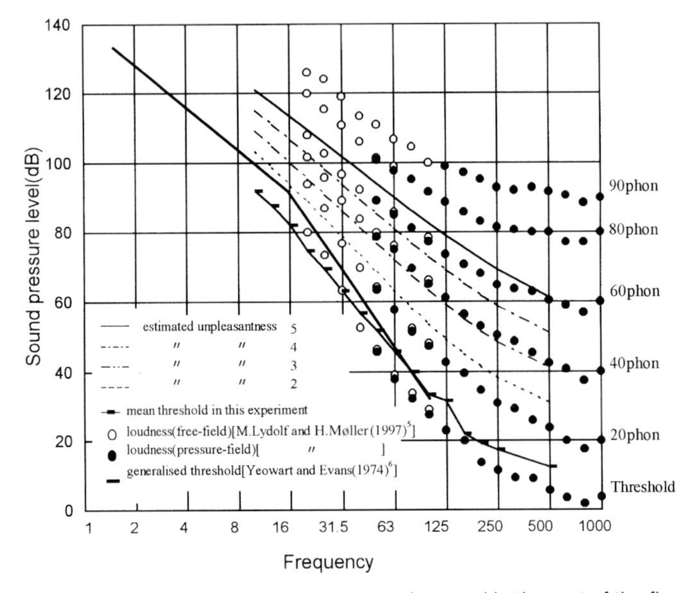

Figure 5. Estimated unpleasantness contours superimposed in the part of the figure cited from Lydolf and Møller's contours[5]

CONCLUSIONS

1) Equal unpleasantness contours of low frequency sounds were successfully estimated by 3rd order polynomials of the log frequencies of the sounds.
2) The gradients of the equal unpleasantness contours were in general very similar to those of the equal loudness contours from 20 to 40 phon in the frequency range below 500 Hz, but the former gradients were less steep than the latter's for the lower frequencies below 100Hz.
3) Acceptable limit levels were different depending on the assumed four living situations.
4) For a given living situation, acceptable limit levels coincided with a particular level of unpleasantness over different frequencies, and

the corresponding levels of unpleasantness were very low in some living situations.

REFERENCES

1. Bryan M. E. (1976). *"Low frequency noise annoyance", In Tempest, W. (Ed.): Infrasound and Low Frequency Vibration, Academic Press, London.*

2. Andresen, J. and Møller, H. (1984). *"Equal annoyance contours for Infrasonic frequencies.", J. Low Freq. Noise & Vibration, 3(3), 1–17.*

3. Kuwano, S., Fastl H and Namba, S. (1999). *"Loudness, annoyance and unpleasantness of amplitude modulated sound", Inter-Noise 99, 1195–1200.*

4. Nakamura, N. and Inukai, Y. (1998). *"Proposal of models which indicate unpleasantness of low frequency noise using exploratory factor analysis and structural covariance analysis", J. Low Freq. Noise, Vib. and Act. Cont. 17(3), 127–131.*

5. Lydolf M. and Møller, H (1997). N. *"New measurements of the threshold of hearing and equal-loudness contours at low frequencies", Proceedings of the 8th International Meeting on Low Frequency Noise & Vibration, Gothenburg, June, 76–84.*

6. Yeowart, N. S. and Evans, M. J. (1974). *"Threshold of audibility for very low-frequency pure tones", J. Acoust. Soc. Am., 55(4), 814–818.*

An investigation of the perception thresholds of band-limited low frequency noises: influence of bandwidth

Yasunao Matsumoto[1], Yukio Takahashi[2], Setsuo Maeda[2], Hiroki Yamaguchi[1], Kazuhiro Yamada[1] and Jishnu K Subedi[1]
[1]Department of Civil and Environmental Engineering, Saitama University, 255 Shimo-Ohkubo, Saitama, 338-8570, JAPAN
[2]Department of Human Engineering, National Institute of Industrial Health, 6-21-1 Nagao, Tama-ku, Kawasaki, 214-8585, JAPAN

ABSTRACT
Perception thresholds of complex low frequency noises have been investigated in a laboratory experiment. Sound pressure levels that were just perceptible by subjects were measured for three complex noises and three pure tones. The complex noises had a flat constant spectrum over the frequency range 2 to 10, 20, or 40 Hz and decreased at 15 dB per octave at higher frequencies. The frequencies of the pure tones used in this study were 10, 20 and 40 Hz. The perception thresholds were obtained using an all-pass filter, one-third octave band filters, and the G frequency weighting defined in ISO 7196. The G-weighted sound pressure levels obtained were compared with 100 dB which is described in ISO 7196 as the G-weighted level corresponding to the threshold of sounds in the frequency range 1 to 20 Hz. The perception thresholds of the pure tones measured in this study were comparable to the results available in various previous studies. The one-third octave sound pressure levels obtained for the thresholds of the complex noises appeared to be lower than the measured thresholds of the pure tones. The G-weighted sound pressure levels obtained for the thresholds of the complex noises appeared to be lower than 100 dB.

1. INTRODUCTION

For the assessment of infrasound and low frequency noise in the daily living environment, an understanding of the characteristics of the human response to those noises is required. It is accepted in some countries that the level of environmental infrasound and low frequency noise must be below the human perception thresholds determined in laboratory experiments so as to avoid the effect of those noises on people, including complaints and adverse effects on health, although this may be a subject of controversy.

There have been previous studies in which the perception thresholds of infrasound and low frequency noise were determined by using pure tones, such as Yeowart et al. [1], Yeowart and Evans [2], and Watanabe and Møller [3]. In the assessment of real-life noises, the characteristics of the perception thresholds of complex low frequency noises are relevant information as well as those of pure tones. Laboratory investigations of the perception thresholds of complex low frequency noises in which the environment and stimuli are controlled are useful so as to understand the characteristics of the thresholds of complex noises. However, there have been few studies of this kind [4, 5] and the characteristics of the thresholds of complex low frequency noises are not yet well understood.

The objective of the present study was to investigate the characteristics of the perception thresholds of complex noises having major frequency components in the infrasound range. In Japan, the Environment Agency (now the Ministry of Environment) has conducted a nation-wide survey to obtain information on environmental infrasound and low frequency noise since 2000 [6]. The survey includes the collection of the G-weighted sound pressure levels defined in ISO 7196 [7] in various environments. It may, therefore, be useful to discuss the relationship between the G-weighted sound pressure levels and the perception thresholds based on the results obtained in this study.

2. METHOD

An experiment involving human subjects was conducted with the infrasound system within the National Institute of Industrial Health, Kawasaki, Japan [8]. A schematic diagram of the plan of the experimental facilities is presented in Figure 1. The capacity of the test chamber was about 25 m^3. Twelve loud speakers, Pioneer TL-1801, having a diameter of 46 cm, were installed in a wall of the test chamber where subjects were exposed to low frequency noises during the experiment. The loud speakers were covered with jersey cloth so that the subject could not detect any movement of the speaker diaphragms visually.

The perception thresholds were measured for three pure tones and three complex noises. The frequencies of the pure tones used in the experiment were 10, 20 and 40 Hz. Source signals for the pure tones were generated by a function generator. A preliminary test was conducted so as to evaluate the quality of the pure tones produced in the experimental system: the sound pressure levels of the fundamental frequency component and its harmonics produced by the experimental system were compared with the perception thresholds of pure tones reported in the previous studies (e.g. [1–3]). It was unlikely that, when the sound pressure level at the fundamental frequency was at the perception threshold, those of the harmonics were also at, or exceeded, the perception thresholds at the corresponding frequencies. The complex noises generated by a computer had a nominally flat spectrum over the frequency range 2 to 10, 20, or 40 Hz and decreased at higher frequencies. The rate of the decrease in the spectra of the complex noises at high frequencies was determined to be 15 dB per octave. This was because, when this rate was greater than 15 dB/oct, audible noises at high frequencies where there were no significant frequency components in the source signals were produced within the sound generation system, which could interfere with the determination of thresholds. The source signals for the pure tones and complex noises were recorded on a DAT that was then used to feed input signals into the sound generation system shown in Figure 2. An audio mixer, Roland BOSS BX-60, was used to adjust the magnitude of stimulus by subjects while a graphic equalizer, SONY MU-E311, was used by the experimenter to adjust the stimulus magnitude. The details of other components of the sound generation system are available in Takahashi et al. [8]

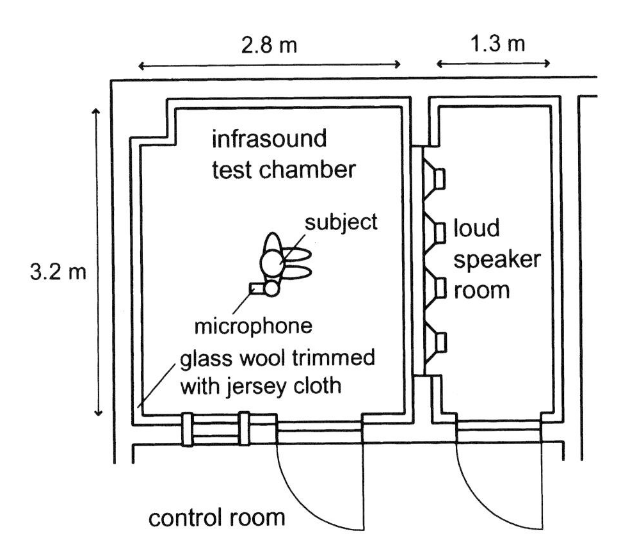

Figure 1. Schematic diagram of the plan of the experimental facilities used in this study

Ten male volunteers, aged from 21 to 24 yrs, took part in the experiment. All subjects appeared to have normal hearing in the hearing test at each octave centre frequency from 125 to 8000 Hz conducted prior to the experiment. The perception thresholds of the six input stimuli were determined by the method of adjustment. Subjects were asked to determine the magnitude of each stimulus by adjusting the audio mixer shown in Figure 2 so that they could just detect the presence of the stimulus. The measurement of perception threshold was repeated four times for each stimulus: the measurement that started with a sound pressure level at which subjects definitely detected the stimulus was repeated twice and the measurement that started with a sound pressure level at which they could not detect the stimulus was repeated twice. The order of the presentation of the six input stimuli were varied between subjects.

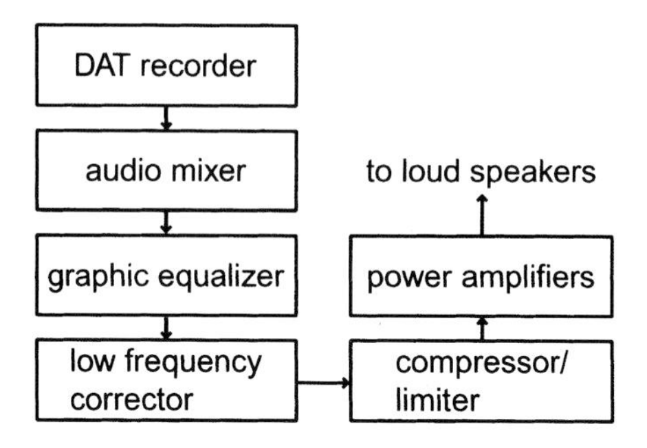

Figure 2. Sound generation system used in the experiment

The measurement of sound pressure was made near the centre of the test chamber at a height of 1.2 m by using a low frequency sound level meter, RION NA-17. Subjects were seated on a flat chair without a backrest located at the centre of the test chamber during the experiment (see Figure 1). The height of the chair was adjusted so that the height of the ears of each subject was 1.2 m from the floor. The position of the microphone was about 0.1 m from the right ear of the subject.

There were no significant frequency components at frequencies above 100 Hz compared to frequency components at lower frequencies in all stimuli used in the experiment. Signals from the low frequency sound level meter corresponding to time histories of the sound pressure were, therefore, acquired in a computer at 1000 samples per second. Overall sound pressure levels, one-third octave band sound pressure levels, and G-weighted sound pressure levels were then calculated for each record. One-third octave band filters used in the calculation were within the margins of error given in JIS C 1513 [9]. For the G frequency weighting, the parameters defined in ISO 7196 [7] were used to calculate weighted values.

3. RESULTS

The perception thresholds for each of the six stimuli were determined in four measurements as described in the preceding section. It was found that the differences in the values obtained in the four measurements were not statistically significant ($p>0.1$, Wilcoxon matched-pairs signed ranks test). The averages of the four measurements for each stimulus are, therefore, presented as the perception thresholds in the following parts of the paper.

Figure 3 shows the median and inter-quartile ranges of the perception thresholds determined for the three pure tones. The reference thresholds of hearing defined for frequencies above 20 Hz in ISO 389-7 [10] are also shown in Figure 3. The median threshold for the pure tone at 20 Hz measured in this experiment was almost equal to the value given in the standard. Although the measured threshold for the pure tone at 40 Hz was greater than the standard value by about 5 dB, it was comparable to the data reported in the previous studies [1–3, 11]. The measured threshold for the pure tone at 10 Hz was also comparable to the previous results. It can be, therefore, concluded that the measurement method used in this study was reasonable to determine the perception thresholds.

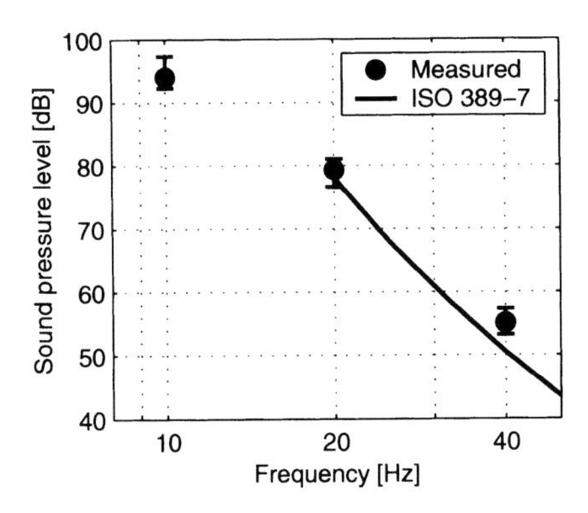

Figure 3. The median and inter-quartile ranges of the perception thresholds determined for pure tones in this experiment. The reference threshold of hearing defined in ISO 389-7 is also presented

The median and inter-quartile ranges of the one-third octave band sound pressure levels obtained for the three complex noises at the perception thresholds are presented in the frequency range between 4 and 200 Hz in Figure 4. The measured thresholds are compared with the reference thresholds of hearing defined in ISO 389-7 [10]. In the course of the experiment, the background noise in the test chamber was measured several times. For the complex noise with a cut-off frequency of 40 Hz, the one-third octave band sound pressure levels at the threshold are shown at frequencies above 8 Hz because the one-third octave band sound pressure levels at the thresholds were almost the same as the background noise levels at lower frequencies. Other characteristics of the background noise measured are discussed in a later section.

The inter-quartile ranges of the thresholds of the complex noises were about 5 dB which were similar to the inter-subject variability observed in the perception thresholds for the pure tones. For all the complex noises used in this experiment, the one-third octave band sound pressure levels at the perception thresholds tended to be greater than the values in ISO 3897 [10] at frequencies above 80 Hz.

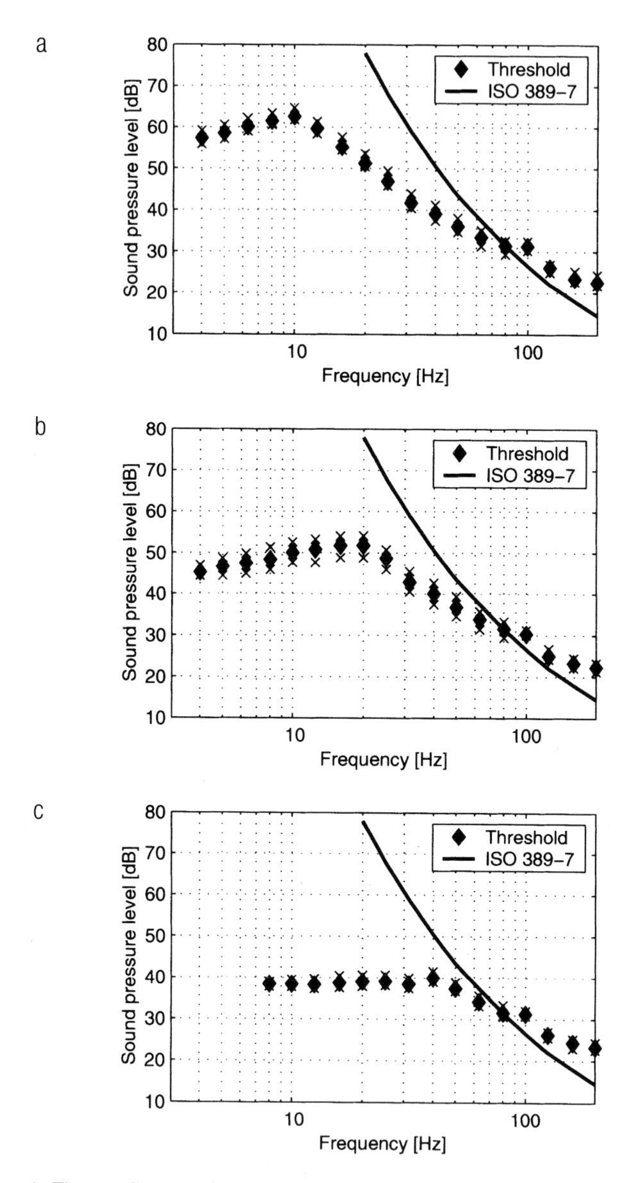

Figure 4. The medians and inter-quartile ranges of the one-third octave band sound pressure levels at the perception threshold determined for the complex noises with cut-off frequencies of (a) 10 Hz, (b) 20 Hz and (c) 40 Hz. The reference threshold of hearing defined in ISO 389-7 is also presented

The median one-third octave band sound pressure levels at the perception thresholds for the complex noises are compared with the median perception thresholds for the pure tones in Figure 5. The reference threshold of hearing defined in ISO 389-7 [10] is also presented in Figure 5. It was found that, for the complex noises, the one-third octave sound pressure levels in the frequency range where the sound pressure level decreases with increasing the frequency were almost the same for the three complex noises: above 20 Hz, the sound pressure levels for the complex noise with a cut-off frequency of 10 Hz were almost the same as those for the complex noise with a cut-off frequency of 20 Hz, and above 40 Hz, the sound pressure levels were almost the same for the three complex noises. At 10, 20 and 40 Hz, the one-third octave band sound pressure levels for the perception thresholds of the complex noises were lower than the perception thresholds for the pure tones.

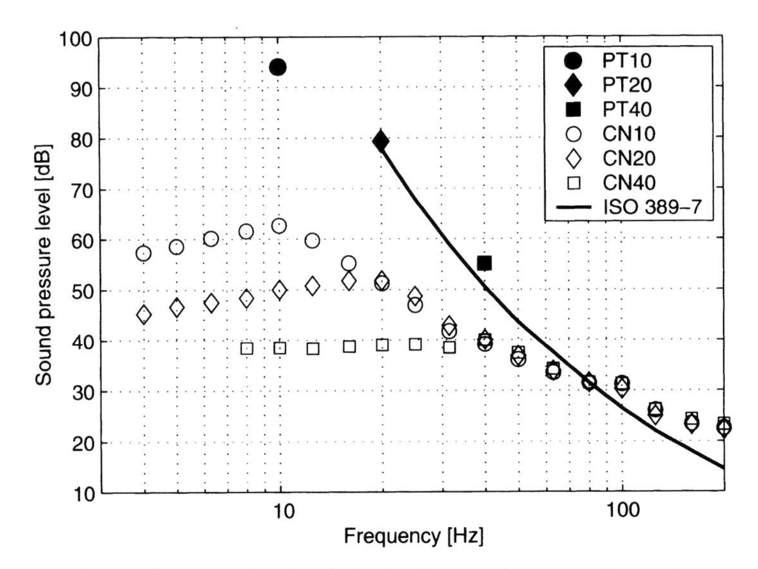

Figure 5. The median perception thresholds for the pure tones and the median one-third octave band sound pressure levels at the perception thresholds for the complex noises. The reference threshold of hearing in ISO 389-7 is also presented. Legend: PT10, pure tone at 10 Hz; PT20, pure tone at 20 Hz; PT40, pure tone at 40 Hz, CN10, complex noise with 10 Hz cut-off; CN20, complex noise with 20 Hz cut-off; CN40, complex noise with 40 Hz cut-off

Table I shows the median and inter-quartile ranges of the G-weighted sound pressure levels calculated for the perception thresholds for all six stimuli. It is stated in ISO 7196 [7] that "in the frequency range 1 Hz to 20 Hz, sounds that are just perceptible to an average listener will yield weighted sound pressure levels close to 100 dB". The median G-weighted sound pressure level of the perception threshold measured with the pure tone at 10 Hz was 94.0 dB, lower than the value given in the standard by 6 dB, and the median G-weighted level for the threshold for the pure tone at 20 Hz was 88.2 dB, lower than the standard value by about 12 dB. The threshold of the pure tone at 40 Hz was out of scope of the standard, so that the measured G-weighted level of the perception threshold was much lower than 100 dB. The G-weighted level of the perception thresholds for the complex noises were lower than 100 dB by more than 30 dB.

Table I. The median and inter-quartile ranges of the G-weighted sound pressure levels at the perception thresholds measured in this study. PT10, pure tone at 10 Hz; PT20, pure tone at 20 Hz; PT40, pure tone at 40 Hz, CN10, complex noise with 10 Hz cut-off; CN20, complex noise with 20 Hz cut-off; CN40, complex noise with 40 Hz cut-off

G-weighted sound pressure level [dB]	Pure tone			Complex noise		
	PT10	**PT20**	**PT40**	**CN10**	**CN20**	**CN40**
25th percentile	92.4	85.4	46.0	67.7	61.3	50.9
Median	94.0	88.2	48.1	68.7	64.3	52.2
75th percentile	97.3	89.9	49.4	70.8	66.6	53.6

4. DISCUSSION

It was found that, for the three complex noises used in this study, the median one-third octave band sound pressure levels at the perception threshold tended to be greater than the reference thresholds of hearing defined in ISO 389-7 [10] at frequencies above 80 Hz. Although the perception threshold of the pure tone at frequencies higher than 80 Hz were not measured with the subjects used in this study, it could be assumed that the thresholds of the subjects might have been similar to those given in ISO 389-7 [10]. The one-third octave band sound pressure levels for the thresholds were similar at frequencies above 40 Hz for all the complex noises and at frequencies above 20 Hz for the noises with a cut-off frequency at 10 Hz and with a cut-off frequency at 20 Hz.

It might be possible to hypothesise that, in the perception of complex noises, the noise component in the frequency range where the one-third octave band sound pressure level is greater than the threshold level obtained with pure tone contributes to the response of the subject. If this hypothesis is valid, the perception of all the complex noises used in this study might be interpreted as the contribution from the frequency components at higher frequencies above about 80 Hz, although there are more significant sound energy inputs at lower frequencies.

The A frequency weighting was applied to the recorded data for the perception thresholds of the complex noises so as to investigate the effect of the frequency components in the audible range to the threshold, although the highest one-third octave band centre frequency used in the calculation was 400 Hz due to the limit of the equipment used in the measurement. The calculated values should not be used in discussion about the absolute A-weighted value but could be used to compare the thresholds of the three complex noises measured because it could be expected that there was no significant difference in the sound pressure level between the three complex noises at higher frequencies where no data were available (see Figure 5). The nominal A-weighted sound pressure levels calculated for the median thresholds were between 23.0 dB and 24.0 dB for the three complex noises: this might support the interpretation of the measured data that the frequency components above 80 Hz may contribute to the perception thresholds of the complex noises used in this study.

Although the data described above supported the hypothesis that the noise components at frequencies above about 80 Hz contributed to the perception of the complex noises, the effect of the noise components at lower frequencies on the perception may not be ignored. It was reported by the subjects and recognised by the experimenters that the three complex noises used in this study was perceived as different noises when the noises were just noticeable and clearly heard: when the cut-off frequency of the complex noise was lower, the noise was perceived as a "lower sound". There were differences in the one-third octave band sound pressure level between the noise with a cut-off frequency of 10 Hz and that with a cutoff frequency of 20 Hz mainly at frequencies below 20 Hz. If these differences were major factors that had an effect on how the subject perceived the noises, the frequency components below 80 Hz, and even below 20 Hz, might be perceived by the subjects, although the one-third octave band sound pressure levels at these frequencies were lower than the thresholds measured with pure tones. It was reported that at the perception thresholds of complex noises consisting of a few tonal components the sound pressure levels of those noises at the frequencies of each tonal component could be lower than the thresholds obtained with pure tones at the corresponding frequency [4, 5]. It might not be, therefore, reasonable to ignore the effect of low frequency components on the perception of the complex noises used in this study. However, it was not possible from the available experimental proofs to understand the mechanism of the influence of low frequency components on the perception.

The measurement of the background noise in the test chamber was made several times during the period of experiment, as mentioned in the preceding section, some variation of the sound pressure levels of the background noise was found. In all measurements made, the one-third octave band sound pressure levels of the background noise were greater than the reference threshold of hearing in ISO 389-7 [10] at higher frequencies. This was observed at frequencies above 100 Hz in some measurements. It is difficult to compare the measured thresholds with those of background noise levels directly because the background noise level varied and was not measured in each experimental session. However, it was found that, in some background noise records, the one-third octave band sound pressure levels of background noise at frequencies above 100 Hz were comparable or greater than the one-third octave band sound pressure levels obtained for the perception thresholds.

The pure tones at 10 and 20 Hz and the complex noises used in this study, particularly those with a cut-off frequency of 10 and 20 Hz, had major frequency components in the frequency range below 20 Hz, so that it was reasonable to assess those stimuli by using the G frequency weighting. For the pure tones, the median G-weighted sound pressure level for the threshold measured at 20 Hz was lower than that measured at 10 Hz by about 6 dB (see Table I, although these were expected to be at almost the same level. The median G-weighted value of 88.2 dB for the threshold at 20 Hz may be inconsistent with the statement in ISO 7196 [7] that "weighted sound pressure levels which fall below about 90 dB will not normally be significant for human perception". The median weighted value obtained for 20 Hz was close to, for example, the "recommended limit for environmental infrasound" in Denmark, 85 dB, which was determined based on "the average hearing threshold for infrasound" of about 96 dB in association with an inter-subject variability of about 10 dB [12]. For the complex noises, the G-weighted sound pressure levels shown in Table I were lower than the just perceptible level, 100 dB, defined in ISO 7196 [7] and even lower than Danish recommended limit, 85 dB. This may imply that there was no contribution from the noise components in the infrasound frequency range to the perception thresholds of the complex noises used in this study. This may be consistent with the hypothesis that the subjects perceived the noise components at frequencies above about 80 Hz only in this experiment, as described above, but inconsistent with the fact that the subjects reported that the three complex noises were perceived differently.

5. CONCLUSIONS

The perception thresholds of pure tones measured at frequencies of 10, 20 and 40 Hz were comparable to those obtained in the previous studies. The G-weighted sound pressure levels obtained for the thresholds of the pure tone at 10 Hz for individuals were lower than the weighted value for the threshold stated in ISO 7196 by about 6 dB, and those at 20 Hz tended to be lower than the standard value by about 12 dB.

At the perception thresholds of the complex noises including significant frequency components in the infrasound range, the one-third octave band sound pressure levels at frequencies above 40 Hz were similar for all the noises used in this study. The one-third octave band sound pressure levels at frequencies above 80 Hz at the perception thresholds tended to be greater than the reference thresholds of hearing defined in ISO 389-7. These data may imply that the perception thresholds of those complex noises were determined by the noise components at those higher frequencies. However, the subjects recognised the differences between the complex noises used in this study by means of their perception, which might suggest that there were some contribution from the low frequency components. Further investigation is required for the interpretation of the results obtained for the complex noises.

REFERENCES

1. Yeowart, N. S., Bryan, M. E., Tempest, W., The monaural M.A.P. threshold of hearing at frequencies from 1.5 to 100 c/s, *Journal of Sound and Vibration*, 1967, 6(3), 335–342.

2. Yeowart, N. S. and Evans, M. J., Thresholds of audibility for very low-frequency pure tones, *Journal of Acoustical Society of America*, 1974, 55(4), 814–818.

3. Watanabe, T. and Møller, H., Low frequency thresholds in pressure field and in free field, *Journal of Low Frequency Noise and Vibration*, 1991, 9(3), 106–115.

4. Watanabe, T. and Yamada, S., Study on perception of complex low frequency sounds, *Proceedings of the 9th International meeting on low frequency noise and vibration*, Aalborg, Denmark, 2000, 199–202.

5. Mirowska, M., Evaluation of low-frequency noise in dwellings. New Polish recommendations, *Journal of Low Frequency Noise. Vibration and Active Control*, 2001, 20(2), 67–74.

6. Ochiai, H., The state of the art of the infra and low frequency noise problem in Japan, *Inter-Noise 2001*, The Hague, The Netherlands, 2001.

7. International Organization for Standardization, Acoustics – frequency-weighting characteristic for infrasound measurements, *ISO 7196*, 1995.

8. Takahashi, Y., Yonekawa, Y., Kanada, K. and Maeda, S., An infrasound experiment system for industrial Hygiene, *Industrial Health*, 1997, 35, 480–488.

9. Japanese Industrial Standards Committee, Octave and third-octave band analyzers for sounds and vibrations, *JIS C 1513*, 1983.

10. International Organization for Standardization, Acoustics – reference zero for the calibration of audiometric equipment – part 7: reference threshold of hearing under freefield and diffuse-field listening conditions, *ISO 389-–7*, 1996.

11. Tokita, Y., On the evaluation of infra and low frequency sound, *Journal of Acoustical Society of Japan*, 1985, 41 (11), 806–812. (in Japanese)

12. Jakobsen, J., Danish guidelines on environmental low frequency noise, infrasound and vibration, *Journal of Low Frequency Noise Vibration and Active Control*, 2001, 20(3), 141 –148.

Masked perception thresholds of low frequency tones under background noises and their estimation by loudness model

Jishnu K. Subedi*, Hiroki Yamaguchi, Yasunao Matsumoto
Department of Civil and Environmental Engineering, Saitama University, 255 Shimo-ohkubo, Saitama, 338-8570, JAPAN
**E-mail: s01d2053@post.saitama-u.ac.jp*

ABSTRACT
This paper presents experimental measurements of masked thresholds of low frequency tones under background noises and application of loudness model to estimate the thresholds. The measurements of thresholds for tones at frequencies 20, 31.5, 40 and 50 Hz were conducted under three background noise conditions: one ambient noise and 60-100 Hz band-pass noises at two levels. The measurements were carried out in an uncontrolled environment in relatively quiet times. The perception thresholds of the same subjects were also measured for frequencies 31.5, 40 and 50 Hz inside a cabin with ambient noise levels well below the average hearing thresholds specified in ISO 389-7. Moore's loudness model has been used to estimate the masked thresholds. The estimated thresholds from the loudness model have been compared with the results obtained in the experiment. The results indicate that the noise above 50 Hz is effective in masking the low frequency tones at 50 Hz and below, and that Moore's loudness model can predict reasonably the average of the measured masked thresholds.

1. INTRODUCTION

Sources of low frequency noise in the environment are growing. Many house appliances, such as ventilation systems and refrigerators, and some civil engineering structures, such as viaducts and railway tunnels, are some of the common sources of low frequency noise. Recent field investigations [1-3] have indicated that increasing numbers of people are complaining about problems arising from low frequency noise. The low frequency noise occurs, normally, as a part of a complex sound containing energies over wide frequency range. As the response of the auditory system to sounds of different frequencies differs, the quantification of the total response from these complex sounds is complicated. It is now understood that the levels obtained from commonly used frequency weighting networks, such as the A-weighting, do not correlate well with the response to complex sounds with audible low frequency components [4-6]. The reason for this is the fact that the rapid change in the auditory sensations, such as loudness and annoyance, with respect to the change in the sound pressure levels at low frequencies is not taken into account in the weighting networks.

In order to calculate the response to complex sounds more accurately, Zwicker and co-workers developed loudness models [7]. Although the model is widely used for practical purposes, low frequency sounds below 50 Hz are not included in it. Moore et al. [8,9] developed revised loudness models based upon the original work by Zwicker, which could be used for frequencies down to 20 Hz.

These loudness models were developed basically from experiments on masking effects, which can be measured quantitatively by measuring the masked threshold of a test sound in the presence of masker sounds. The measurements of masked thresholds show that low frequency sounds below 50 Hz can also produce masking effects. Finck [10] used 100, 115 and 130 dB sound pressure levels of 10, 15, 25 and 50 Hz tones as masker sounds and measured the masked thresholds of test sounds in the frequency range of 50 to 4800 Hz. His results showed that the maskers could produce constant masking effects up to 500 Hz. Watanabe et al. [11,12] used pure tones and complex tones at frequencies of 10 and 20 Hz as masker sounds to measure the masked thresholds of tones at frequencies from 4 Hz to 50 Hz. They also used band-pass noises with different widths centered on 20 Hz as masker sounds to measure the masked thresholds of pure tone at 20 Hz. Their results varied greatly among the subjects, and in some cases the masked threshold appeared lower than the threshold in quiet. In a similar study, Fidell et al. [13] showed that sound at 40 Hz is masked by a masker with band limits of 11-400 Hz.

Although these results indicate that masking effects are present in the low frequency regions below 50 Hz, they cannot be used directly to construct a loudness model because of the limited available data and the large variations among the data. Therefore, Moore's loudness model below 100 Hz is based on the extrapolation of data above 100 Hz [9]. The application of the model to estimate the threshold of complex sounds [9] and the loudness of complex sounds [14] for high frequencies showed that the results are accurate enough within subjective variability. However, the applicability of the model for low frequencies has yet to be verified.

Furthermore, direct application of the model based on the auditory mechanism is questionable in the low frequency region below 50 Hz, as there are reports suggesting a presence of other mechanism of the perception besides auditory at these frequencies. From the experiments with components of sound below 50 Hz and noise above 50 Hz, Inukai et al. [15, 16] indicated that other factors of perception such as vibration and feelings of pressure are also associated with sounds below 50 Hz. However, results from a survey of complaints about infrasound and low frequency noise showed that 93% of the complainants perceive the sound through the ears [17].

In order to understand the mechanism of perception of low frequency sounds more accurately, the measurement of masked thresholds of low frequency sounds masked by high frequency sounds is useful. The present study, therefore, has been carried out to measure masked thresholds of low frequency tones under different levels of background noise and to investigate the applicability of Moore's loudness model to estimate the thresholds. The perception thresholds of low frequency tones at 50 Hz and below are measured under different background noise conditions in controlled and uncontrolled environments and the results are compared with the estimated results from the loudness model.

2. EXPERIMENTAL METHOD

2.1 Threshold measurement under background noises

In order to investigate the masked thresholds of human subjects for low frequency tones under masker sounds, a room environment with ambient sound was selected. A room (6.5 × 3.75 × 5.3 m), as shown in Figure 1(a), was used for the measurement, and test sounds were produced from a low frequency speaker (YAMAHA, YST 800) placed in the middle of the room at a height of 1.0 m above the ground. An infrasound microphone (RION NA-18) was used to measure the sound. In order to keep the ambient sound at constant levels, the experiments were conducted during night hours at a relatively quiet time.

A function generator (NF ELECTRONIC INSTRUMENTS, E-1011A) was used as the source for the pure tone test sounds, and the frequencies of the test sounds used in the experiment were 20, 31.5, 40 and 50 Hz. Masked thresholds of the test sounds were measured under three different masker sounds: one ambient noise (Ambient Noise) and two band-pass noises from frequency 60 to 100 Hz at different levels (Noise 1 and Noise 2). The band-pass noises were generated from a PC operated by the experimenter. In later discussion, the test sounds are referred as "signals" and the masker sounds in the background are referred as "background noises".

Four male and one female subjects, aged between 26 and 29 yrs, participated in the experiment. The subjects were placed in front of the speaker at a distance of one meter for all cases except for signals at 20 Hz, where the subjects were placed at 30 cm from the speaker. The change in the position was necessary to achieve sufficient level of 20 Hz signal without any significant higher harmonics. During the measurements, the subjects were seated in an upright position with the height of their ear adjusted at 1.2 m as shown in Figure 1(b). The microphone was placed at 0.2 m from center of the subjects' head. The noises were produced from another speaker of same type. The speaker was placed below the speaker for the signal, as shown in Figure 1b.

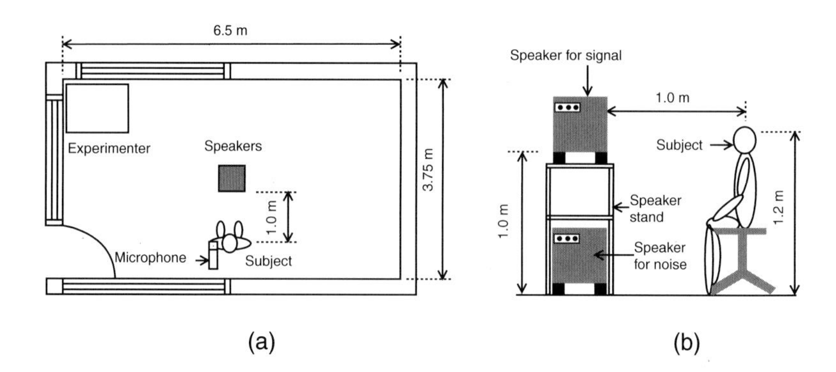

Figure 1. Schematic diagram of the experimental setup for the uncontrolled environment. (a) Plan view of the experimental setup; (b) Elevation showing placement of speakers and subject.

The subjects could adjust the level of the signals from the function generator, and the thresholds were measured by the method of adjustment with four repetitions - two starting below audibility and two starting above. Before starting the measurements, the subjects were given a sufficient time for practice so that they could distinguish between the signals and the background noises.

The 1/3 octave band sound pressure levels of the three background noises are shown in Figures 2(a) and 2(b) by continuous lines. As seen in the figures, the background noises exceed the ISO hearing threshold above 50 Hz. As the ambient noise was not controlled during the experiments, the reproducibility of these noise conditions during the measurement of the perception thresholds was investigated.

The average 1/3 octave band sound pressure levels in the frequency range 50-200 Hz, which were measured for signals of frequency 31.5 Hz at its masked threshold under Ambient Noise, are shown in Fig. 2(a) by filled circles. Their comparison with the 1/3 octave band sound pressure levels of Ambient Noise only showed that the difference is about 1 dB at 63 Hz, 2 dB at 100 Hz and no difference at other frequencies. Similar comparison for measurements under Noise 1 (shown by crosses in the figure) with Noise 1 only and Noise 2 (shown by filled triangles in the figure) with Noise 2 only showed that the differences are within a similar range. The results were similar for measurements of perception threshold at frequencies 40 and 50 Hz. Therefore, the background noises were considered reproducible for the measurement of perception thresholds at frequencies 31.5, 40 and 50 Hz and they are represented by the measured 1/3 octave band sound pressure levels of the noise only conditions in further discussion.

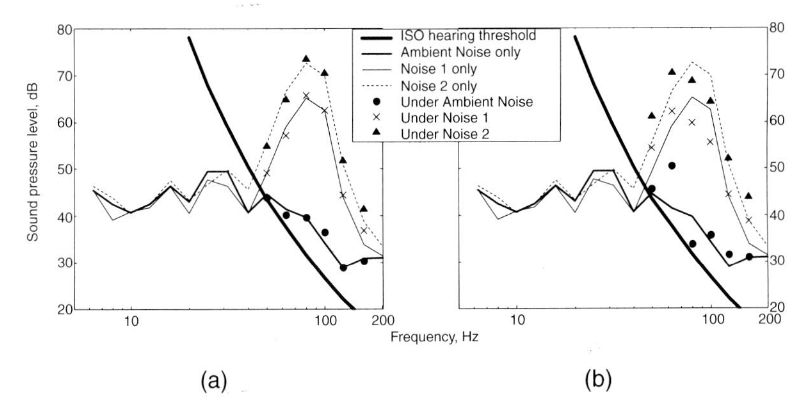

Figure 2. Three levels of background noises used in the experiment for the signals at frequencies (a) 31.5 Hz; and (b) 20 Hz. Measurement of noises only conditions (continuous lines) and the average of measured 1/3 octave band sound pressure levels of five subjects measured at frequencies from 50 to 200 Hz for both the noise and the signals (symbols) are shown. The reference hearing threshold curve specified in ISO 389-7 is also shown.

However, as seen in Fig. 2(b), the average 1/3 octave band sound pressure levels of signal at 20 Hz under background noises, shown by symbols in the figure, differ significantly from the measurements for noise only cases. The measurements showed that higher harmonics at 60 Hz exceeded the ISO threshold by about 5 dB when the sound pressure level of 20 Hz signal was at 80 dB. Although the levels of background noises were at sufficiently high to mask these 60 Hz harmonics, there was an effect of varying the sound pressure levels of the background noises in some sets of measurements. Therefore, the three background noise conditions (Ambient Noise, Noise 1 and Noise 2) for the signal at 20 Hz are represented by the average 1/3 octave band sound pressure levels measured for all the subjects.

2.2 Threshold measurement in quiet

The measurements of the thresholds in quiet for the same subjects were conducted in a cabin of size 1.8 × 1.2 × 2.3 m (Fig. 3) designed for experiments on low frequency noise. Four speakers (YST 800) placed in two horizontal lines were used, and the subjects were placed in front of the speakers at a distance of 1.0 m. The microphone (RION NA-18) was placed 0.2 m from center of the subjects' head position near their right ear. The measured background noise in the cabin at the location of the microphone is shown in Fig. 4 along with the average hearing threshold level specified in ISO 389:7 [18]. The sound pressure level of the background noise crosses the ISO hearing threshold curve above 160 Hz. The noise conditions above 100 Hz in the cabin were similar to another experiment room for low frequency noise [19].

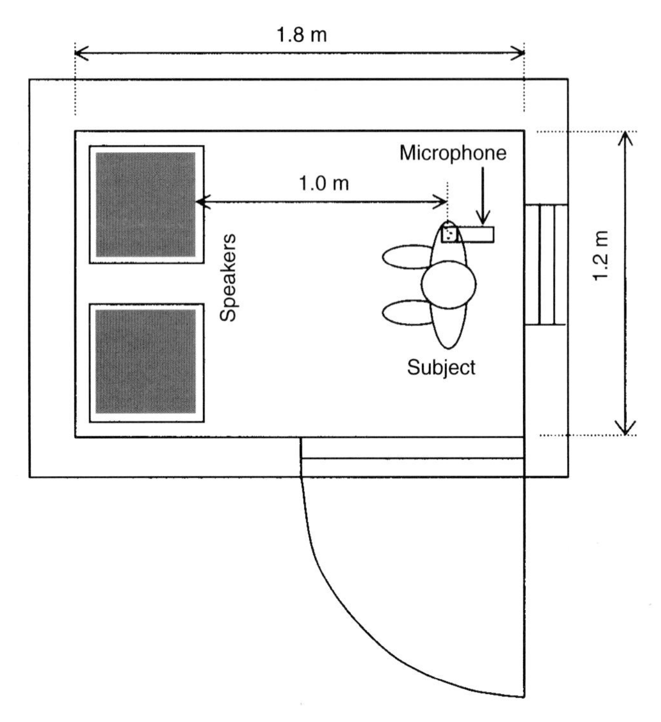

Figure 3. Schematic diagram of the plan view of the experimental facilities inside the cabin.

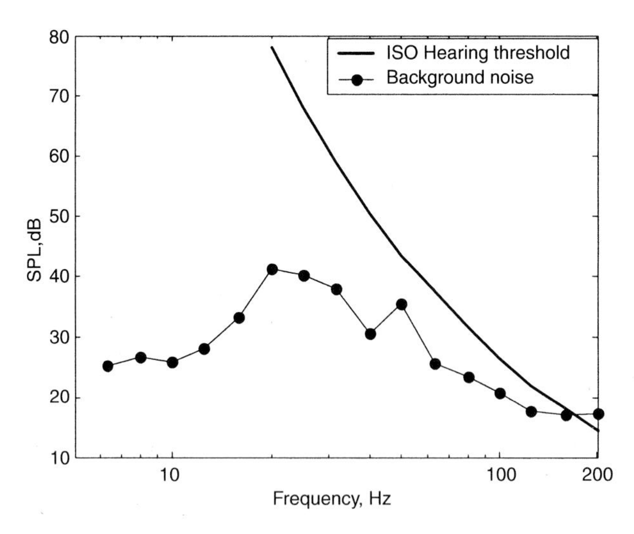

Figure 4. Background noise measured inside the cabin at the position of the microphone shown in Figure 3.

Table I. Summary of the experimental conditions

Background condition	Noise type	Distance (m) from speaker to the subjects for test frequencies				Environment	Method	Number of repetitions
		20 Hz	31.5 Hz	40 Hz	50 Hz			
Quiet	-	-	1	1	1	Controlled environment inside cabin	Indirect method of adjustment	2*
Ambient Noise	Ambient noise in relatively quiet time	0.3	1	1	1	Uncontrolled environment	Method of adjustment	4
Noise 1	60-100 Hz band-pass noise	0.3	1	1	1			
Noise 2	60-100 Hz band-pass noise	0.3	1	1	1			

*4 repetitions were made for one subject

The threshold measurement method was the indirect method of adjustment with UP and DOWN sequence, where the subjects did not have direct control over the sound pressure level. During the measurements, the subjects and the experimenter could not see each other and specially designed buttons and indicators were used for communication. In the UP sequence, the subjects were presented continuous signals well below their hearing threshold and they were asked to press the 'UP' button until the sound was just noticeable to them. The experimenter would increase the level until the subject responded by pressing the 'DECISION' button. In case the signal became sufficiently high and the subjects asked to decrease the level, the signal was decreased to the level below audibility and the process started again. The process was similar for the 'DOWN' sequence, but the starting sound pressure level was well above the audible level of the subjects. The subjects were then asked to press DOWN button to decrease the level of the sound. The higher harmonics produced during the measurement of perception threshold at 20 Hz in quiet were significant in the absence of the background noises. Because of this limitation, the measurement of thresholds for 20 Hz tone was not carried out. Two repetitions at each frequency were made for four subjects, and four repetitions were made for one subject. The summary of all the experimental conditions is given in Table I.

3. EXPERIMENTALLY MEASURED THRESHOLDS

3.1 Thresholds for pure tones in quiet

Results of threshold measurements for the five subjects in the quiet are shown by the filled symbols in Fig. 5(a). As can be seen in the figure, the average threshold in quiet is 6.6, 8.9 and 8.0 dB above the average threshold of hearing defined in ISO 389-7 [18] for frequencies 31.5, 40 and 50 Hz, respectively. The large difference in the two thresholds could not be due to the presence of the noise in Fig. 4, because the sound pressure level of the noise higher than ISO hearing threshold only above 160 Hz should not affect the results at 50 Hz and below adversely. Although the recommended age limit of the subjects for ISO hearing threshold is from 18 to 25 years inclusive, all of the subjects of this study were of age above 26 years. Hence, it is possible that the subjects' thresholds were higher than average thresholds specified in the ISO. As separate audiometric tests were not conducted for the subjects, this could not be verified, while the average thresholds obtained in this study are similar to the average thresholds obtained by Inukai et al. [20] for subjects aged between 19 and 62 years. For further discussion in this study, the perception thresholds measured inside the cabin are considered as the threshold in quiet for the subjects.

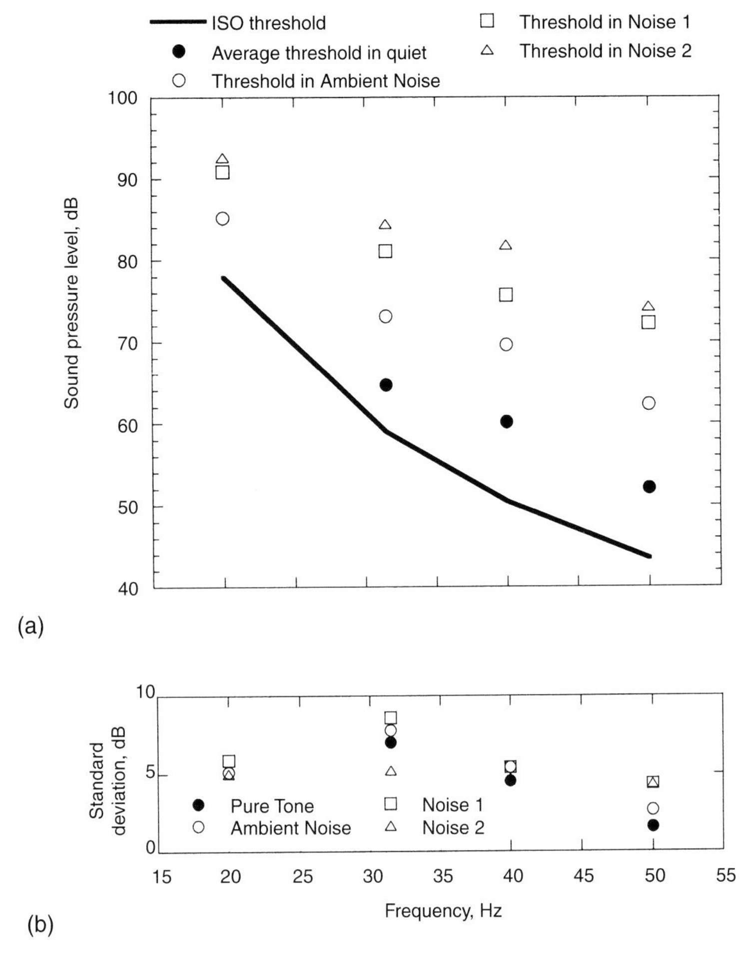

Figure 5. Perception thresholds in quiet and under background noises. (a) Average thresholds (The reference hearing threshold curve specified in ISO 389-7 is also shown.); and (b) Standard deviation.

3.2 Masked threshold

The average masked thresholds with different levels of background noises are also shown in Fig. 5(a). The standard deviations of the thresholds among five subjects for all the cases are shown in Fig. 5(b). The detailed experimental results for individual subjects are given in Subedi et al. [21]. It can be seen in the figure that increases in the level of background noise cause an increase in the perception threshold level. As the sound pressure levels of background noises vary among the subjects for the case of test sound at 20 Hz and average sound pressure levels of the noises are different from those at other frequencies, the results cannot be compared directly with results from other frequencies. However, it can be seen from the results that the increase in the perception threshold with increase in the level of background noise is observed at 20 Hz also. The results suggest that masker sounds at frequencies 50 Hz and above can produce masking effects to sounds down to 20 Hz. However, the increase in the perception threshold decreased with decreases in the frequency for all background noises. This tendency suggests that the masking effect decreased with increases in the frequency separation between the noise and the signal.

The possible indication of these results would be that the main mechanism of perception of sounds at 50 Hz and below is an auditory mechanism. If the masker sounds in the frequencies above 50 Hz are perceived only by auditory mechanism, these sounds should be able to produce masking effects mainly on the auditory mechanism, and perception by other mechanisms, such as vibration, pressure feeling and vibro-tactile perception, would not be masked by these masker sounds. As the masker sounds caused increase in the perception thresholds of the test sounds at 50 Hz and below by more than 20 dB, it could be possible that the main mechanism of perception of these test sounds is an auditory mechanism. However, other mechanisms of perceptions, such as vibro-tactile perception, are also involved at sufficiently higher sound pressure levels as suggested by Landstrom et al. [22].

4. APPLICATION OF MOORE'S LOUDNESS MODEL

4.1 Auditory system in the model

Moore's loudness model [9] is an empirical approach to estimate the loudness from the sound stimulus. The model takes into account the processing of the sound stimulus in different parts of the auditory system at different stages to calculate the loudness. The sub-systems of the model are shown schematically in Fig. 6. The sound stimulus 6(a) passes through outer and middle ear, and the processing in the outer and middle ear is achieved in the model by fixed transfer functions 6(b). The stimulus after the processing in the outer and middle ear reaches the inner ear and excites the inner ear where the impedance increases with decrease in the frequency. The increase in the impedance suggests that the auditory system has less efficiency at lower frequencies. This impedance in the inner ear is represented in the model by the "excitation at the threshold" 6(c), which is a threshold expressed as the sound pressure level reaching the inner ear. The inner ear is modeled as bank of overlapping auditory filters, and the "excitation pattern" 6(d) is calculated as an output of the filters for the corrected stimulus reaching the inner ear. The excitation pattern corresponds directly to the specific loudness 6(e), and the summation of the specific loudness across the ERB scale gives the loudness 6(f) for that sound. The ERB stands for the "equivalent rectangular bandwidth" of the auditory filter at certain frequency and is a function of frequency [23]. Besides the loudness of pure tones and complex sounds, the model can also be used to estimate their thresholds.

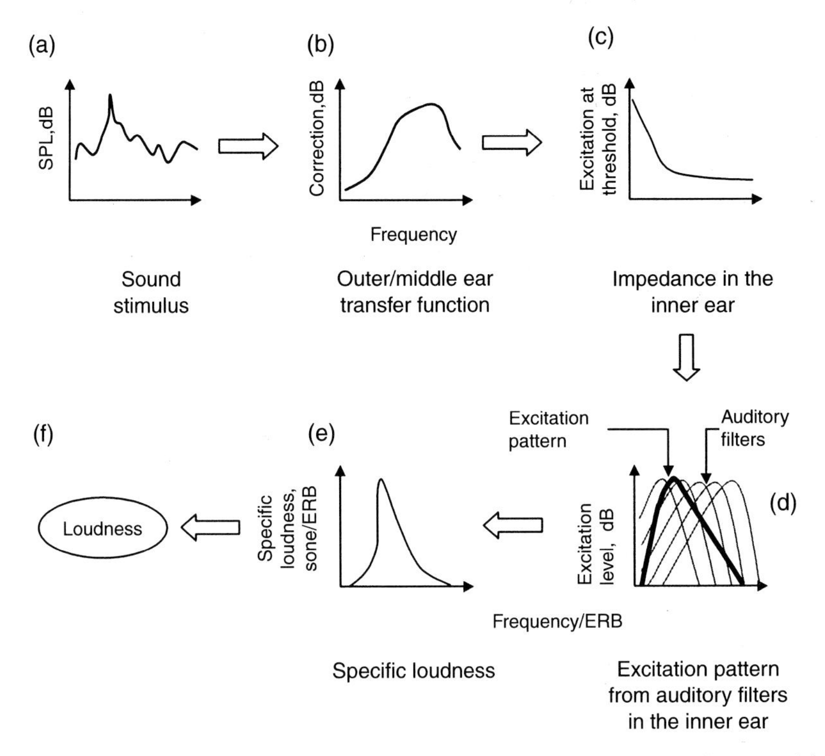

Figure 6. Schematic diagram showing different stages in the Moore's loudness model (based upon Moore et al. [9]).

4.2 Input of masker sounds to the model

The sound pressure levels of the 1/3 octave band spectrum up to 200 Hz, which were measured under the noise only conditions, were taken as inputs representing the masker sounds in the model to estimate masked thresholds of pure tones at frequencies 31.5, 40 and 50 Hz. As the sound pressure levels measured under the noise only conditions differed from the average sound pressure levels measured in the presence of the noise and signal of 20 Hz, the model was not applied to estimate the masked thresholds of 20 Hz.

Because the input for the masker sounds, i.e. the background noises, was limited for frequencies up to 200 Hz, it was assumed that the sounds above 200 Hz did not have any masking effects for signals below 50 Hz. In order to verify this assumption, the specific loudness for a 50 Hz signal under Noise 2 at its masked threshold is shown in Fig. 7. As seen in the figure, the specific loudness for the 50 Hz signal approaches an insignificant level above 80 Hz. Hence, the contribution to the masked threshold is only from the frequency below 80 Hz. Although the maximum contributing frequency changes with the level of masker sound, the choice to limit the frequency up to 200 Hz for all three levels of background noises might seem appropriate.

The unit of specific loudness in Fig. 7 is sone/ERB and loudness is obtained by summing specific loudness across the ERB scale. The ERB scale is also shown on the upper axis in the figure. The obtained loudness in sones can be converted to loudness level in phons from the relation given by Moore et al. [9]. The total estimated loudnesses from Moore's model for Ambient Noise, Noise 1 and Noise 2 are 25.4, 46.5 and 55.0 phon, respectively.

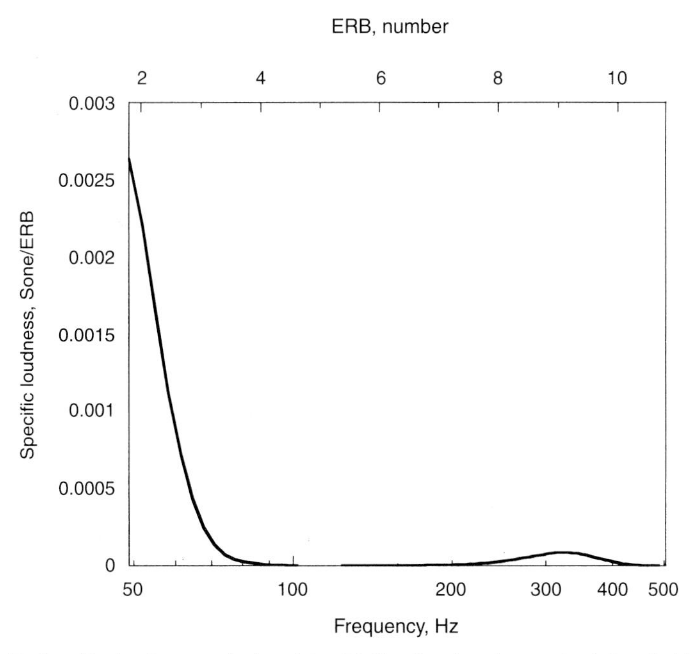

Figure 7. Specific loudness calculated for 50 Hz signal at its masked threshold under Noise 2. Upper scale is equivalent rectangular bandwidth (ERB).

5. COMPARISON OF ESTIMATED AND MEASURED MASKED THRESHOLDS

Moore's loudness model was applied to estimate the masked thresholds for the low frequency tones at frequencies 31.5, 40 and 50 Hz under the background noises. Comparison of the average of the measured masked thresholds for the five subjects under the three background noises with the estimated results from the model is shown in Fig. 8.

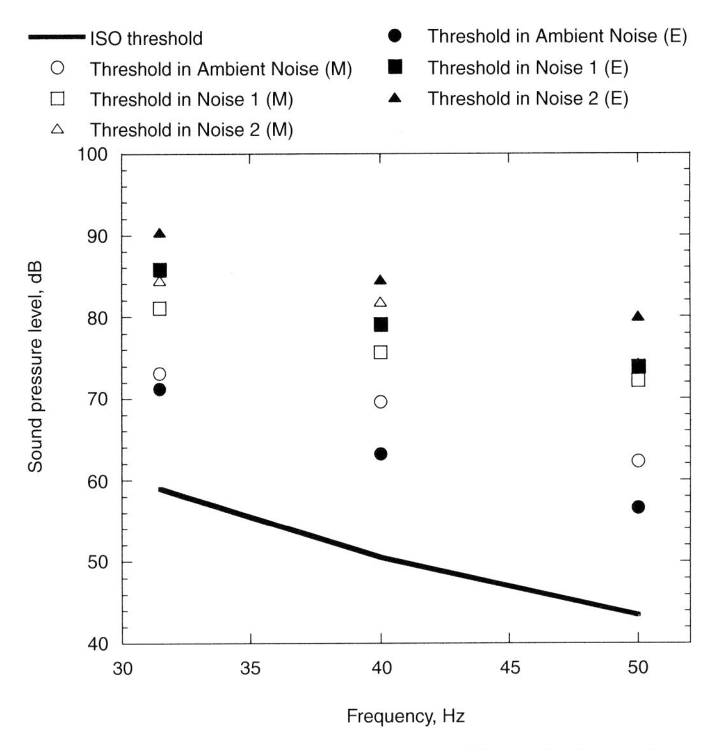

Figure 8. Average measured thresholds under three different background noises and estimated results from Moore's loudness model (E – Estimated values, M – Measured values). The reference hearing threshold curve specified in ISO 389-7 is also shown.

As shown in the figure, the estimated results in Ambient Noise are lower than the measured masked thresholds, while the estimated results are higher than the measured masked thresholds for Noise 1 and Noise 2. The differences in the estimated and average of measured results are within 6 dB for these frequencies. The differences are larger for the frequencies 40 and 50 Hz in the case of Ambient Noise, while they are larger for 31.5 Hz in the case of Noise 1 and Noise 2.

In order to further investigate the estimated results of the model, the estimated loudness levels from the model were compared with the loudness levels specified in ISO 226 [24]. The comparison of the equal loudness level contours calculated from the model and the ISO is shown in Fig. 9. It can be observed in the figure that at lower loudness levels the estimated contours are lower than the contours specified in ISO 226, and with increase in the loudness levels the estimated contours gradually become higher than the contours specified in the ISO. As shown in Fig. 8, a similar change in the trend is observed also in the case of the measurements of masked thresholds, where the estimated thresholds are lower than the measured thresholds in Ambient Noise and higher in Noise 1 and Noise 2. However, the change in the trend occurred at lower sound pressure level in case of the thresholds compared to the case of loudness contours.

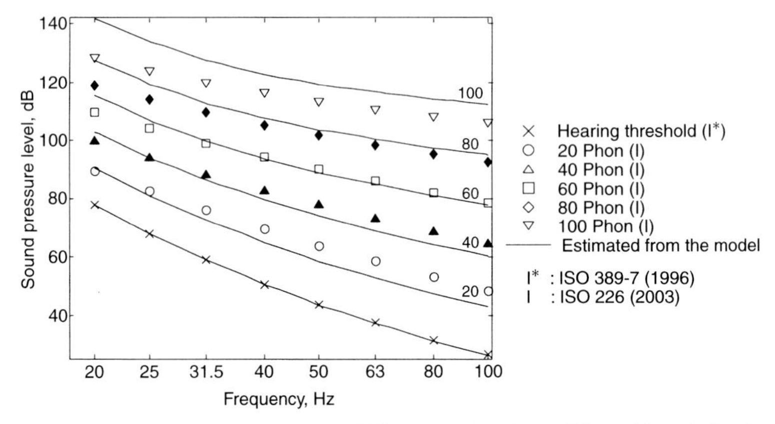

Figure 9. Equal loudness contours from ISO 226 and estimated from Moore's loudness model. Loudness levels in phon are shown for each estimated contours. The lowest solid line is hearing threshold estimated from the model.

Furthermore, it can be pointed out that the two contours from estimated results and ISO 226 are not parallel to each other. At 20 phons loudness level the two contours cross each other roughly at 25 Hz. Although this trend is observed at all levels, the crossing point is shifting towards higher frequencies with increase in the loudness level. The possible reason for this is that the assumed efficiency of the auditory system in the loudness model at lower frequencies is lower with respect to the assumed efficiency at higher frequencies in comparison to the real behavior, and this tendency increases with increase in the loudness level. The comparatively large differences in estimated and measured thresholds at 31.5 Hz for Noise 1 and Noise 2 can also be explained by this phenomenon.

The possible reason for the relatively lower efficiency in the low frequencies is because of the assumption in the model that the assumed auditory filters in the inner ear are limited only down to 50 Hz [9]. It is considered in the model that the frequencies below 50 Hz are detected because they produce outputs from the auditory filters tuned above 50 Hz. This contradicts the experimental study on low frequencies, which suggest that tonal behavior of the sound is present down to 20 Hz [25]. Furthermore, it has been observed that the change in the slope of hearing threshold from approximately 20 dB/octave at higher frequencies to 12 dB/octave at lower frequencies occurs at about 15-20 Hz [26]. These studies suggest that the change in the hearing mechanism occurs at 20 Hz not at 50 Hz. Hence, it is recommended that the lower limit of the assumed auditory filters should be extended at least to 20 Hz.

6. CONCLUSIONS

The perception thresholds of low frequency pure tones at 50 Hz and below were measured under four background conditions in controlled and uncontrolled environments. Moore's loudness model was applied to estimate the thresholds and its performance in the low frequency region was investigated. The findings of this study are summarized as follows:

1. The masked perception thresholds of low frequency tonal sounds at 50 Hz and below increased with increases in the level of the background noises above 50 Hz. The thresholds increased by more than 20 dB by adding higher levels of background noises. The increase in the threshold indicates that sounds below 50 Hz can be masked by masker sounds above 50 Hz.
2. The estimated thresholds from Moore's model and the average of measured masked thresholds match reasonably well within the subjective variability. However, it can be pointed out that the gradients of the equal loudness-level contours for frequencies below 50 Hz specified in ISO 226 are steeper than the contours obtained from the loudness model.

ACKNOWLEDGEMENTS

The authors are grateful to Prof. Brian R. Glasberg and his co-workers at University of Cambridge for providing data tables of figures from their Journal of Audio Engineering Society paper. The authors also gratefully acknowledge the help of Mr. Mitsutaka Ishihara in the experiments. The research was supported by the Japan Society for the Promotion of Science, the Grant-in-Aid for Scientific Research (15656108, 15760343), and the TOSTEM Foundation for Construction Materials Industry Promotion.

REFERENCES

1. Kamigawara, K., Tokita, Y., Yamada, S. and Ochiai, H., Community responses to low frequency noise and administrative actions in Japan, *Proceedings of 32nd Internoise*, Korea, 2003, 1221-1226.

2. Persson-Waye K. and Rylander, R., The prevalence of annoyance and effects after long-term exposure to low-frequency noise, *Journal of Sound and Vibration*, 2001, 240(3), 483-497.

3. van den Berg, F., Low frequency sounds in dwellings: A case control study,

Journal of Low Frequency Noise, Vibration and Active Control, 2000, 19(2), 59-71.

4. Bryan, M.E., Low frequency noise annoyance, in: Tempest, W., ed., *Infrasound and Low Frequency Vibration,* 1976, Academic press, London, 65-96.

5. Kjellberg, A., Goldstein, M. and Gamberale, F., An assessment of dB(A) for predicting loudness and annoyance of noise containing low frequency components, *Journal of Low Frequency Noise and Vibration,* 1984, 3(3), 10-16.

6. Møller, H., Comments to: Infrasound in residential area – case study, *Journal of Low Frequency Noise and Vibration,* 1995, 14(2), 105-107.

7. Zwicker, E. and Fastl, H., *Pyscho-acoustics - Facts and Models,* Second updated edn., Springer-Verlag, Germany, 1999.

8. Moore B.C.J. and Glasberg B.R., A revision of Zwicker's loudness model. *Acustica - acta acustica,* 1996, 82, 335-345.

9. Moore, B.C.J., Glasberg, B.R. and Baer, T., Model for the prediction of thresholds, loudness, and partial loudness, *Journal of Audio Engineering Society,* 1997, 45(4), 224-239.

10. Finck, A., Low frequency pure tone masking, *Journal of the Acoustical Society of America,* 1961, 33(8), 1140-1141.

11. Watanabe, T. and Yamada, S., Study on the masking of low frequency sound, *Proceeding of 10th International Meeting on Low Frequency Noise and Vibration and its Control,* 2002, UK, 41-46.

12. Watanabe, T. and Yamada, S., Study on perception of complex low frequency sounds, *Proceeding of Ninth International Meeting on Low Frequency Noise and Vibration,* 2000, Denmark, 199-202.

13. Fidell, S., Horonjeff, R., Teffeteller, S. and Green, D. M., Effective masking bandwidths at low frequencies, *Journal of the Acoustical Society of America,* 1983, 73(2), 628-638.

14. Meunier, S., Marchioni, A. and Rabau, G., Subjective evaluation of loudness models using synthesized and environmental sounds, *Proceeding of 29th Internoise,* France, 2000.

15. Inukai, Y., Taya, H., Miyano, H. and Kuriyama, H., An evaluation method of combined effects of infrasound and audible noise, *Journal of Low Frequency Noise and Vibration,* 1987, 6(3), 119-125.

16. Inukai, Y., Taya, H., Miyano, H. and Kuriyama, H., A multidimensional evaluation method for the psychological effects of pure tones at low and infrasonic frequencies, *Journal of Low Frequency Noise and Vibration,* 1986, 5(3), 104-112.

17. Møller, H. and Lydolf, M., A questionnaire survey of complaints of infrasound and low-frequency noise, *Journal of low frequency noise, vibration and active control,* 2002, 21(2), 53-64.

18. International Organization for Standardization, Acoustics – Reference zero for the calibration of audiometric equipment – part 7: Reference threshold of hearing under free field and diffuse-field listening conditions, ISO 389-7, 1996.

19. Takahashi,Y., Yonekawa, Y., Kanada, K. and Maeda, S., An infrasound experiment system for industrial hygiene, *Industrial Health,* 1997, 35, 480-488.

20. Inukai, Y., Nakamura, N. and Taya, H., Unpleasantness and acceptable limits of low frequency sound, *Journal of Low Frequency Noise, Vibration and Active Control,* 2000, 19(3), 135-140.

21. Subedi, J. K., Yamaguchi, H. and Matsumoto, Y., Application of psycho-acoustic model to determine perception threshold of low frequency sound in the presence of background noise, *Proceeding of 32nd Internoise,* Korea, 2003, 2768-2775.

22. Landstrom, U., Lundstrom, R. and Bystrom, M., Exposure to infrasound – Perception and changes in wakefulness, *Journal of Low Frequency Noise and Vibration,* 1983, 2(1), 1-11.

23. Moore, B.C.J. and Glasberg, B.R., Suggested formulae for calculating auditory-filter bandwidths and excitation patterns, *Journal of the Acoustical Society of America,* 1983, 74(3), 750-753.

24. International Organization for Standardization, Acoustics-Normal equal-loudness-level contrours, ISO 226, 2003.

25. Yeowart, Y., Bryan, M.E. and Tempest, W., The monaural M.A.P. threshold of hearing at frequencies from 1.5 to 100 c/s, *Journal of Sound and Vibration,* 1967, 6(3), 335-342.

26. Leventhall, G., *A review of published research on low frequency noise and its effects,* Report for Defra, 2003.

Thresholds and acceptability of low frequency pure tones by sufferers

Yukio Inukai, Hideto Taya[1] and Shinji Yamada[2]
[1]Institute for Human Science and Biomedical Engineering, National Institute of Advanced Industrial Science and Technology, AIST Tsukuba Central 6, Higashi 1-1-1, Tsukuba, Ibaraki 305-8566, Japan, inukai-yukio@aist.go.jp
[2]Dept of Eng., Yamanashi University, Takeda 4, Kofu, Yamanashi 400-8511, Japan, yamada@ccn.yamanashi.ac.jp

ABSTRACT
In order to investigate sensory thresholds and to make subjective evaluations of low frequency pure tones in noise sufferers who complain of annoying environments in their everyday life, sound pressure levels of sensory thresholds and subjectively acceptable maximum SPL levels for a living room were measured in a low frequency chamber. These measurements involved a psychophysical experiment using eleven pure tones at low frequencies from 10Hz to 100 Hz as stimuli, and the psychophysical method of subject adjustment was used for the measurements. Twelve members of the noise-sufferer's society in Japan participated as subjects (referred to as participants in the measurement experiment). The results show that all the participants' acceptable maximum sound pressure levels were relatively low, and nearly equal to their sensory thresholds. These results are characteristic of the participants and differ from the previous results obtained from the other adults.

1. INTRODUCTION

Our previous paper [1] reports sensation thresholds, equal unpleasantness contours of low frequency sound and the acceptable limits of SPL for some living situations of 39 adults aged from 20 to 55. Pure tones at frequencies from 10 Hz to 500 Hz were used and the means obtained were well predicted by a polynomial function. In addition, similar but somewhat different results were obtained from older adults aged 60-75; the results were reported in another paper [2]. However, there are fairly large individual differences in the sensations and in the evaluations of low frequency sound. In particular when participants are exposed to and complain of annoying environments in their daily lives, their responses in the form of sensations and in evaluations of low frequency sound are expected to differ from those of other adults. We recently measured sensation thresholds and acceptable limits of SPL for some members of the noise-sufferer's society in Japan [3]. This paper investigates the characteristics of the noise sufferer's thresholds and the acceptability of low frequency pure tones at frequencies from 10 Hz to 100 Hz, and discusses the similarities and differences of their results in comparison with the previously reported results of other adults.

2. METHODS

Stimuli

Eleven low frequency pure tones were used as stimulus sounds at frequencies from 10Hz to 100Hz with 1/3-octave intervals.

Participants

Nine female and 3 male noise sufferers aged from 39 to 71 voluntarily participated in this study. The participants were members of the noise-sufferer's society in Japan. They complained about and seriously suffered from their annoying environments in their daily lives. Most of them had unresolved noise problems with their neighbours and suspected low frequency noise as the cause. However, they were not necessarily exposed to low frequency noise. The gender, age, noise generators and/or complaints are shown in Table I

Table I Self-reported causal systems or contents of complaints.

Par.	Age	Causal system or Complaints	Par.	Age	Causal system or Complaints
M1	63	compressor, pump, machinery sound	F5	60	unkown, low level buzzing
M2	54	outside machine of air conditioner			
M3	48	dryer's ventilator	F6	39	air conditioner and electric generator on the roof of the hospital neighbourhood
F1	71	air conditioner			
F2	60	air conditioner, motor	F7	43	37 air conditioners on the roof of the hospital
F3	68	unknown	F8	48	power cable?, unknown continuous sound
F4	65	air conditioner, pump	F9	56	trucks, hitting sound of tennis ball

Par: participant, M: male, F: Female

Apparatus

The experiment was conducted in a 22.75 cubic metre pressure-field chamber at AIST. The internal chamber dimensions were 3.5 x 2.5 m base with a height of 2.6 m. Sixteen 46-cm diameter loudspeakers were mounted in a wall of the chamber and were driven by sixteen 150 w amplifiers. The background noise level of the chamber was less than 17 dB (A). There was a small chair for participants. The stimulus sounds were generated by a sine/noise generator (B&K 1049) and presented to participants through the loudspeakers in the chamber. The sound levels could be adjusted at steps of less than 1dB by a subject using a remote volume controller.

Measurement items

The following two items were measured at each stimulus frequency for each participant.

(1) Sensation threshold of SPL.
(2) Acceptable limit of SPL.

Method of measurement

A method of subject adjustment was used for the both items of threshold and acceptable limit. The participant adjusted the sound pressure levels to subjectively appropriate levels by manipulating the remote volume controller by hand.

Procedure

Sensation thresholds and acceptable-limit levels were measured in separate sessions. One participant entered in the chamber for each session. The stimuli were presented in random order once at each frequency per session. To measure acceptable limits, participants were required to produce a sound pressure level of a given stimulus reflecting the acceptable limit of SPL for her or his living room using the remote volume controller. The produced sound pressure levels were measured by a microphone in the chamber and calibrated at the position of the participant's head. Each session took about 15 minutes and about the same resting time was taken between sessions.

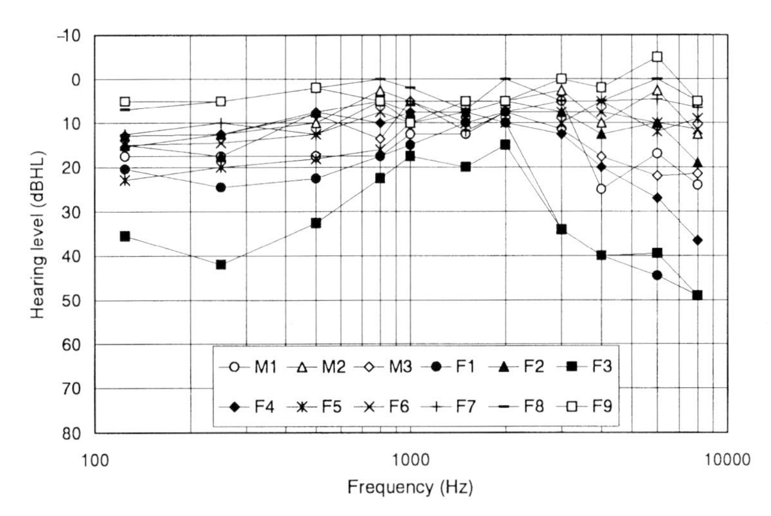

Figure 1. Hearing levels of 12 participants at higher frequencies than 100 Hz measured by the audiometer.

3. RESULTS AND DISCUSSION

Hearing level

Hearing levels of the participants were measured by the audiometer (Rion AA-79S or AA79) to check their hearing levels at higher frequencies than 100 Hz before and after the low frequency measurement sessions. Figure 3 shows the hearing levels obtained from the 12 participants. The hearing levels were represented by those of the better right or left ear of each participant. If both ears had the same hearing level, 3 dB were subtracted from the monaural level. The results show some individual differences in hearing levels of the participants but most except F4 have normal hearing for their age.

Threshold

Observed thresholds were averaged for each participant and are shown in Figure 2. The measurements were repeated 2 or 3 times. The first trials were regarded as practice and were excluded. For reference, the mean thresholds and estimated equal unpleasantness curves, U2 (a little unpleasant) to U5 (very unpleasant), of 39 other adults from our previous paper [1] are also shown in the same figure. The sound pressure level (L) of an equal unpleasantness curve for a given unpleasantness rating grade (y) was estimated by the following equation (1).

$$L=(y+43.5 -37.4x+10.7x^2 -0.988x^3) / (0.312 -0.176x + 0.0360x^2) \quad (1)$$

Where L means a sound pressure level in dB (L) at frequency h and x = log h. There are rather large individual differences in the thresholds, but the dispersion was almost the same as that of other adults. The thresholds of sufferers were rather less sensitive than those of other adults, and no one showed an especially more sensitive threshold.

SPL of acceptable limits

The measurements were normally made once and were sometimes repeated when the observed values were unstable. Figure 3 shows the sound pressure levels of acceptable limits for each participant. Acceptable limits were measured assuming the situation that participants were reading a newspaper quietly in their living room. For reference, the thresholds and equal unpleasant curves of other adults are shown in this figure. The acceptable limit levels of the participants in Figure 3 are commonly very similar to the threshold levels in Figure 2.

Reliability of the Measured values

The method of adjustment for measuring threshold is considered more difficult than the other psychophysical methods for inexperienced participants. Fortunately, five participants, M1, M2, F2, F8, F9, also

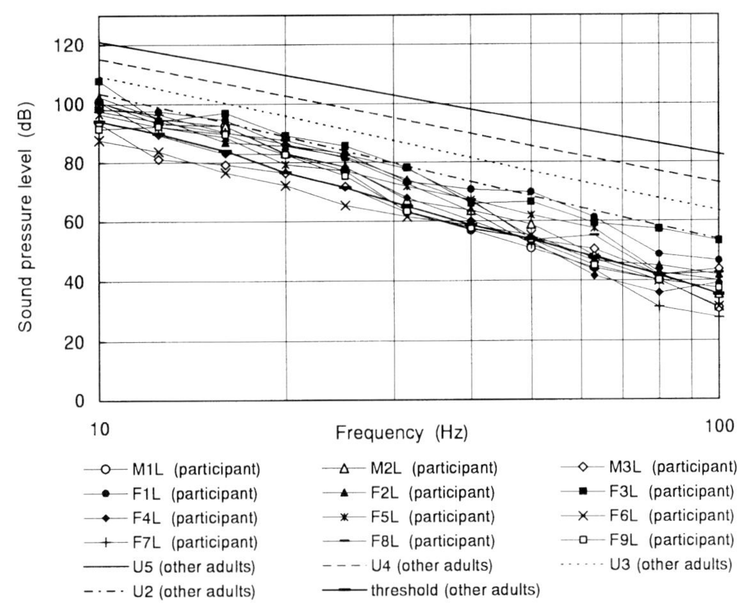

Figure 2. Thresholds of 12 participants (M1L-F9L) in this experiment, and mean threshold, estimated the umpleasantness (U2-U5) of other adults from our previous paper [1]

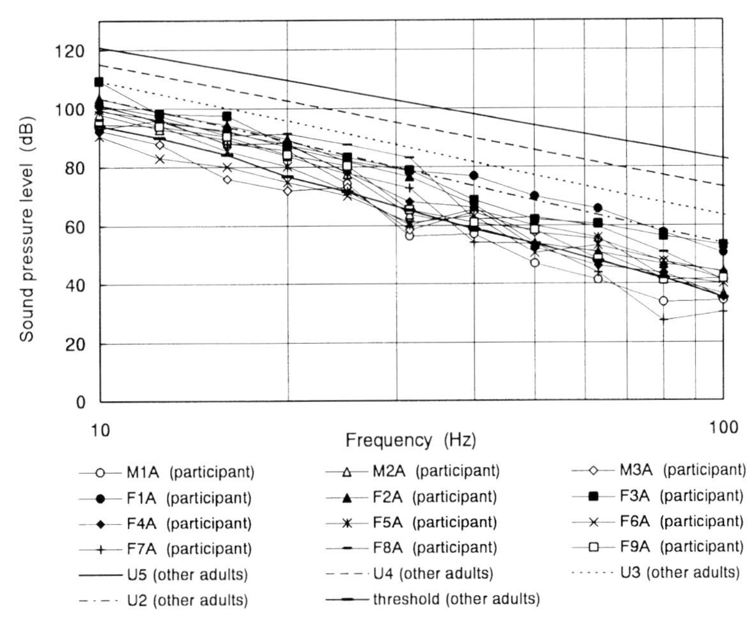

Figure 3. SPL of acceptable limits of low frequency pure tones obtained from 12 participants (M1A-F9A) in this experiment, and mean threshold, estimated unpleasantness (U2 -U5) of other adults from our previous paper [1] .

participated in the measuring threshold experiments by the method of limits at Yamanashi University. So, the results obtained from the two methods were compared. The threshold curves averaged over the five participants for each method at different locations, AIST or Y.U. are shown in Figure 4. The means and standard deviations of the differences between the thresholds obtained by the two different methods for each participant are also shown. The results of the method of adjustments were 1.5dB higher than those of the method of limits on average. The reliability of the mean data by both methods was considered very high, although there was a small response-bias particular to the measurement methods. The standard deviations of the differences were rather large, but this was probably due to the small number of replications.

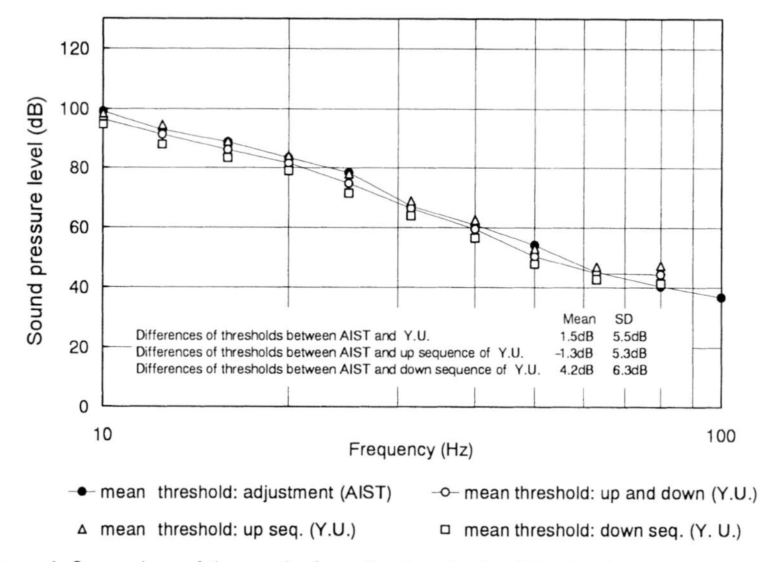

Figure 4. Comparison of the results from the 2 methods of threshold measurement.

Mean threshold curve and mean accessible limits of SPL

The thresholds of all participants were averaged and the mean threshold values obtained were plotted by circles with solid line segments in Figure 5. At the same time, mean acceptable limits of SPL were plotted by filled dots with dotted line segments. The results show that there was almost no difference between the two groups, and their acceptable limit levels were nearly equal to their threshold levels. This implies that the participants judged low frequency sounds as being at an acceptable limit level as soon as they heard them. This tendency was clearly different from that of the other adults shown in the same figure. In the other adults, there was an approximately 10 dB difference between their threshold curve and their acceptable limit curve which were estimated by equation (1) by substituting y=1.6 for the same situation of a living room [1]. Although the cause of the difference between the 2 groups of participants was not obvious, it is conjectured that some response mechanism which had developed during the long-term experiences of sufferers to their annoying environments might be related to their unique criteria in their evaluations of low frequency noise, otherwise the sufferer's strict criteria might be due to the fact that the subjects belong to the category of people who are most sensitive to noise annoyance. That such a group of persons exist has been well established in the literature (the authors are grateful to an anonymous referee for his valuable comments on this point) [4,5], however, to the authors knowledge such relationship between sufferers and their personality has not been reported. It is thought that the cause of a complaint is often a much more complicated one, i.e. to be a result of the interaction of various factors which may include friction in human relations with persons in charge of noise sources.

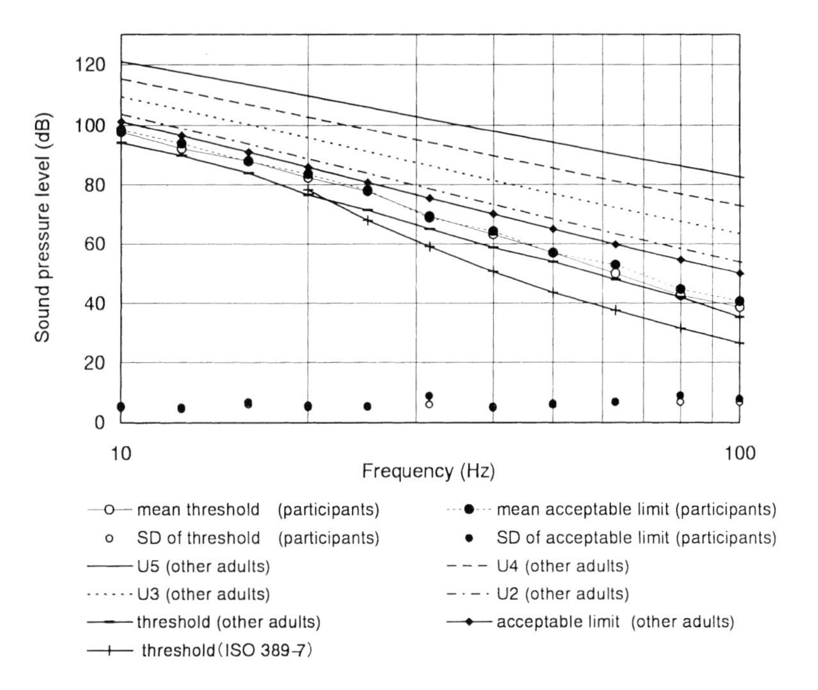

Figure 5. Mean threshold curve and mean acceptable limits of SPL for the participants on the assumption of reading newspapers in their living rooms.

The sound pressure levels of the vertical axes of Figure 5 were transformed to A-weighted noise levels and are shown in Figure 6. This figure shows that the noise levels of acceptable limits for their living rooms were rather low levels from about 21 to 34 dB(A). It is supposed that noise at lower frequencies might be more troublesome because noise at lower frequencies is not easily masked by the surrounding noise in their living environments.

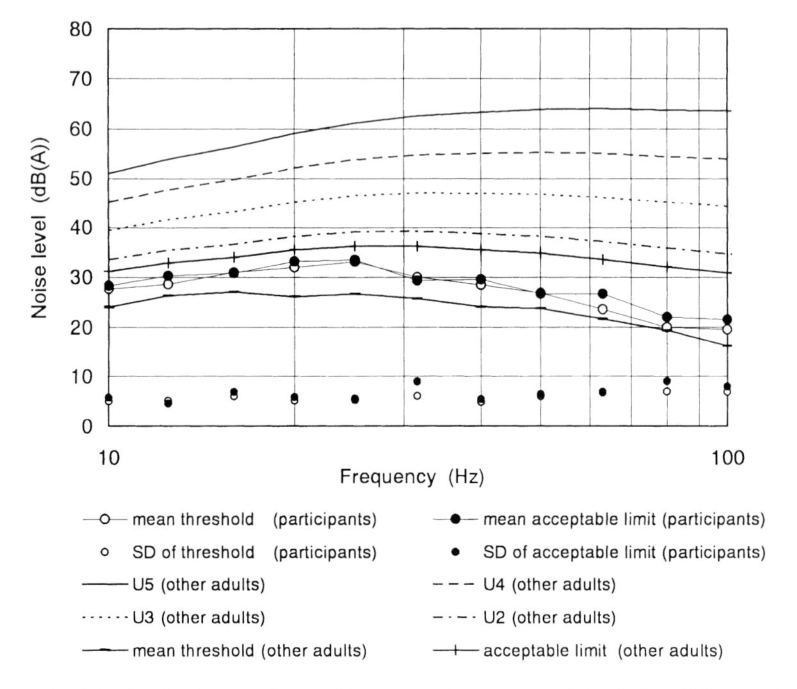

Figure 6. Noise levels of mean threshold curve and mean acceptable limits of SPL for the participants on the assumption of reading newspapers in their living rooms.

CONCLUSIONS

1) The thresholds of the participants (noise sufferers) were not necessarily lower than the thresholds of other adults.
2) The sound pressure levels of acceptable limits judged by the participants were nearly equal to their hearing threshold levels, and the noise levels were as low as from 21 dB (A) to 34dB (A).
3) The causes of their strict criteria for the evaluation of low frequency sound were not obvious, and it was conjectured that some of them were probably subjects belonging to the category of people most sensitive to noise annoyance, and some might be caused by sufferers' experiences of long-term exposures to their annoying environments.

REFERENCES

1. Inukai, Y., Nakamura, N. and Taya, H. (2000), Unpleasantness and acceptable limits of low frequency sound, *J. Low Frequency Noise, Vibration and Active Control*, 19(3), pp. 135-140 .

2. Inukai, Y., Nakamura, N. and Taya, H. (2001), Threshold, unpleasantness and acceptable limits of low frequency pure tones for elderly people, *Proceedings of 17th International Congress on Acoustics*, Rome, 4(8-11), pp. 30-31.

3. Inukai, Y., Taya, H. and Yamada, S. (2002), Sufferers' sensation threshold and acceptable SPL for pure tones at low frequencies. *Proceedings of the 2002 Spring Meeting of the Institute of Noise Control Engineering of Japan*, April, pp. 23-26 (In Japanese).

4. Langdon, F.J. (1985), "Noise Annoyance" in *The Noise Handbook* Ed. Tempest, W. Academic Press, London.

[5] Bryan, M.E. (2002), Personality and annoyance by low & mid frequency environmental noise, *Proceedings of the 10th International Meeting on Low Frequency Noise and Vibration and its Control*, York, England, pp. 91-102.

Chapter 2: Effect of low frequency noise on people in terms of annoyance and sleep deprivation

Most of the research effort on low frequency noise and vibration is focused on assessing annoyance and sleep disturbance. All of the following papers focus on one or both of these aspects.

1. Unpleasantness and acceptable limits of low frequency sound.

The unpleasantness of tonal sound at frequencies from 10 Hz to 500 Hz was evaluated using twenty seven female and twelve male test subjects and acceptable levels of sound for various living spaces was also evaluated by the same subjects. In some cases it was found that acceptable limits were close to very low levels of unpleasantness in some situations. The paper is included in Chapter 1.

2. Low frequency noise annoyance in classroom

Noise in class rooms in Sweden was found to contain both low and high level low frequency components. The authors found that the level of low frequency noise was not correlated to the level of annoyance when the difference between dB(C) and dB(A) levels were used to assess the level of low frequency noise. The authors conclude that the above method for assessing the level of low frequency noise is only reliable at levels higher than 30 dB(A) and that the method may not be reliable at levels as low as or lower than 30 dB(A).

3. A structured approach to LFS-complaints in the Rotterdam region of the Netherlands.

A structured approach for handling low frequency sound complaints has been developed and implemented and reported here. The author has proposed licences curves which are presented in Figure 3 and suggested as regulation limits for low frequency noise.

4. Annoyance of low frequency noise and traffic noise

The annoyance of a number of low frequency noised sources was compared to the annoyance of traffic noise using eighteen normal hearing test subjects and four subjects with special low frequency problems. As expected the special group found the low frequency noises (which ranged between 10 Hz and 200 Hz) more annoying than the others found them. Traffic noise was among the least annoying noises.

5. Perceptions of the public of low frequency noise

This paper introduces a range of systems for measuring both noise and vibration at low frequencies. In addition, this paper discusses the correlation of both noise and vibration at low levels and low frequencies with sleep disturbance as well as suggesting that long-term exposure sensitizes people making the problem worse with as time elapses.

6. A questionnaire survey of complaints of infrasound and low-frequency noise

A survey was carried out on 198 people who had complained about infrasound reporting perception of the sound with their ears as well as their bodies. No conclusive evidence was found that the infrasound originated from an external source in any of the cases and this resulted in the recommendation for more tests and measurements.

7. Assessment of low frequency noise complaints among the local environmental health authorities and a follow-up study 14 years later.

Swedish local environmental health authorities were interviewed to assess the occurrence of low frequency noise. Most complaints originated from fans, amplified music, compressors and laundry rooms. It was also found that guidelines of acceptability based on 1/3 octave band measurements performed much better than A-weighted guidelines (as expected).

8. Low frequency noise annoyance: the behavioural challenge

This paper is concerned with how the quality of the interpersonal relationship between the person making the complaint and the Environment Officer to which the complaint is being made affects the resolution of the low frequency noise problem.

9. Comparison of objective methods for assessment of annoyance of low frequency noise with the results of a laboratory listening test.

Subjective assessments made by 9 male and 9 female test subjects of a number of different types of low frequency noise and the results were compared with a number of objective measurement and calculation methods. The results showed that the Danish objective measuring method correlates better with subjective annoyance than methods used in other countries.

10. Blast densification method: sound propagation and estimation of psychological and physical effects

Blast noise equates to impulsive low frequency noise for which frequencies below 10 Hz are the most important in terms of affecting communities at distances larger than 1 km. However at distances greater than 300 m from the blast site, the effects of the blast sound are small and generally limited to rattling windows and building fittings. Closer than 300 m, complaints of sleep disturbance will occur and at distances closer than 50 m people report oppressive or oscillatory feelings.

11. Annoyance of low frequency noise (LFN) in the laboratory assessed by LFN-sufferers and non-sufferers.

T. Poulsen (2003) 177

Eight different low frequency noise environments were tested using eighteen normal hearing test subjects and four who has previously reported being annoyed by low frequency noise. The environments included continuous noise with and without tones, intermittent noise, music, traffic noise and impulsive LFN. The special group assessed the annoyance of the noises much higher than the "normal" group, especially at night.

12. Psychological analysis of complaints on noise/low frequency noise and the relation between psychological response and brain structure.

T. Kitamura, M. Hasebe and S. Yamada (2005) 191

The conclusion from this paper is that part of the cure for complaints arising from low frequency noise problems is to counsel the sufferer because it was found that sometimes complaints continued after the noise ceased.

13. Annoyance of low frequency tones and objective evaluation methods

J.K. Subedi, H. Yamaguchi, Y. Matsumoto and M. Ishihara (2005) 199

Annoyance of low frequency pure (31.5, 50 and 80 Hz) and combined tones (31.5 + 40 Hz, 50 + 40 Hz and 80 + 40 Hz) was measured using fourteen male and two female test subjects. It was found that the rate of increase in annoyance as the level increased was greater for the lower frequencies and the annoyance of a complex sound was dependent on the relative level of each component as well as the separation in frequency. They also tested the correlation of Moore's loudness model, A-weighting and total energy summation with annoyance and found the A-weighted level was the most accurate predictor. However, they did not test the dB(C) -dB(A) model.

14. LFN and the A-weighting

P. Sloven (2005) 219

This paper considers the extent to which conventional dB(A) assessment underestimates the annoyance of low frequency music noise (with a heavy beat) and proposes a new rating procedure based on the dB(C) measurement.

15. Thresholds and acceptability of low frequency pure tones by sufferers

Twelve members of the Japanese noise-sufferers society were used to assess the sensory thresholds and subjectively maximum acceptable SPLs for a living room. It was found that the thresholds of the sufferers were not necessarily any lower than other adults, but the levels of acceptability for sufferers were very close to their threshold levels, which is not the same result for studies on other adult subjects. The full paper is included in Chapter 1.

16. Low frequency noise annoyance and the negotiation challenge for environmental officers and sufferers.

This paper is concerned with issues surrounding the application of inappropriate assessment procedures to the assessment of low frequency noise problems and how this influences the resolution of a complaint by a sufferer.

Low frequency noise and annoyance in classroom

Pär Axelsson, Kjell Holmberg and Ulf Landström
National Institute for Working Life, P.O. Box 7654, S-907 13 Umeå, Sweden

ABSTRACT
The most common method for noise assessment is the A-weighted sound preasure level. The question has been raised whether the frequency weighting with an A-filter gives a correct result when assessing the annoyance response of noise containing strong low frequency noise (LFN) components. One method sugested to identify LFN is the dB(C) – dB(A) difference. The aims for this study is to investigate if background noise in Swedish elementary school is to be considered as LFN, further to test the hypothesis that students exposed to audible LFN at high levels are more annoyed than students exposed to LFN at lower level. The results indicates that the noise in 16 of 22 classrooms are to be considered as LFN. The analysis did not show any difference in rated annoyance between students exposed to high LFN levels and students exposed to low LFN levels.

INTRODUCTION
In 1997 The National Board of Occupational Safety and Health in Sweden carried out a survey on the working environment in Swedish elementary school (7-15 years) (Swedish National Board of Occupational Safety and Health 1997). Headteachers in Sweden were asked about the working environment at their school. Findings in this survey were that the most common environmental problems at school were related to indoor climate and "noise, sound and acoustic problems". The daily activities in the classroom are based on communication and concentration. The working methods in schools in Sweden today differs a lot from the traditional method of lecturing. The teaching methods are nowadays more focused on problem solving. The students are more interactive, eg. discussing with each other and working in groups and projects. The teacher has more and more become a supervisor, guiding not lecturing. The acoustic conditions in these environments are very complex for the constitution of the framework for classroom organisation, management and the learning process. Noise exposure problems vary in different school environments due to the presence of different noise sources, speech and other noises related to the student activity. However, the risk for hearing damaging noise exposure during classroom activities is low. On the other hand there is a compelling body of research showing adverse effects of noise on speech communication, performance, short time memory and learning (e.g.

Pekkarinen & Viljainen 1991, Jones 1990, Enmarker et al. 1998).

A great number of external noise sources as ventilation systems, activities in the school building, air- and road traffic noise interfere with the school work. Road traffic noise may deteriorate attention (Enmarker et al. 1998). Bronzaft and McCharty (1975) compares reading ability and sound pressure levels for schools close to railroads. Students in classroom with the highest sound levels were found to be late in their development of reading. Moreover, Cohen et al. (1980) have reported that air traffic noise can affect reading comprehension and mathematic proficiency. Inside the classroom noise sources such as projectors and other technical equipment are present. However, at school as in many other working environments such as offices, shops, service and hospitals, the main part of the noise originates from human activity.

An effect of noise that has to be considered in the school environment is annoyance. Holmberg (1997) points out that in occupational environments noise annoyance as well as negative effects on performance will increase with increasing sound level, tonal character and variability of the noise. Differences in responses seem to exist between high- and low frequency noise exposures. Despite the complexity of noise annoyance, the most common method for noise assessment is still the A-weighted sound preasure level. The question has been raised whether the frequency weighting with an A-filter gives a correct result when assessing the annoyance response of noise containing strong low frequency noise (LFN) components. The interval between the hearing threshold and an unacceptable level is much smaller for LFN than noise with higher frequencies (ISO 1985). At low frequencies annoyance may appear just above the hearing threshold level. Persson Waye (1995) presents results indicating that annoyance experienced from low frequency noise is higher than annoyance from noise without dominant LFN components at the same level. There are also several observations that point out that the A-weighting underestimates the annoyance from LFN (Kjellberg et al. 1984, Persson & Björkman 1988, Leventhall 1980). The definition of LFN varies and there is no international agreed definition. Castelo Branco et al. (1999) suggests frequencies up to 500 Hz and Berglund et al. (1996) suggests that frequencies up to 250 Hz should be considered as LFN. Dealing with annoyance, an appropriate definition of LFN could be "noise with a dominant frequency content of 20 to 200 Hz" (Persson Waye 1995). One method that has been taken into use in some Swedish recommendations to identify LFN is the dB(C) – dB(A) difference (Swedish National board of Health and Welfare 1995, Swedish Royal Board of Building 1992). This will constitute an estimate of how much energy that is to be found in the low frequency part. A limit of 15-20 dB is also given over which the noise is to be considered as LFN (Swedish National Board of Health and Welfare 1995). Persson Waye

(1999) suggests that this method should only be used when the level is above 30 dB(A). One aim for this study is to investigate if background noise in Swedish elementary school is to be considered as LFN. Further to test the hypothesis that students exposed to audible LFN at high levels are more annoyed than students exposed to LFN at lower level.

METHOD

Sound levels were recorded in 22 unoccupied classrooms at three typical schools in Sweden under similar conditions with normal activity in the surroundings. The noise was recorded using a sound level meter (Brüel & Kjær 2237) with a ½" microphone (B&K 4189) and a digital tape recorder (TEAK DA-P20). The sound level meter was placed in a representative, asymmetrically situated position in the classroom corresponding to the ear height of the students in order to measure the percieved noise. The measurements were made for 10 minutes.

The recordings were analysed according to A- and C- weighted levels. The $L_{Ceq} - L_{Aeq}$ difference was calculated from the equivalent sound levels.The classrooms with a $L_{Ceq} - L_{Aeq} < 15$ dB were categorised as low LFN level exposure and classrooms with $L_{Ceq} - L_{Aeq} > 20$ dB were categorised as high LFN exposure. In the extent of the study A- and C-weighted levels were calculated for a limited frequency range, i.e. 63-20kHz. This was made to obtain a $L_{Ceq} - L_{Aeq}$ level difference based on the frequency range in which the sound pressure level exceeded the hearing threshold level. The result is a $L_{Ceq} - L_{Aeq}$ difference based on the frequency range with levels above the hearing threshold level for frequencies in the lower part of the spectra, $L_{Ceq(63\text{-}20kHz)} - L_{Aeq(63\text{-}20kHz)}$. The classrooms with a $L_{Ceq(63\text{-}20kHz)} - L_{Aeq(63\text{-}20kHz)} < 10$ dB were categorised as low LFN level exposure and classrooms with $L_{Ceq(63\text{-}20kHz)} - L_{Aeq(63\text{-}20kHz)} > 13$ dB were categorised as high LFN exposure. The 337 students working in these 22 classrooms were asked to report their annoyance in a multiple-choice question. Five alternatives were given. "Not at all annoyed ", "Somewhat annoyed", "Quite annoyed", "Much annoyed" and "Very much annoyed".

To test the hypothesis that students working in classrooms with high LFN levels are reported to be more annoyed than students working in classrooms with low LFN levels, a Mann-Whitney Utest was used. The same statistical tool was used to provide with sufficient founds that there was no difference between the groups in L_{Aeq}. The level of significance was chosen to be $p<0.05$.

RESULTS

Table 1 shows that eight classrooms have a $L_{Ceq} - L_{Aeq}$ difference of 15 dB or below. These classrooms are categorised as low LFN exposure. Nine classrooms have a $L_{Ceq} - L_{Aeq}$ difference of 20 dB or above. These classrooms are categorised as high LFN exposure. The statistical analys shows that there is no difference in L_{Aeq} between the groups low LFN exposure and high LFN exposure ($Z= -.93$ $p> .05$).

Table 1. LAeq, LCeq , LCeq – LAeq difference for unoccupied classrooms.

Class room	L_{Ceq}	L_{Aeq}	$L_{Ceq} - L_{Aeq}$
9	52	43	10
15	50	40	10
1	48	37	11
6	53	40	12
11	50	38	12
10	54	40	14
5	52	37	15
2	51	36	15
19	52	35	17
18	53	35	18
22	56	37	18
20	53	34	19
21	55	36	19
3	58	39	20
17	60	40	20
13	54	34	20
16	60	40	21
14	60	39	21
8	60	39	21
7	56	34	22
4	61	38	23
12	58	35	23

Figure 1 displays the distribution of how students reported their annoyance during the lesson. The comparison of reported annoyance between the high- and low LFN-exposed groups, based on a $L_{Ceq} - L_{Aeq}$ difference to identify LFN, shows that there is no difference in rated annoyance between these groups ($Z= -.84$ $p> .05$). The arithmetical means of the L_{Aeq} for the LFN exposure groups is 39 dB(A) for the low LFN exposure and 38 dB(A) for the high LFN exposure.

The median, max and mean of sound pressure levels in 1/3-octave bands for the 22 classrooms are shown in figure 2. The spectra is related to the normal hearing threshold (ISO 1985). The analysis shows that for the frequencies 25, 31, 40 and 50 Hz all levels are close or below the hearing threshold level.

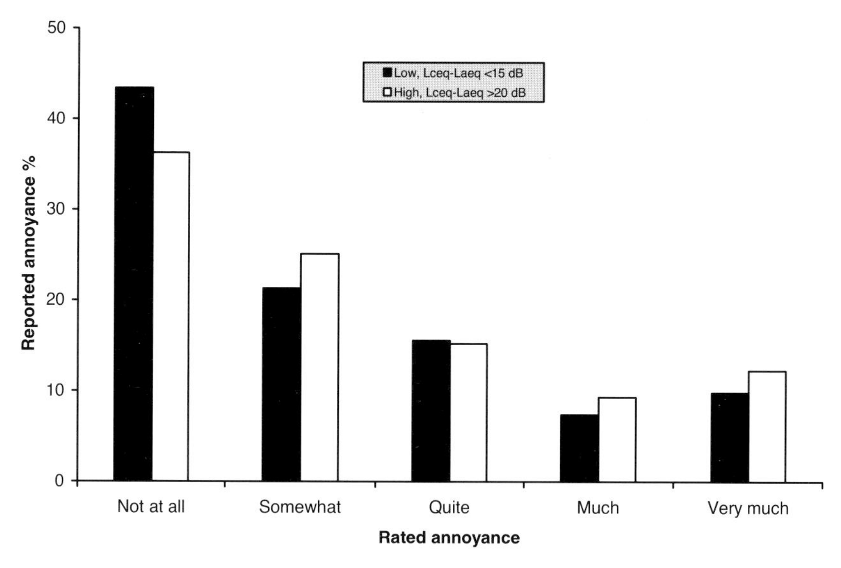

Figure 1. The distribution of reported annoyance between the high- and low LFN-exposed groups based on a $L_{Ceq} - L_{Aeq}$ difference to identify LFN.

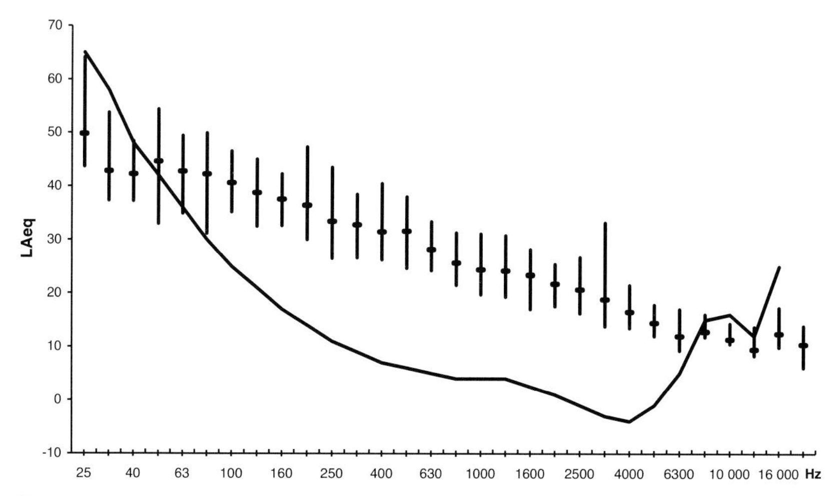

Figure 2. Median, maximum and minimum of sound preasure level of 1/3 octavebands for the background noise in the 22 clasrooms. The full line shows the normal hearing treshold (ISO 226-1 1985).

Table 2 shows the $L_{Ceq} - L_{Aeq}$ difference based on the frequency range in which the sound pressure level exceeded the hearing threshold level (Figure 2), $L_{Ceq(63\text{-}20kHz)} - L_{Aeq(63\text{-}20kHz)}$. Nine classrooms, that has a $L_{Ceq(63\text{-}20kHz)} - L_{Aeq(63\text{-}20kHz)}$ difference of 10 dB or below are categorised as LFN exposures. Ten classrooms, that has a $L_{Ceq(63\text{-}20kHz)} - L_{Aeq(63\text{-}20kHz)}$ difference of 13 dB or above are categorised as high LFN exposures.

Table 2. $L_{Aeq(63\text{-}20kHz)}$, $L_{Ce(63\text{-}20kHz)}$, $L_{Ce(63\text{-}20kHz)} - L_{Aeq(63\text{-}20kHz)}$ difference for unoccupied classrooms.

Class room	$L_{Ceq(63\text{-}20kHz)}$	$L_{Aeq(63\text{-}20kHz)}$	$L_{Ceq(63\text{-}20kHz)} - L_{Aeq(63\text{-}20kHz)}$
1	44	37	7
17	47	40	7
9	50	43	7
15	47	40	8
11	47	40	8
6	49	41	9
19	45	35	10
18	45	36	10
2	46	37	10
5	48	38	11
10	52	40	11
21	48	36	12
8	52	39	13
3	52	39	13
22	51	38	13
12	48	35	13
14	52	39	13
7	48	34	14
4	52	38	14
13	48	34	14
16	54	40	15
20	49	34	15

Figure 3 displays the distribution of how the students reported their annoyance during the lesson. The comparison of the reported annoyance between the high- and low LFN-exposed students, based on a $L_{Ceq(63\text{-}20kHz)} - L_{Aeq(63\text{-}20kHz)}$ difference to identify LFN, shows that there is no difference in rated annoyance between the groups ($Z= -.57$, $p> .05$). The arithmetical means of the $L_{Aeq(63\text{-}20kHz)}$ for the LFN exposure groups is 39 dB(A) for the low LFN exposure and 37 dB(A) for the high LFN exposure.

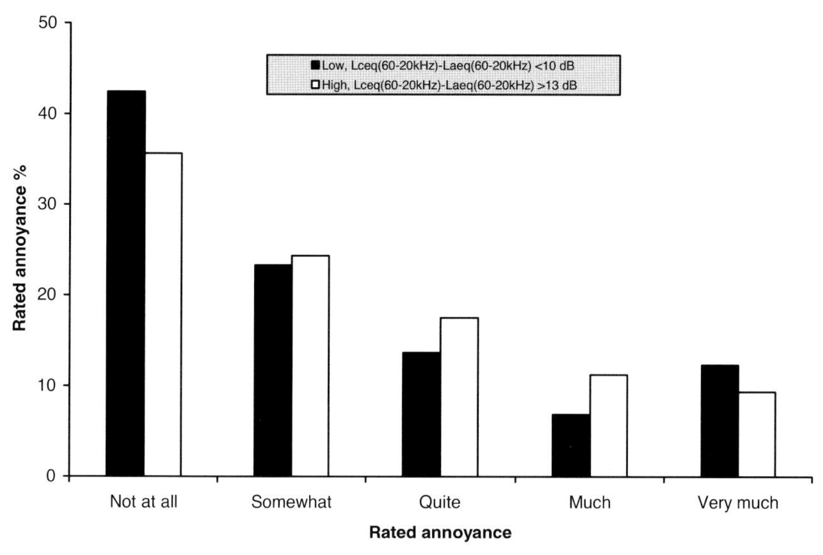

Figure 3. The distribution of reported annoyance between the high- and low LFN-exposed groups based on a $L_{Ceq(63\text{-}20kHz)} - L_{Aeq(63\text{-}20kHz)}$ difference to identify LFN.

DISCUSSION

This field study shows that the noise in 16 of the 22 classrooms investigated are to be considered as LFN using the method cited by the Swedish National Board of Health and Welfare (1995). Further the statistical analysis did not show that students exposed to high LFN levels are reported to be more annoyed than students exposed to low LFN levels when using the $L_{Ceq} - L_{Aeq}$ difference as identify LFN. In the extent of the study, A- and C-weighted levels were calculated for a limited frequency range, 63-20kHz. The idea is a classification of high/ low LFN exposure in a frequency range in which the sound pressure level exceeds the hearing threshold. The theoretical basis is that an annoyance response is based on audible sound. The statistical analysis did not show any difference in reported annoyance between the group exposed to higher LFN levels exposure and the group exposed to lower LFN levels using the $L_{Ceq(63\text{-}20kHz)} - L_{Aeq(63\text{-}20kHz)}$ difference to identify LFN. There si no difference in LAeq exposure between high and low LFN level exposure groups.

On the basis of earlier research there is a problem using this method at low levels. High levels at low frequencies may contribute to a high C-weighted level, yet below the hearing threshold level. If so, the method to identify LFN may overestimate energy in the lower part of the spectra. According to this, Persson Waye (1999) suggests that this method should only be used when the level is above 30 dB(A). There are reasons to believe that a recommendation for the use of this method should be even higher than 30 dB(A). The dominant source of the background sound in these classrooms is ventilation. Yet, it is not a correct assumption that ventilation noise is LFN at all times. The National Board of Health and Welfare (1995) also points out the necessity to move on to the use of other methods as 1/3-octave band analyses to confirm a LFN exposure. The results in this study do not contradict the proposed method for defining the LFN component, neither that this definition is a relevant method for the risk assessment of LFN annoyance. Earlier research raise the question whether the frequency weighting with an A-filter is a correct method when assessing the annoyance response of noise especially when containing strong low frequency noise (LFN) components. There is a continous discussion about the complexity of noise annoyance, the concept of noise reactions and interactions with other factors. Maybe this problem is more complex than a matter of measuring methods. The ratings is based on the total stimuli, noise load as well as personal and social factors. The school environment is very complex and LFN components are just one of many factors that may influence the annoyance response. The possibility can not be excluded that the complexity of the school environment obscured the detection of a LFN – annoyance relationship. Noise exposure problems vary in different school environments due to the presence of different noise sources, speech and other noises related to the student activity. These factors have to be considered asessing the effects of noise annoyance in school environments. The problem with noise annoyance must be looked upon as a whole.

ACKNOWLEDGEMENT

Thanks to students participating and making this study possible. This study was financed by grant from Swedish Council for Work Life Research, Dnr 1997-0070.

REFERENCES

Berglund, B., Hassmén, P. and Job, S. (1996) Sources and effects of low frequency noise. *Journal of Acoustical Society of America*, 99, 2985-3002.

Bronzaft, A. L. and McCarthy, D. P. (1975) The effect of elevated train noise on reading ability. *Environment and Behavior*, 7, 517-527.

Castelo Branco, N. A. A. and Rodriguez, E. (1999) The Vibriacoustic disease – an emerging pathology. *Aviat Space Environ Med*, 70, suppl A1-6.

Cohen, S., Evans, G. W., Krantz, D. S. and Stokols, D. (1980) Psychological, motivational and cognitive effects of aircraft noise on children: Moving from the laboratory to the field. *American Psychologist*, 35, 231-243.

Enmarker, I., Boman, E., Hygge, S. (1998) The effects of noise on memory. In Carter, N. & Job, R.F.S. eds. Noise Effects 98 – Proceeding of the 7:th International Congress on Noise as a Public Health Problem. **1**. 353-356. Sydney, Australia: National Capital Printing ACT.

Holmberg, K. (1997) *Critical noise factors and their relation to annoyance in working environments*. Doctoral Thesis. Division of Environment Technology. Luleå University of technology.

International Standard Organisation (1985) Acoustic. Normal equal-loudness contours for pure tones and normal treshold of hearing under free field listening condition. ISO 226-1. Second edition. Switzerland.

Jones, D.M., Miles, C., Page, J. (1990) Disruption of proofreading by irrelevant speech: Effects of attention, arousal or memory? *Applied Cognitive Psychology*. **4**. 89-108.

Kjellberg, A., Goldstein, M. and Gamberale, F. (1984) An assessment of dB(A) for predicting loudness and annoyance of noise containing low frequency noise. *Journal of Low Frequenzy Noise and Vibration* **3**, 10-16.

Leventhall, H.G. (1980) Annoyance caused by low frequency/ low level noise. In Moller, H. Rubak, P. eds, *Proceedings of the Conference of Low Frequency Noise and Hearing,* Aalborg, Denmark, Pp 113-120.

Pekkarinen, E., Viljainen, V. (1991) Acoustic conditions for speech communication in classrooms. *Journal of Scandinavian Audiology*. **20,** 257-263.

Person Waye, K. and Björkman, M. (1988) Annoyance due to low frequency noise and the use of the dB(A) scale. *Journal of Sound and Vibration*. **127**, 491-497.

Persson Waye, K. (1995) *On the effects of environmental low frequency noise.* Doctorial thesis. University of Gothenburg. Sweden.

Persson Waye, K. (1999) In: *Annoying noise; An outline for criteria documentation.* In Swedish. Summary in English. Arbete och Hälsa 1999:27. National Institute for Working Life. Solna. Sweden.

Swedish National Board of Health and Welfare. (1996) General advise for indoor noise and high sound levels, (SOSFS 1996:7) Stockholm.

Swedish Royal Board of Building. (1992) *Low Frequency noise from ventilation installations.* In Swedish.

Swedish Royal Board of Building. Solna. Sweden. ISBN 91-540-5533-4 Swedish National Board of Occupational Safety and Health. (1997) *Working environment at school; Astudy of comparison 1992-1997*. In swedish. Solna. Swedish National Board of Occupational Safety and Health. Solna. Sweden.

The Swedish National Board of Occupational Safety and Health (1992) *Noise, Statute book of the Swedish National Board of Occupational Safety and Health*. ASF 1992:10. Stockholm.

A Structured approach to LFS-complaints in the Rotterdam region of the Netherlands

Ing. Piet Sloven
DCMR Environmental Protection Agency, Box 843, NL-3100AV Schiedam

ABSTRACT
In the working area of the DCMR Environmental Protection Agency (EPA) an increasing number of people have registered complaints because of low frequency sound (LFS). A 'protocol of low frequency sound' for use in the Rotterdam-area has been developed. The purpose of the approach on LFS is threefold: (1) to offer the several organisations involved a handle to work with, in order to clarify the nature of the LFS complaints and structure them in a procedure; (2) to register their experiences with LFS; (3) to possibly develop a policy on LFS. Conclusions are as follows. (a) The indications lead us to assume that LFS is a growing problem. (b) If the source is not found directly there are complex factors contributing to the LFS-annoyance. (c) Those whose quality of life and health are threatened by LFS need help. (d) The system introduced is a good start. (e) Periodic modifications should lead to an improvement in efficiency. (f) A necessary condition is a consistent, well-equipped and well balanced team of people with experience in handling LFS-complaints. (g) To improve the approach it is worthwhile gathering information from complaint handlers elsewhere.

1. FRAMEWORK

The word 'sound' is used intentionally because most of the complainants do not use the word 'noise'. Hence, LFS instead of LFN. This paper clarifies the backgrounds, the systems used with the protocol, together with a report of experiences and a look into the future.

The Rotterdam region (Rijnmond) and its acoustic climate

The figures between brackets are a comparison with the Netherlands. Rijnmond is about 2700 km^2 (3%), with 1.2 million inhabitants (7%), about 25,000 companies, 200 industrial sites, and some 30% of the inhabitants are noise-annoyed. There are 18 municipalities in the Rotterdam region. Noise is coming from various sources: traffic (road, rail, water, air) and main sites (factories/harbours).

The Environmental Protection Agency (EPA; in Dutch: DCMR) has about 450 employees. One of the responsibilities is the registration and investigation of complaints, reports and incidents. Specialists are on duty 24 hours a day as telephone operators and field inspectors. Most of their incoming information is passed on to the enforcement-sections. They catalogue the information (in most cases on the next day) and arrange follow-up proceedings. The follow-up can consist of a visit to both the companies responsible for the noise annoyance and the complainants. Reports are made of all the activities following a complaint and this is added to the system MIRR [EPA97].

2. LFS-HISTORY IN THE NETHERLANDS, ESPECIALLY IN THE ROTTERDAM REGION

LFS-annoyance is not predicted by the A-weighted levels [VRO, Per] and is not included in legislation, as this aspect was not important enough in relation to other environmental problems. Since then, justice sees the LFS problem as a case with 'probable special sensitivity of the receiver' which means that aggrieved conditions cannot yet be linked to a licence.

In 1995 and 1996, national meetings of the 'interdisciplinary panel LFS' took place, based on experiences of the Monitoring Network for Health and Environment (a non-governmental organisation of citizens/volunteers). The product was a bundle of 125 pieces of work [MGM96]. Several members of the panel have published their summaries.

In 1997, EPA was host to a National LFS-workshop [NSG97]. This was the starting point for a Dutch guideline on LFS. In the same period questions were asked in the Dutch Parliament by persistent LFS complainants. Together with MHS (the Environmental Health department of the Municipal Health Service, in Dutch: GGD/MMk), supported by the Municipality of Rotterdam, EPA started an ad-hoc co-operation to handle the LFS-complaints. Afterwards the City of Rotterdam ordered the EPA to officially handle LFS-complaints and to gather knowledge.

In the meantime, EPA participated in national LFS-studies for health symptoms and the indoor environment, resulting in the NSG-Guideline {NSG99]. In November 1999, a start was made with the establishment of a Dutch platform of approximately 60 LFS-sufferers.

Earlier, in December 1998, the EPA-protocol 'Handling of LFS-complaints in the Rijnmond region' was completed. The LFS-complaints persisted, sometimes accompanied with serious health and social problems, while similar difficulties were recognised in the rest of the Netherlands. Following from our experience, this paper is a clarification of the situation.

3. REASONS TO DEAL WITH LFS-COMPLAINTS

These are the following.

More attention is being given to the effect of nuisance on the environment [e.g. MGM99].

Social-economical motivation

- A district turns into a poor district when people with higher incomes are leaving the area. If people are leaving because of pollution, including LFS, it is bad advertising!
- Health complaints occur after a certain period of time. Waiting for epidemiological results is an unnecessary delay.
- To prevent mentally exhausting juridical procedures.

Social interest

- Sufferers who cannot find an ear for their problems, will seek medical help. Paying special attention to LFS-complainants, reduces their feeling of isolation.

Aid

- Minimise insecurity, dissatisfaction and fear of the sufferers. Complainants get peace of mind, feeling their problem is seriously handled.

Acoustic climate

- From 1980 the spontaneous increase of levels of LAeq has stopped. However, the growing social activities within society are blocking things. Other kinds of noise pollution such as that related to Lmax, LFS, are going to play a bigger role as a result of a growing number of 'special events'.
- Complaints are indicators. In general: the higher the quantity of complaints, the better the quality of the indicator.
- Raising of consciousness regarding LFS in those who are responsible for sources of sound, and civil policy makers as well.

Spatial planning

- Most noise-problems are in fact environmental planning-problems. Keeping a sufficient distance from sources can reduce the problem.
- Concern about the growing use of time and land with noisy activities.

4. DEALING WITH LFS-COMPLAINTS

The EPA-protocol 'Approach to LFS-complaints'

The number of times that most of the acoustical or social workers come into contact with complaints arising from LFS is in fact minimal. In those cases they need support, such as that provided by a protocol, which is outlined in Fig 1.

- receipt of lfs complaint

- administration / forming a team (ad hoc)

---

objectivating complaint

- intake-interview with complainant
- geographical-historical complaints-investigation
- making a survey

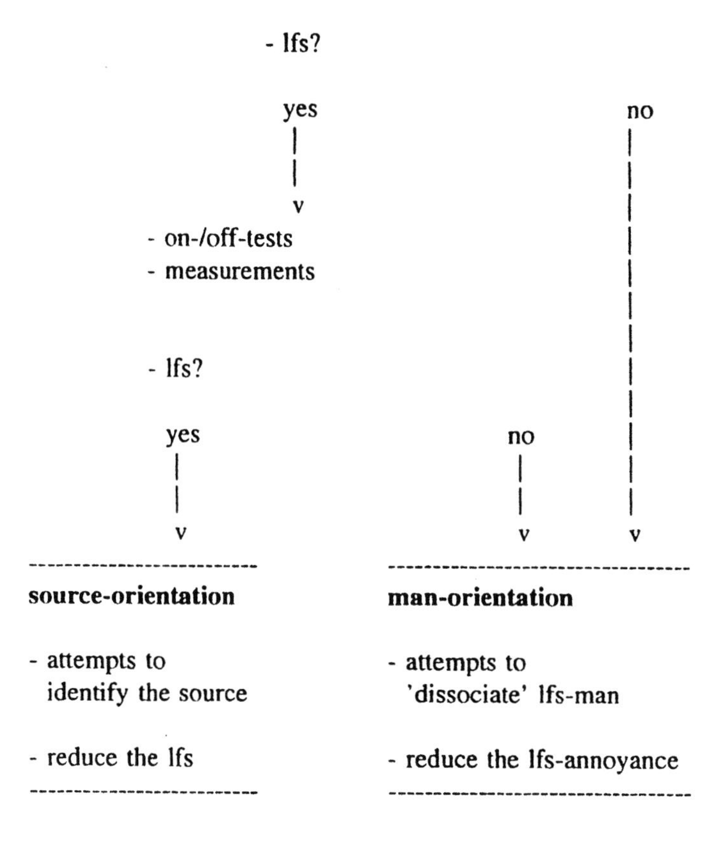

- message to complainant and teammembers
- close file
- complaint dispatched

Figure 1. Main outline of the Protocol 'Dealing with lfs complaints'

Parties involved

The complaints possibly related to LFS come in through different channels, so it is probable that more than one person handles the problem.

- Environmental Protection Agency Rijnmond: complaint-centre employee, enforcement co-ordinator, co-ordinator and manager of the Noise Section are involved. There are 6 EPA-enforcement departments, most of them divided into two groups.
- Environmental Health Department of the Municipal Health Service (MHS). There are four MHS's in the working area 'Rijnmond' of the EPA.
- Police: interim agents (reporters), environmental co-ordinator.
- Community or province (especially in case of written complaints): handling manager. In the working area of the EPA there are 18 municipalities. The municipality of Rotterdam is divided into 10 parts, each with a certain independence.
- Housing association: managing employee, caretaker, social worker. There are many of them. An enormous amount of reorganising and merging is going on.
- Monitoring Network for Health and Environment: the provincial agent. - Dutch Noise Annoyance Foundation (NSG): information-agent.

The collaboration with the Municipal health Service (MHS, psychological and social aspects) and EPA (technical aspects and management) is part of that protocol, together with a streamlining between the different branches of the EPA.

Possible steps to take

The course of action a complainant usually follows is:

EPA-complaint centre, one of the enforcement inspectors, assistant section Noise. That section will contact one of the departments of the MHS's in the area. The protocol guides the different partners in the process of assistance. The protocol is not a plan, but a description of experiences until 1998. It is also not a blueprint, but a guide. To find LFS-sources, the EPA-protocol may help, but it does not specify methods by which to identify them.

Ending a case

It depends upon the degree of involvement whether EPA or MHS end a procedure on their own. Examples of such cases are:

- source found (EPA starts negotiations with the one who is responsible),
- sufferer is a confused person (MHS guides to another kind of help). In many cases EPA and HMS deliberate.

A case may also be ended if:

- the measured sound levels are extremely low, to find a specific source would take very much effort,
- there is no possibility of reducing the sound (e.g. traffic),
- in similar cases there is no prospect of solution,
- LFS is one of the (many) problems of the complainant, but not one of the important ones,
- The complainant refuses (further) co-operation or wants to sell his house.

5. OTHER AIDS BESIDES THE PROTOCOL

The protocol is a good basis but more aid is welcome. We developed the following.

Forms

Several of them support the procedure as a help to gather information (about situation and circumstances) for filing.

- The one-page communication form. Every involved party gets a copy of a communication form, to enable them to see working procedures and who has to be informed. The form gives the complainant, e-mail addresses, telephone numbers/fax numbers and the addresses of all parties.
- The four questionnaires A, B, C and D give a clear picture of the situation. Form A focuses on the basic circumstances. When the EPA-inspector pays the informant a visit after he has investigated the environment 'B is used. Form 'C focuses on the acoustical aspects. Form 'D is personal, meant for the MHS-assistant.

Files of complaints, factories and acoustic climate

To start the process the EPA-complaint centre does geographic-historical investigations. The register of all complaints enables one to find similarities. The EPA also has the vast system 'MIRR' [EPA97] in which most of the information about institutes is filed. It is a help to find sources of noise. The section Noise has rough indications of the acoustic climate of most of the dwellings and what kind of noise (traffic, airport, factories) is dominant. But field research is always necessary, the information mentioned above is just a help.

Complaint follow-up system (CFS) on LFS

About two years ago, the growing number of complaints became too many to be memorized, so a simple system was constructed to keep those who were involved informed. Administrative details are: name and place of informant/complainant, dates of complaining, first action (e.g. house call), in put MHS date of measurements and report, case-code, provision of the evaluation forms. The dates immediately give a survey of the time-scale involved. The characters of the case-code are the entrance to a subsystem, a log in which the main points of the process are kept.

The CFS requires the user to keep it up to date. Otherwise the consequence will be that one has to ask around and to search through files. The CFS contains four categories: A= active, processing (34), W = waiting, for new facts or a sign from the complainant (9), P = predicted (3), X = finished (66). The figures between brackets are the numbers for February 2000. In the year 1999 the section Noise handled about 40 cases. In a quarter of them measurements have taken place. From the LFS-complaints reported through the complaint centre, about a half of them will be put through to the Noise section. Those are inserted in CFS and are probably the most difficult ones.

Measurements

The preparations for measuring and the making of the reports take a lot of time. This is done only if necessary. Sometimes, for certain cases in which measurements are combined with on-off tests, a measuring-scheme is devised. Semi-automatic measurements are also made by the complainer. The circumstances and the facts as to what exactly should be measured are not always clear. If possible we allow time for the measurements and sample exclusively the sounds of interest. Experience is that it takes much more time in the cases where disturbances are traced and sifted after the event. Measurements are assessed in relation to Figs 2 and 3.

We use a list to memorise what machines should be turned off in the house, and also to remind us what to put on again afterwards (refrigerator, alarms, heating system-pump etc.).

The measurements are made at the time that the complainant hears the sound best, very often at night and in the complainer's bedroom. But to gather more information, the measurements also take place at other places in the house.

The files of complainants

The Noise section of the EPA makes work files, titled by the main complainant's name. The work files are ordered geographically. The official aim of this is because the way EPA works geographically. The acoustical aim is to recognize similar sources and cases in the neighbourhood, even if the time between cases is several years.

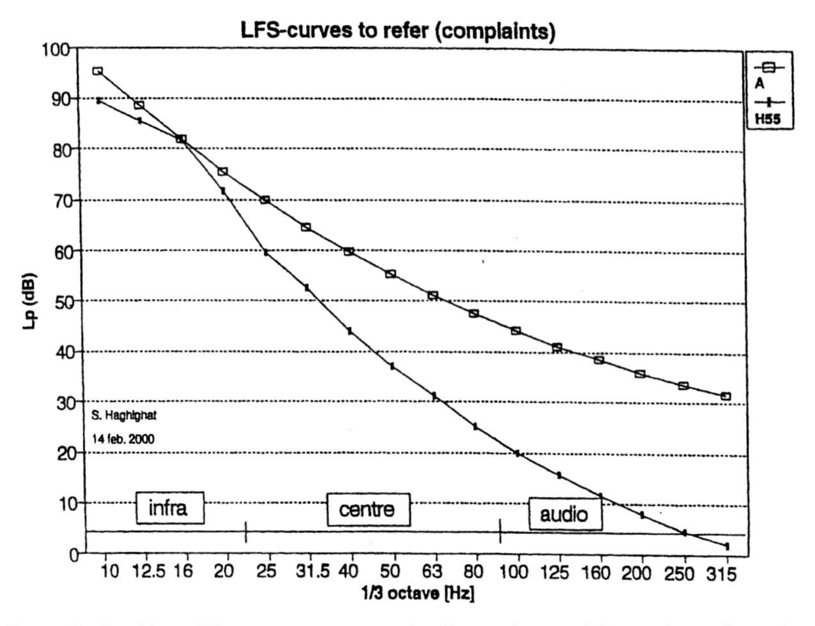

Figure 2. Checking of lfs measurement results (Protocol curves). Legend: see figure 3

Communication

We have developed several ways to communicate with the people dealing with the LFS-problems: complainants, EPA + MHS, other relief workers.

With complainants. Ambition: sincerity, honesty, and - last but not least - clarity. In many cases there is dialogue with MHS. Sometimes it is found that there are personal problems not caused by LFS. To draw attention to the sub-problem and to enlarge it that way is not appropriate. That is a reason to end the complaint handling with a letter, written by EPA and/or MHS, which explains why the investigations have been stopped and what can be done next or instead. In 1999 a brochure was prepared and distributed through town halls, libraries, pharmacies, hospitals and medical doctors in the Rotterdam-region.

Between EPA and MHS. The EPA-section Noise and the Environment Health Department of the MHS have almost daily contacts to deal with one or more LFS-cases. In order to discuss general aspects, lines of policy, changes in the way we work and to keep each other well-informed, the LFS assistants of the Noise-section and assistants of MHS's have periodical meetings on LFS and meet about three times a year.

With other relief workers. The experiences, results of research in literature, information and possible ways to prevent LFS or to reduce the nuisance, and the quantitative state of LFS case-handling are gathered in quarterly reports. This is a way of sharing knowledge, to keep people informed and to use with the general quarterly reports made by EPA.

6. RESULTS

Although in the acoustical view not many cases lead to a solution, there is considerable success in fulfilling the aims formulated under 'Reasons to deal with LFS-complaints'.

The annoyed, the society and the EPA.

Nowadays there is a new attitude, showing respect, and that the attitude, the experiences and the knowledge of the sufferers is an important part in case-handling. It gives them back their self-confidence and enables them to gain perspective on the situation. The EPA really works for the empowerment of citizens. We give them our expert support. The internal co-operation at EPA has improved; other authorities have learned where to find us.

Experiences.

Thanks to the intensive way of dealing with the LFS-problems much useful experience has been gained [e.g. SLO99]. The most important aspects are the following.

Sources of noise

- In many cases no obvious sources were found. Further investigation is not likely.
- Most of the cases don't have specific 'major source'.

Measurements

- Of particular importance in assessing the complaint are tests in which the suspected source is turned on and off. The complainant has to tell the investigator when and what differences are heard. Other simple aids to get any impression about the hearing of the complainant are whispered speaking whilst watching reactions, turning the volume knob of a television or tuner, using sound generators.
- In most cases analyses of third-octave bands from 40Hz upwards are enough. Maybe in future a simple rough quick scan, in octaves, starting with the 31Hz octave, under ideal circumstances, will give enough information to make further decisions.
- Up to now, we judged vibration-measurements necessary in only three LFS cases.

Acoustic LFS-references (indoors, at night)

- The (normally-used) 25 dB(A)-limit, is not sufficient to recognize complaints.
- Nor is the general rule 'LFS, if L(C) - L(A) > 20 dB (German)' [DIN]. If that indicator is exceeded, then the LFS-nuisance is almost certain.

- In cases where the investigators also experience some of the LFS, the A-weighted level was usually 21 dB(A) or higher.
- There is no indication that many cases are dealing with audible tonal sound [PSI].
- The threshold-curves do not explain the LFS-nuisance felt by all the sufferers, but are a help in comparing the measurement results.
- Telling the complainer the percentage of people that can hear his or her measured sound is valuable.

Attitude in way of working

- Involving citizens in the process of watching has the advantage that the dealing-process goes faster. Besides such citizens are keeping you awake.
- Do not trust the certainties of the sufferer, nor your own preconceptions.
- Work systematically.
- Provide documentation: for your own good, and for authorities, other parties involved and sufferers.
- Although the numbers are too small for statistical use, it is obvious that the reporting of complaints until now is not uniform. Per quarter, in which the months of May + June + July are the first quarter, the relation is 6 : 3 : 2 : 3.
- EPA- assistants, are inclined to forget the step 'calling in MHS'. Very often, we made that mistake which then needs to be corrected.

Status of 'assistance' (relief worker)

- Be prepared for resistance. Dealing with LFS-complainants might be seen as dealing with 'losers' and with non-quantifiable results.
- The LFS principle of open dealing results in many and unexpected contacts with both complainants and colleagues. Take enough time to communicate, every day.
- In many cases a combination of aspects complicates the investigation.

LFS-sufferers

- Sufferers are not different from the ordinary-Hollander. Mostly they are normal-hearing alert people; women and the elderly are over-represented. [compare MGM96]
- The EPA-LFS files give following results: (a) average age 55; half of them living alone, (b) 2/3 female, 1/3 male (exactly in accordance with [Gie98]). But: note that a small group in the complaint centre of EPA learns that for all kinds of complaints coming in (20,000/year) also 70% are female and the estimated average age is 55 as well! In this respect the pattern of LFS-complainants is not very special.

- The impression is that the wishes of the LFS-sufferers about quiet are high. They seem to respect and to like silence; in many cases there is a lack of indoor sound.
- The results of other investigators [e.g. Per] are confirmed that in many cases there is a relation between the personal expression of susceptibility and the reactions to LFS.

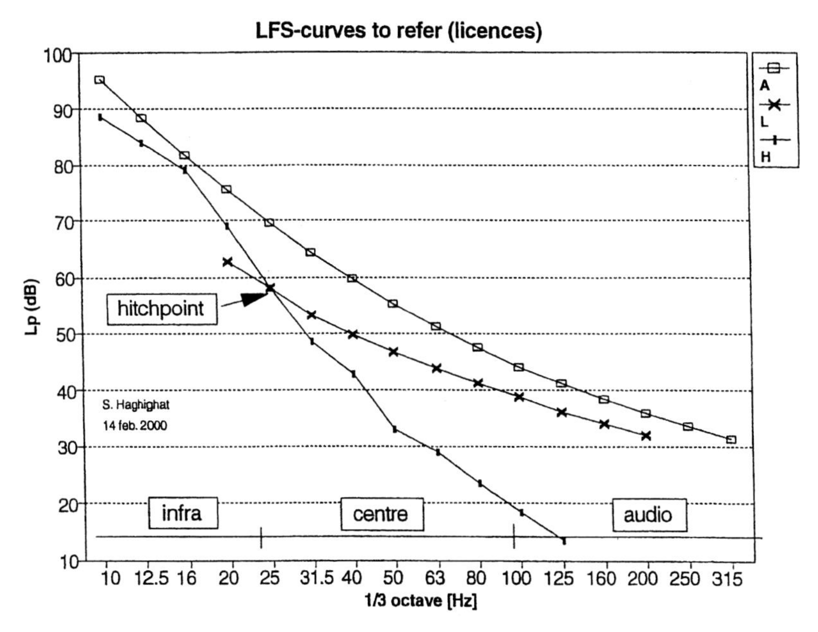

Figure 3. Proposal to use in licences (P. Sloven. 2.2.2000 to be discussed)

- Just in the most aggravating circumstances.
- For the present: situations within homes at night.
- More are less continuous and normal lfs limiting extra annoyance (tonal banging characteristics combination with tangible vibrations).
- To be judged more severe in case of combination with tangible vibrations and in case of lfs with a clear banging character.

Above the x-axis the parts of the lfs-frequencies.
The y-axis: equivalent unweighted 1/3 octave band sound levels.

Lfs-curves intended for use by permit granters [Slo00] and the EPA-Protocol.

A = 25 dB(A); all sound-energy concentrated in one 1/3-octave.

L = Lfs-limit. Above that curve there is excessive annoyance. To use as license-limit from 25 Hz.

H = HTL5 = Hearing Threshold Level that can be experienced by 5% of the average Dutch. To use as license-limit up to 25 Hz.

H55 (figure 2) = Hearing threshold level to be heard by 5% of the most important group of complainants: those who are aged about 55 years. [Pas].

The public spread of information about LFS

- Due to the complexity of complaint handling and our own lack of experience, it is useful to keep the tools used like questionnaires, letters and procedures, up to date.
- In several cases EPA and MHS 'took a risk' and asked publicly for help. There were no resulting floods of complaints.

7. A LOOK IN THE FUTURE

Expectations

In relation to the handling of LFS-complaints, gradual changes will take place.

- In 15 years 1 /3 (1/2 in all Holland) of the Rotterdam inhabitants will be over 55 years of age. At the moment most of the LFS-complainants are in this age group.
- The shared use of buildings for 'industrial' and 'living' purposes is increasing.
- Ventilation systems, heating systems and air-conditioning etc. can cause problems.
- At the moment only 3% of Dutch homes are provided with climate control installations. Those are sources of LFS and the numbers will grow.
- Due to the lack of building space more underground infrastructure is being developed. Vibrations caused by traffic result in LFS via the foundations of residential buildings.
- Within a few years in the Netherlands there will be new legislation for noise and urban planning [Wig97, Wig98]. The municipalities will then have the right to more independence regarding the setting of noise level limits. Due to the prohibitive cost of space this will in some cases result in excessive noise taxation, particularly LFS.

To achieve a higher return from complaint handlers in relation to the resources used, a shortened procedure is in some cases required. The original procedure will only be adhered to if: 1) the informant can substantiate the complaint (eg. witness testimony), or if a quick-scan of the complaint by one of the researchers provides a clear cause in the first instance. 2) a justifiable request by a member of a social assistance agency is received (from housing association to general practitioner).

8. THINGS TO DO IN THE FUTURE TO IMPROVE THE APPROACH

- Continually improving the data on LFS will make it possible to handle the gradually increasing number of complaints more efficiently and shorten processing time without compromising the original aims.
- Development of source-detection methods: sound intensity,

microphone-array, intelligent signal analysis, on-off tests, panel of LFS-sensitive people?
- To find a way to deal with licenses that is acceptable to most of those involved, (authorities, judicial reviewers, companies and permit granters). Such rules must be communicable (relatively simple) and not too severe (business activities must be possible).

More knowledge about house-front insulation, noise-transfer, spread in thresholds of audibility/sensitivity, stress aspects eg serotonin, degrees of annoyance.

ACKNOWLEDGEMENTS
I express my gratitude to my colleagues of the EPA: Jose van Reede and Sian Jones for helping me with the English language and Shahram Haghighat for his production of the figures and his advice in the past.

REFERENCES

DIN(1997). *Messung and Bewertung tieffrequenter Gerauschimmissionen in der Nachbarschaft (Measurement and assessment of residential low frequency sound immission)*. DCMR Environmental Protection Agency, Schiedam, Holland.

EPA97 *MIRR = Milieu-informatiesysteem Regio Rijnmond (Environmental Information system). Contains information of about 25,000 factories in the region of Rotterdam*. DCMR Environmental Protection Agency, Schiedam, Holland.

EPA98 (1998). *Protocol aanpak laagfrequentgeluid* (*Protocol dealing with LFS*. Composers P. Sloven, F. Houtkamp, In Dutch). An example of a case investigation and of the use of the protocol is in Sloven 1999. DCMR Environmental Protection Agency, Schiedam, Holland.

EPA99 (version 13 Sep 1999). *Deltaplan Geluid, figures*. DCMR Environmental Protection Agency, Schiedam, Holland.

Gie98 Gielkens-Sijstermans, C., Coltijn, T. H., Jongmans-Liedekerken, A. W. (1998). *Gevoeligheid (...) (Sensitivity for LFS; a study into possible factors)*. GGD Limburg; in Dutch. Jongmans:@knmg.nl

MGM96 Meldpuntennetwerk Gezondheid en Milieu: lfg-coded papers of the interdisciplinary LFS-group 1995+1996. Summary: G.P. van den Berg (1996). (...) *informatiebundel* (*LFS and nuisance*; in Dutch) Health and Environmental Monitoring Network Foundation, www.ecomarkt.nl/sgm.

MGM99 Monitoring Network for Health and Environment. (1999). Statement. NGO preparatory WHO. Health and Environmental Monitoring Network Foundation. www.ecomarkt.nl/sgm.

NSG97 NSG. (1997). *Laagfrequent geluid; verslag van een workshop (LFS; report of a workshop)* Nov 13 1996 Netherlands Foundation for Noise Abatement (NSG) Postbox 381 2600 AJ Delft, Netherlands.

NSG99 NSG. (1999). *NSG-Richtlijn laagfrequent geluid (guide LFS)*. Also: J. Kramer, in Geluid April 1999; in Dutch. Netherlands Foundation for Noise Abatement (NSG) Postbox 381, 2600 AJ Delft, Netherlands.

Pas Passchier-Vermeer, W. (1998). *Beoordeling laagfrequent geluid in woningen* (Assessment of lfs in dwellings). TNO-rapport 98.028; in Dutch.

Per Persson Waye, K. (1995). *On the effects of environmental low frequency noise*. Abstract of thesis. Goteborg University.

PSI This is not in accordance with the declaration of G.P van den Berg c.s., 1999: 'Stil geluid': lfg in Woningen Report NWU-83 in Dutch.

Slo95 Sloven, P.A. (1995). Milieu en RO op lokaal niveau: houd afstand! (Environment and spatial planning: keep distance). In : *ROM-magazine* (Sept); in Dutch.

S1o99 Sloven, P.A., Soede, W. (1999). *Toetsing in praktijk. Meer duidelijkheid door Protocol en Richtlijn Lfg? (A test in practice. More clearness by Protocol and Guideline LFS?)*. In "Geluid" (June); in Dutch.

S1o00 Sloven, P.A. (2000). *LRA, LFS in rooms, assessment.* Proposal to the Dutch (license) SESAM-system. (Curve 'L' goes at higher frequencies in the direction of curve 'A', in dependence of the loudness). DCMR Environmental Protection Agency, Schiedam, Holland.

VRO Min. *Housing, Spatial Planning and the Environment.* (1988). Laagfrequent geluid; een literatuurstudie (LFS; literatur). GF-HR-01-04 (writers Heringa, Vercammen (Peutz)).

Wig97 Wiggers N.K.J., Sloven, P.A. (1997). Geluid gedereguleerd en de hinder ontspoord (Noise deregulated and the nuisance derailed). In: *Geluid* (March); in Dutch.

Wig98 Wiggers N.K.J., Sloven, P.A. (1998). MIG mag. (contributions to the deregulation of noise) In: *Geluid* (December); in Dutch. Discussion on new national legislation. Municipalities will become more free to decide what is permitted.

Annoyance of low frequency noise and traffic noise

Frank Rysgaard Qistdorff and Torben Poulsen
Department of Acoustic Technology, Building 352, Technical University of Denmark, DK-2800 Lyngby, Denmark, E-mail dat@dat.dtu.dk Web: www.dat.dtu.dk

ABSTRACT
The annoyance of different low frequency noise sources was determined and compared to the annoyance from traffic noise. Twenty-two subjects participated in laboratory listening tests. The sounds were presented by loudspeakers in a listening room and the spectra of the low frequency noises were dominated by the frequency range 10 Hz to 200 Hz. Pure tone hearing thresholds down to 31 Hz were also measured. Eighteen normal hearing subjects and four subjects with special low-frequency problems participated in the tests.

INTRODUCTION
Complaints against low frequency noise sources in the environment have increased in recent years and in some cases the complaints cannot be verified by objective noise measurements. Typical low frequency noise sources are power plants, high-speed ferries, ventilating noise, etc.

In order to investigate the annoyance from low frequency noise sources, and compare this annoyance to the annoyance from traffic noise, The Danish Environmental Protection Agency initiated the preset investigation. At the time of writing the investigation is not finalized and the data presented here should thus be seen as preliminary results.

LISTENING TESTS
Listening tests were performed in a listening room where digitally recorded signals were presented over loudspeakers. The test subjects listened one at a time to eight different sound signals, each presented at three different levels. After each presentation a questionnaire was presented to the test subject. A written instruction was given to the subjects and the subjects could ask questions about the procedure throughout the tests. Information about the sound signals was only given after all the tests were finalized. A full training session was performed before the final tests were initiated.

Test signals
Seven low frequency test signals and one traffic noise signal were used: The low frequency signals were sounds from a gas turbine, a high speed

ferry, a steel factory, a generator, a compressor, a drop forge (transmitted through the ground) a music from a discotheque. The traffic noise was a recording from a busy motorway. The signals were recorded on disk and filtered to simulate an indoor listening situation. The duration of all the signals was two minutes. The signals were presented at L_{Aeq} levels of 20 dB, 27.5 dB and 35 dB. The sound signals were presented in a random order and were presented twice to the test subjects.

Test subjects

Eighteen normal hearing test subjects participated. Pure tone audiometry was performed in the frequency range 125 Hz to 8000 Hz with a Madsen Midimate 602 audiometer equipped with Sennheiser HAD 200 earphones. The audiometer was calibrated according to the values given in Han and Poulsen (1998). Hearing threshold levels at or below 15 dB HL were accepted in the frequency range 125 Hz to 4 kHz and a hearing threshold level of 20 dB at a single frequency (incl. 8 kHz) was also accepted. Besides the normal hearing subjects, four test subjects with special low frequency problems participated in the investigation. Pure tone audiometry for these subjects was also performed but no specific limits were set for these persons.

Listening room

The dimensions of the listening room were 7.52 x 4.75 x 2.76 m. (L x W x H). The room fulfilled the recommendations given in IEC 268-13 (1983). A detailed description of the equipment and the presentation system may be found in Mortensen (1999). The sounds were reproduced by means of two KEF 105 loudspeakers and two Amadeus Sub subwoofers. The loudspeakers were hidden by a light curtain. The test subject was seated at a position where only one natural room-mode is dominant, i.e. a 45.5 Hz resonance. All signals were filtered to compensate for the resonance at this frequency.

The subject's task

After each presentation the subject filled in a questionnaire with four questions: 1) How loud is the sound? 2) How annoying is the sound if it should be heard at home during the day and in the evening? 3) How annoying is the sound if it should be heard at home during the night? Below each of these questions a horizontal line was given (length 10 cm, left end named 'not annoying', right end named 'very annoying') and the subject responded by making a mark on the line. The fourth question was 4) Is this noise annoying? And the response was given by a mark in either a 'yes' or a 'no' box.

Low frequency hearing threshold

A determination of the pure tone hearing threshold was performed with a Two Alternative Forced Choice method with 800 ms sound signal durations. The procedure determines the 79.4% point of the psychometric function. A detailed description of the procedure may be found in Buus et al. (1997). The threshold was determined at 31 Hz, 50 Hz, 80 Hz and 125 Hz. A computer controlled Tucker Davis system with Sennheiser HAD 200 earphones was used for these threshold determinations.

RESULTS AND DISCUSSION

The results presented here are the raw average data and should be regarded as preliminary. No analysis has been performed yet. The average annoyance as a function of presentation level is shown in figure 1. Each point represents the average annoyance from eighteen test subjects. The left panel shows the response from the question about the day and evening situation. The right panel shows the response from the question about the night situation. It is seen that the annoyance increases with increasing L_{Aeq} level. It is also seen that the annoyance generally is evaluated somewhat higher at night than during day-evening. It is also seen that the Drop Forge is the most annoying sound followed by Music and that these two sounds together with the Compressor may form one group during day/evening and the other sounds form a less annoying group. A detailed statistical analysis will show whether this grouping can be confirmed.

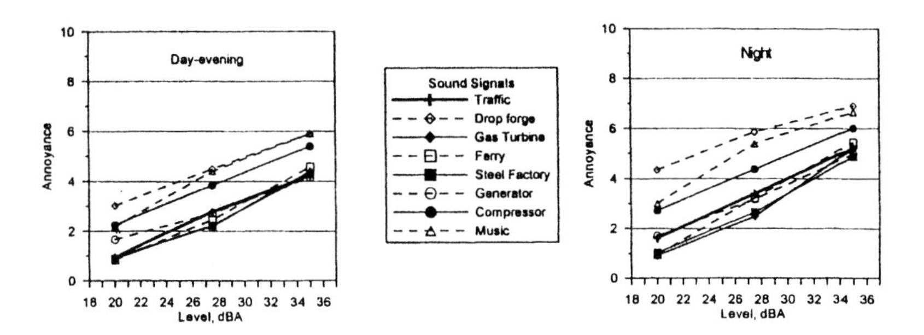

Figure 1. Perceived annoyance of seven different low frequency sounds. Each point represents the average over eighteen normal hearing subjects. Left panel: Annoyance during day and evening. Right panel: Annoyance during night. The annoyance is given as the position of the subjects' mark on the 10 cm questionnaire line. Annoyance of traffic noise is represented by the heavy solid line.

In Figure 2 the average annoyance perceived by the two subjects groups are shown for the day-evening and for the night situation. It is seen that the special group generally evaluates the sounds more annoying than the normal hearing group. This difference between the groups is more pronounced at night than during day/evening. At the time of writing a statistical analysis has not yet been performed.

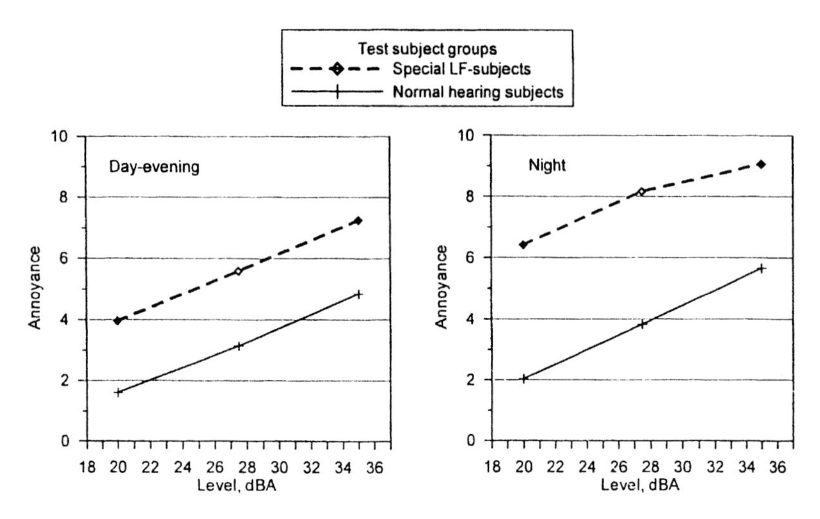

Figure 2. Comparison of annoyance perceived by eighteen normal hearing subjects and four subjects with special low frequency problems. Average over subjects and over all sound signals.

Hearing thresholds

The average audiogram (125 Hz to 8 kHz) is shown in Figure 3 for both subject groups. It is seen that a high frequency hearing loss is found in the special group. The low frequency hearing threshold (31 Hz to 125 Hz) was measured for both subject groups, but as no audiometric calibration values exist in this frequency range, the measured values have been compared to the hearing threshold values given in ISO 389-7

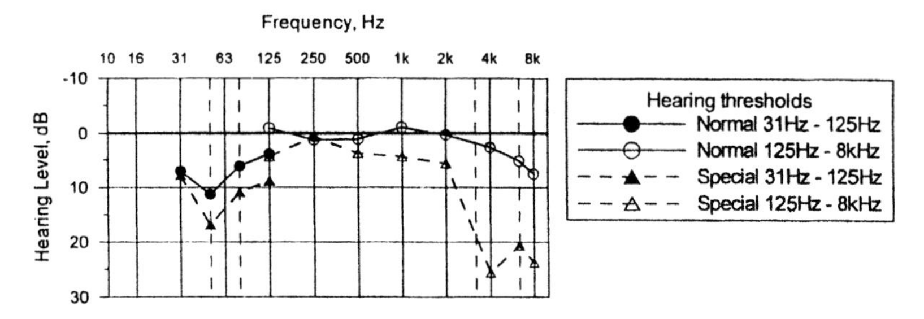

Figure 3. Average pure tone threshold for the group of normal hearing subjects and for the group of subjects with special low frequency problems. The curves in the frequency range 125 Hz to 8 kHz is calculated from audiometric data. The curves in the frequency range 31 Hz to 125 Hz is based on measurements in dB SPL and referenced to ISO 389-7.

(1996). This 'audiogram' (based on sound pressure levels) is also shown in Figure 3. It is seen that the low frequency hearing threshold for the special group is not better than for the normal group.

Due to limitations in the equipment it was not possible to measure the hearing threshold at frequencies below 31 Hz. At the lower frequencies a hissing noise became audible at high presentation levels and this would result in erroneous results.

ACKNOWLEDGEMENTS

The present investigation is supported by The Danish Environmental Protection Agency. Mr. Jorgen Jakobsen is greatly acknowledged for ideas and discussions in the planning phase of the investigation. The sound signals were made available by DELTA Acoustics and Vibration, DK-2800 Lyngby. Tina Hjordt helped in providing the group of test subjects with special low frequency problems.

REFERENCES

Buus, S., Florentine, M. and Poulsen, T. (1997) Temporal integration of loudness and the form of the loudness function, *J. Acoust. Soc. Am.* 1997; 101: 669–680.

Han A.L. and Poulsen T. (1998) Equivalent Threshold Sound Pressure Levels (ETSPL) for Sennheiser HDA 200 earphone and Etymotic Research ER-2 insert earphone in the frequency range 125 Hz to 16 kHz. *Scand Audiol* 1998; 27: 105–112.

IEC 268-13 (1985): *Sound system equipment.* Part 13: Listening tests on loudspeakers.

ISO 389-7 (1996): *Acoustics-Reference zero for the calibration of audiometric equipment-Part 7*: Reference threshold of hearing under free-field and diffuse-field listening conditions. International Organization for Standardization. Geneva, Switzerland.

Mortensen, F.R. (1999) *Subjective evaluation of noise from neighbours with focus on low frequencies.* Main project. Publication no. 53, Department of Acoustic Technology, Technical University of Denmark, DK-2800 Lyngby. ISSN 1395–5985.

Perceptions of the public of low frequency noise

D.M.J.P. Manley[1], P. Styles[2] and J. Scott[3]
[1]*Independent Consultant, 58 Godiva Road, Leominster, Herefordshire HR6 8UQ, UK*
[2]*School of Earth Sciences and Geography, Keele University, Staffordshire ST5 SBG, UK*
[3]*5 Garthdee Terrace, Aberdeen, ABIO 7JE, UK*

ABSTRACT
This paper describes the results of a series of measurements using a variety of systems comprising vibration sensors and sound-level measuring devices. These were carried out at the homes of people experiencing the effects of low frequency noise and in other environments where low-frequency noise was likely to be generated. Suggestions are made as to the possible reasons for the perception of low-frequency noise by sufferers.

INTRODUCTION
About 500 people have complained to Local Authorities concerning Low Frequency Noise (LFN). Many times more than this figure are suspected to be being annoyed by LFN and often do not know who to complain to or when seeking medical advice are not recognised as sufferers of the effects of LFN. Often the effects they complain of are mistaken for tinnitus, but it has been established that tinnitus is clearly definable and is an unrelated issue.

A programme of research was started in the early 1990s to find out:

1 why people are affected by LFN;
2 the capabilities of state-of-the-art instrumentation for measuring LFN;
3 what can be done, if anything, for LFN sufferers.

PERCEPTION OF LFN
Sound Level Meters (SLM) have been designed to measure noise in the range 20 – 20,000 Hertz. In order to arrive at a criterion of equal loudness, one has to compensate for the variable sensitivity of the human ear to different frequencies.

Figure 1 shows equal loudness curves. It can be seen that at low frequencies the ear becomes less sensitive and so an SLM has to compensate for this drop in sensitivity. This compensation is called "weighting" and in the United Kingdom A-weighting is extensively used. Figure 2 shows the weighting curves and compares 'A' weighting with 'C' weighting, a scale considered by many international workers in the field as being more appropriate for use in evaluating the level of Low frequency sound. Following the 'C'-weighting curve, it can be seen that the human being is recognised as being more sensitive to LFN than under "A"-weighted evaluation. Experience of sufferers confirm this sensitivity and yet the "C"-weighting formula which would more properly recognise this is not being adopted.

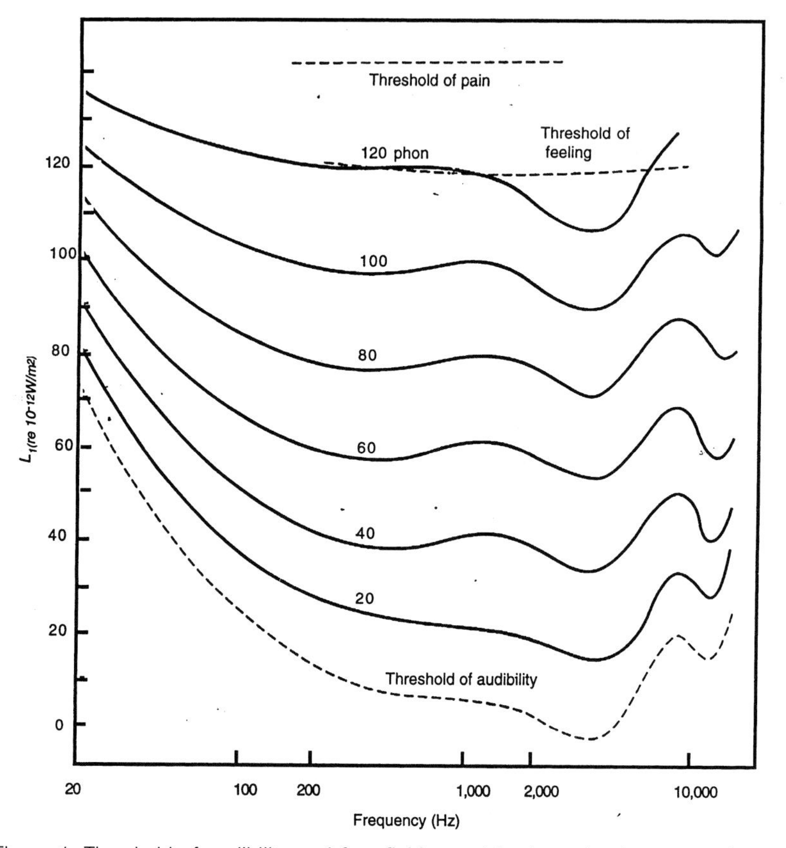

Figure 1. Threshold of audibility and free field, equal loudness level contours for pure tones with subject facing source.

MEASURING LFN

Most people can gain access to SLMs that are now designed to work through computers. Although there are general models that will provide frequency analysis, it appears there are considerable variations in the lowest frequencies that can be detected. This paper describes and discusses systems that the authors have used in the presence of LFN sufferers and which have shown a correlation with inception of irritation and change in level.

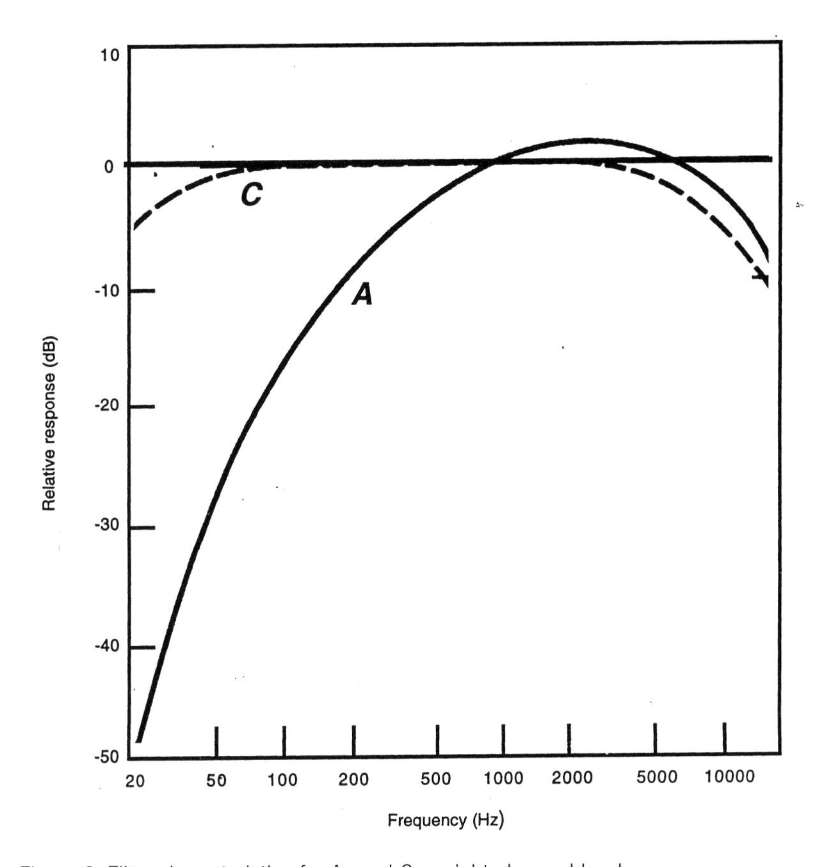

Figure 2. Filter characteristics for A- and C- weighted sound levels

THE SYSTEMS

A number of measuring systems were used during this study. They are described below and significant points about their design and operation are noted. Examples of the results obtained from each of the systems are also presented.

System A

Geosense PS 1 Geophone.

This system only detects ground-borne vibration and has been used by the authors in many private dwellings. It appears that sufferers of LF noise pick up peaks in the 5-20Hz range at levels of about 10 dB below the annoyance criterion in the work-place. This is not surprising when we consider that it is night-time when the subject is trying to sleep that the LF annoyance factors persist.

A graphic example of the variation in results between quiet and noisy occasions as perceived by a sufferer who lived near a busy port and who

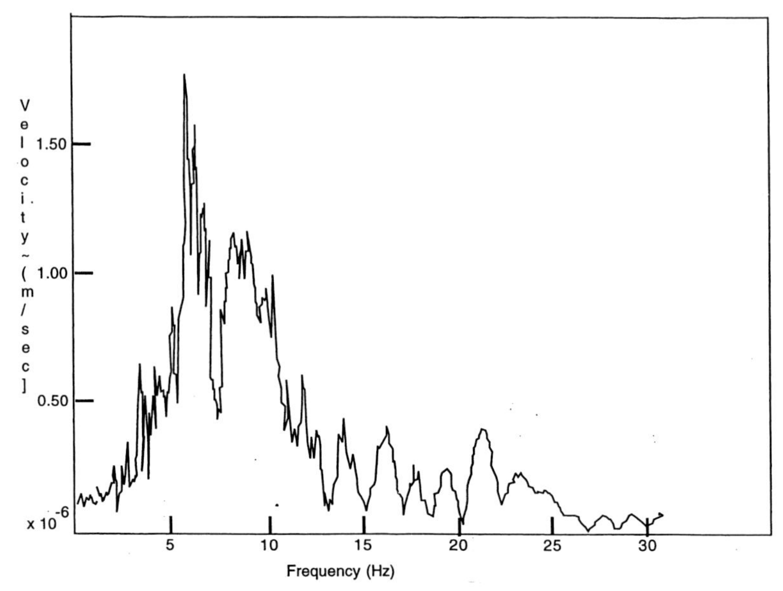

Figure 3a. Quiet occasion as perceived by sufferer – Frequency spectrum

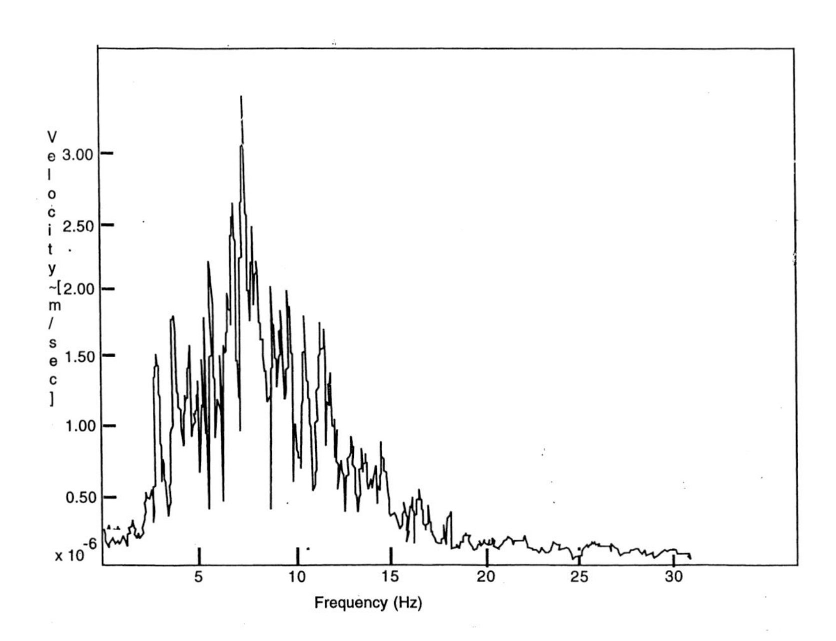

Figure 3b. Noisy occasion as perceived by sufferer – Frequency spectrum

complained of being awakened in the early hours of the morning is shown in figure 3. Here, the PS 1 Geophone equipment was linked to a lap-top computer, set up to enable the equipment to be activated by the sufferer:

a when he went to bed; and
b when he woke up in the night.

The results show quiet peaks of $1.6x10^{-6}$ ms^{-1} at 6 and 6.5Hz in figs 3a and 3b, with noisy peaks of 4 x 10^{-6} ms^{-1} at 7.5 Hz. The noisy peaks are thus double the amplitude of the quiet ones, and whilst it is accepted that these values are in themselves not especially high, they nevertheless correlate with the experience of the sufferer. This experiment was repeated some 12 times with similar results.

System B

An airborne analyser using the microphone channel of a VIBROSOUND 6-channel, 24-bit digital recording system (made by MAGUS Electronics, Sandbach, UK) fed from the output of a Bruel & Kjaer 2231 Sound Level Meter fitted with a microphone that has a flat response down to 1Hz.

Although designed for seismic work, the VIBROSOUND can perform FFT analysis on output (unweighted) signals from the Bruel & Kjaer 2231. Particularly good results using this system have been obtained by the University of Liverpool. Those shown in figure 4 were recorded close to a wind farm where there had been reports from sufferers. The results show a large number of peaks below 20Hz ranging from 20 to 70 dB in the linear unweighted setting. If "C" weighting is applied, peaks at about 4Hz would be 53 dB, some 8 dB above the recommended limit of 45 dB.

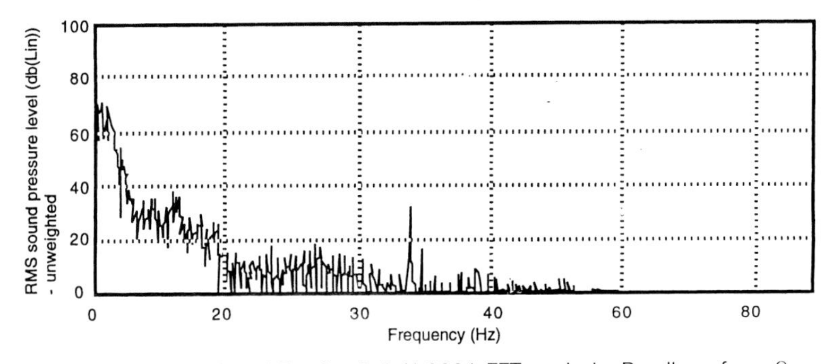

Figure 4. 200 metres from LF noise B & K 2231 FFT analysis. Readings from Caravan Site.

System C

LENNARTZ low-frequency, three component digital trial seismometer to detect vibration with the output fed to the VIBROSOUND instrument for recording and later analysis and display.

This very sensitive seismometer shows LF levels with much greater resolution in tones than System A. A typical output from it is shown in figure 5. Here a peak at 5Hz of nearly 200 nm/sec velocity can be seen.

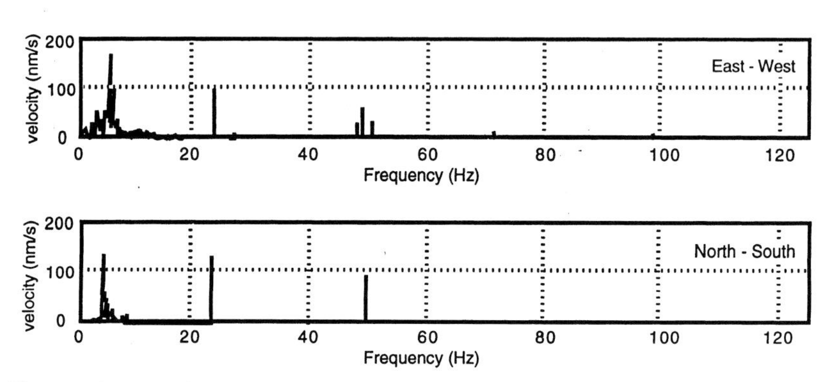

Figure 5. Lennartz Seismometer University of Liverpool

SYSTEM D

CEL 441 Sound Level Meter, working directly into an ONO SOKKI FFT analyser.

It is generally accepted that a tonal sinusoidal signal must have about 100 cycles before it can be recognised by an analogue detector as a clear peak with a "Q" factor of 40dB. For a 4 Hz signal, this represents a sample time of 25 seconds. The ONO SOKKI system used only has a sampling time of 1 second and therefore, as seen, has a resolving power acceptable only above 20 – 30 Hz, whereas for 100 repetitions the frequency would have to be greater than 100 Hz. Therefore, the ONO SOKKI can only adequately resolve part of the total spectrum of the incident low-frequency signals.

Figure 6 shows a typical graph from an investigation of a factory in a rural residential area in Shropshire on a clear, windless winter's day. The factory used a number of compressors and other equipment producing LFN in an environment where there was not any other such noise. There are peaks between 20 and 90Hz that are similar in both graphs.

From the character of the two curves, there is no doubt that the noise from these machines can be detected more than a kilometre away and indeed this caused disturbance to the sufferer.

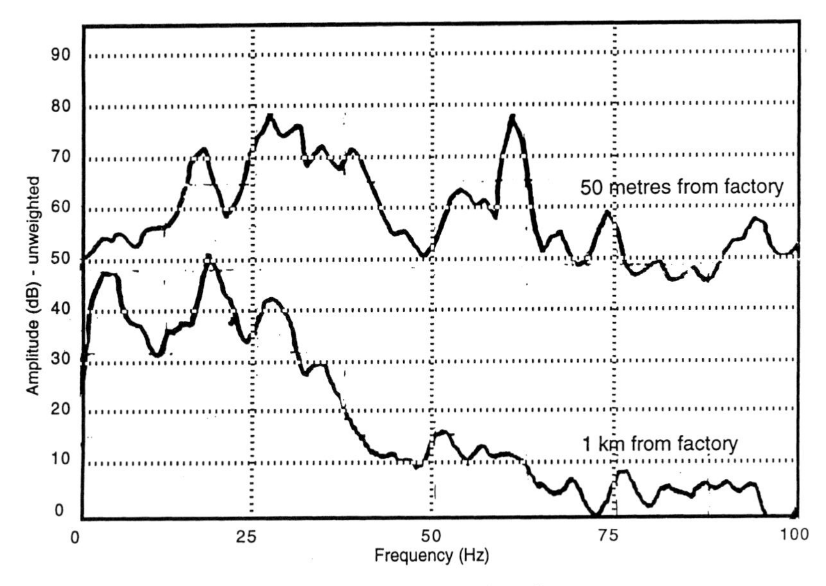

Figure 6. Average of 32 readings airborne LF noise from factory

SYSTEM E

FFT analyser developed for the British Aircraft Corporation and the Ministry of Defence (UK) which displays the analysis in bands of 0.55 Hz width.

This system is unique, in that it:

- uses a fixed band-width of 0.55 Hz from 1 Hz upwards;
- records changes every seven seconds (being the sample time).

Observations over a long period of time showed that through this detection system the peaks varied in amplitude, changing every few seconds. This may well be due to the lack of narrowing of discrete frequency detection bands causing adjacent signals to interfere with each other.

Indeed, in figure 7, the results show that there is much interference between tonal signals at LF and seem to indicate multi-path propagation from a distance and phase changes when the signals mix. It also suggests that the signals are air-borne and that the changes in phase are due to temperature effects on sound velocity and through refraction.

System F

Small accelerometer fixed to a window pane which acts as a vibration sensor for signals induced in the pane by the airborne signals. Its output is recorded on the VIBROSOUND for subsequent processing and analysis.

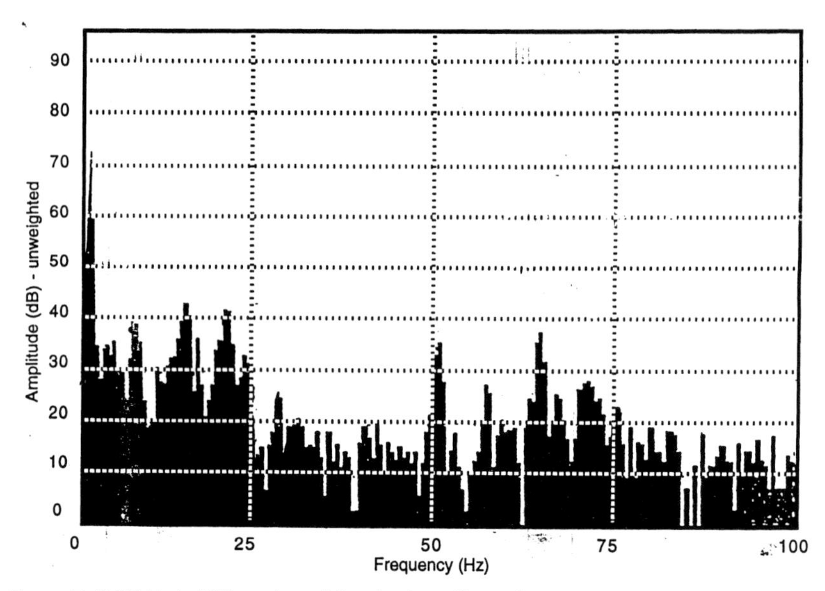

Figure 7. 0.55 Hertz FFT analyser LF noise in sufferers house

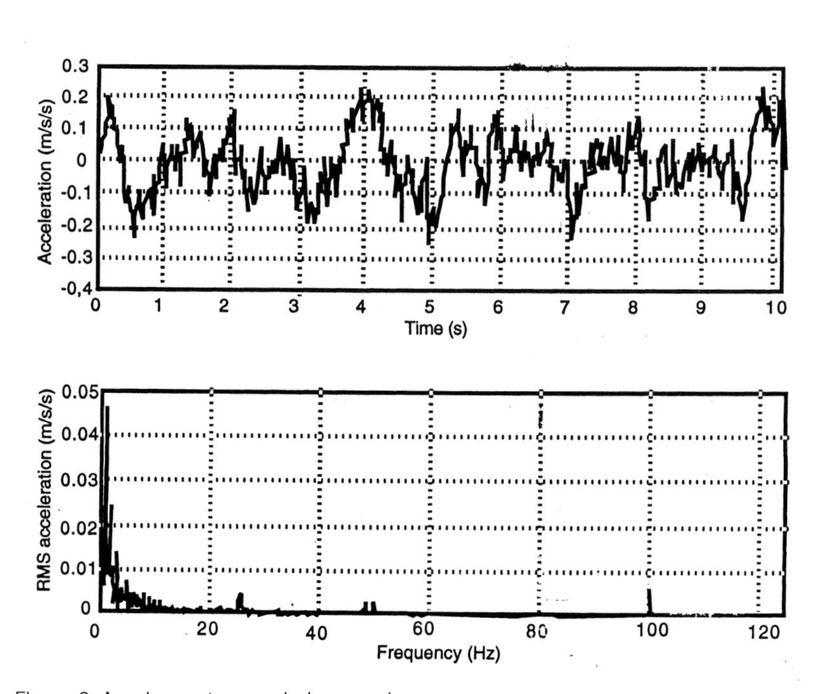

Figure 8. Accelerometer on window – noisy area

The results presented in figure 8 show an acceleration of 0.01m/sec^2 which corresponds to a sound pressure level of about 60 dB (unweighted). This corresponds to readings taken by the SLMs. However, it is unlikely, because the acoustic power is in the order of microwatts, that this would give enough energy to produce the sound pressure levels measured in the room. What is also significant is that there is little variation in sound pressure levels throughout the rooms. This therefore suggests that vibrations of the whole house may be occurring. These could be set up by a low-frequency airborne wave, or by vibration through the ground.

APPRAISAL OF THE SYSTEMS AND THEIR RESULTS

All of the systems show activity in the 5 – 20 Hz region, which is below the usual tonal LF threshold of the human ear. From correlation of the readings with the experience of sufferers there is no doubt that there is a link between tonal activity above a certain level and the effects experienced by some sufferers. It also seems clear that this is not necessarily connected with absolute amplitude, rather more the tonal character of the sound and its relationship with other frequencies in the ambient sound levels at the time.

From the research undertaken it is evident that the levels of both ground-borne and air-borne sound which cause disturbance are lower in amplitude than originally thought to be troublesome or to be detectable by people. In part this seems to confirm that it is not simply the direct effect on the body of the amplitude of the LF sound present, but the way the brain perceives and interprets it and the meaning and reaction which it attaches to it.

The difficulty is that while the cause may be real, the precise frequencies and levels at which individuals are affected may vary from person to person, with perhaps only a few percent of the population able to detect them. Nevertheless with rising levels of LF sound it is not unreasonable to speculate that in future a greater proportion of the population will be troubled as the sounds increasingly breach the thresholds of detection and perception and intrude into the individuals consciousness.

RESOLVING POWER OF THE BRAIN

Anecdotal evidence and reports collected from LFN sufferers would appear to suggest that the most efficient device for detecting Low Frequency noise may well be the human brain. This is borne out be careful questioning of LFN sufferers. A very important point which emerges from discussion with those experiencing the effects, is that, often the sufferer is unaware of LFN until that person has been in the acoustic field for several hours or days. In fact, the sufferer more than often becomes sensitised and hence cannot voluntarily exclude or ignore the noise source. The output of the brain's analysis of infrasound is usually through the generation of heard effects, but visual disturbances have also been reported and it has been suggested that some apparently paranormal sightings are in fact caused by infrasound.

For future work, some thought should be given to the use of devices which have very long sampling times, say many hours, so as to simulate the brain. Based on the discussions we have had with Sufferers, and the ability of the brain to store, process and produce subjective Psych-Acoustic effects, we have no doubt that the memory of the brain is able to hold acoustic information for several hours. The expansion of this issue is proposed to be the subject of a future paper.

CONCLUSIONS

Several different experimental systems have been used variably to detect, process and display LF signals presented as:

- vibrations in the structures, or in the foundations of various buildings, due to the impact of airborne or ground-borne waves, respectively;
- sound level pressure changes due to incident airborne waves, or secondary airborne waves generated by the vibration of the structures of the buildings.

From this work, it is clear that further investigations are needed to:

- find out why there are multiple LF sources;
- better understand the respective roles of ground-borne waves and air-borne waves.
- the significance of particular tonal frequencies, such as 8 Hz and 16Hz present in many of the investigations.

ACKNOWLEDGEMENTS

The authors would like to thank members of Liverpool University Staff, especially I. Rushforth and Dr A. Moorhouse. Thanks are also due to everyone who took part in the experiments and for allowing their homes to be used for such purposes.

REFERENCES

1. *Measurement of Low-frequency Sound, Environmental Law*, DIN 45680 ICS D.140.01.Briefing sheet, Low-frequency noise.

2. UK Noise Association, *Measurements of Environmental Low frequency noise*, Dec 2001.

3. Prof. V.V. Krylov, *Proc. Inst. Acoustics*, Vol. 18, Part 1, pp 21-28, 1986.

4. Andersen and Møller, Equal Annoyance Contours for Infrasonic Frequencies, *Journal of Low Frequency Noise and Vibration*, Vol. 3, No. 3, 1984.

5. N. Broner and H. G. Leventhall, The Annoyance and Unacceptability of Lower Level Low-Frequency, Noise, *Journal of Low Frequency Noise and Vibration*, Vol. 3, Nr. 4, 1984.

6. Henrik Møller, Annoyance of Audible Infrasound, *Journal of Low Frequency Noise and Vibration*, Vol. 6, Nr. 1, 1987.

7. Hinton, J., *Birmingham Noise Map* DETR Report, February 2000.

8. Rushforth, I., Styles. P., and Moorhouse, A., Industrially induced Low-Frequency Noise & Vibration in residential buildings, in *Proc of 'Acoustics 2000'*, University of Liverpool, April 2000.

A questionnaire survey of complaints of infrasound and low-frequency noise

Henrik Møller and Morten Lydolf*
Department of Acoustics, Aalborg University, Fredrik Bajers Vej 7-B4, 9220 Aalborg Ø, Denmark. hm@acoustics.aau.dk
**now at Bang and Olufsen, Struer, Denmark. mlf@bang-olufsen.dk*

ABSTRACT
A survey of complaints about infrasound and low frequency noise has been carried out. 198 persons reported about their troubles in a questionnaire. Their verbal reports often describe *a "deep and humming or rumbling sound, like coming from a distant idling engine of a truck or pump"*. Nearly all respondents report of a sensory perception of a sound. In general they report that they perceive the sound with their ears, but many mention also a perception of vibrations, either in their body or of external objects. The sound disturbs and irritates during most activities, and many consider its mere presence as a torment to them. Many of the respondents report on secondary effects, such as insomnia, headache and palpitation, which they associate with the sound mainly because it occurs at the same place. In a majority of the cases, only a single or few persons can hear the sound, but there are also examples, where it is claimed to be audible to everybody. Typically, measurements have shown that existing limits (and hearing thresholds) are not exceeded. The investigation leaves the key question: Are the troubles induced by an external sound or not, and if they are, which frequencies and levels are involved? The feasibility of a study of this is supported by the results.

INTRODUCTION

For many years there have occasionally been cases where people complain about infrasound or low-frequency noise. This is the case in Denmark, and the situation seems to be comparable in many other countries. Most descriptions mention a deep humming sound in the home of the complainant, which annoys and disturbs sleep, rest and concentration. In addition, the sound is often claimed to cause an impaired quality of life due to headache, pain, stress, and other kinds of trouble, including severe worries of being exposed to a 'mysterious sound'.

Typically, the sound is only perceived by a single person and not the entire household. For this reason, it is often taken for granted that the trouble cannot be induced by an external, physical sound. As a consequence, in most cases no action is taken, and the complainant is left alone with his or her problem. Many of the annoyed persons find this situation unacceptable, and in Denmark some of these have organized themselves in a society, "Infralydens Fjender" ("Enemies of Infrasound"). The society puts a constant pressure on the authorities by repeatedly bringing up their problem, e.g., in the daily press.

A disturbing issue is the widespread misunderstanding that infrasound is inaudible for humans, because the frequency components are placed below the claimed *'audible frequency range'* from 20 Hz to 20 kHz. Although it was shown at least as early as in the 1930'es that infrasound can be perceived, when only the sound pressure level is sufficiently high ([1], [2], [3]), this misunderstanding still exists, even among professionals. As a consequence, the mere mentioning of the word *infrasound* brings up associations to 'inaudible sound' that can hardly be taken seriously.

Official initiatives in Denmark

In 1995 the Danish Environmental Protection Agency arranged noise measurements in some selected cases. The measurements usually showed sound pressure levels well below or, at the highest, around the normal hearing threshold for low and infrasonic frequencies, a fact that added to the skepticism toward the complainants. The hypothesis was put forward that they might suffer from a special low frequency tinnitus, but this was never tested, thus neither confirmed.

In 1997 the Environmental Protection Agency issued an information report on low frequency noise, infrasound and vibrations [4]. The report recommends that the indoor noise in dwellings should not exceed 85 dB(G) for the infrasound and 20 dB(A) for the low frequency noise (10-160 Hz).

In Figure 1 the recommended limits are shown together with the hearing threshold standardized in ISO 389-7 [5], and for the lowest frequencies as measured by Watanabe and Møller [6]. For frequencies below 20 Hz the limits ensure a sound pressure level approximately 10 dB lower than the average hearing threshold. Going toward higher frequencies, the limit passes the average threshold around 30 Hz, and a level 10 dB above the average threshold is reached around 70 Hz.[1]

These limits appear quite reasonable, provided that they are used with measurements that truly represent the human exposure. On the other hand, it seems that in most of those cases that initiated the information report, measured levels are below the limits, and the report apparently stopped further examination of these cases.

[1]The information report [4] states that the limit is 10 dB below the average hearing threshold up to 40 Hz. As seen in the figure, this is not correct. The reason for the disagreement is that the report uses an 'average hearing threshold', which deviates significantly from the standardized hearing threshold in the 25-50 Hz frequency range

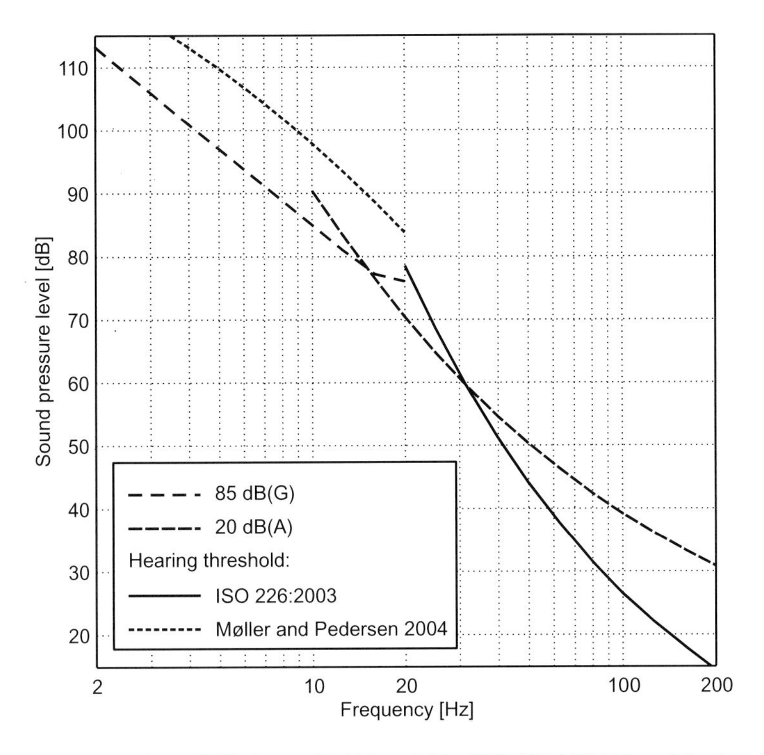

Figure 1. Limits of 85 dB(G) (up to 20 Hz) and 20 dB(A) (10-160 Hz) and the hearing threshold (as standardized in ISO 389-7 [5] and as proposed by Møller and Pedersen [6])

Present study

The survey presented in this chapter was meant to give a better understanding of the trouble experienced by individual complainants. In addition, and – depending on the results – it was thought to possibly serve as background material for planning and seeking of funds for a more thorough investigation of a group of complainants. It was the intention to collect in a systematical way information from a number of cases, and in this way to clarify, 1) whether the troubles experienced by different people are similar and 2) what they are, 3) whether there are reasons to believe that the troubles are induced by physical noise or not, 4) where and when the troubles occur, 5) whether there are problems all over the country, and 6) what has been done to solve the problems. It was not the intention to investigate systematically the extent of claimed low frequency noise problems – except that the survey would reveal, if there were only few isolated cases in Denmark. The present chapter summarizes some important results of the survey. Results have been updated with data from questionnaires received after the initial conference publication [7]. A full report is available in Danish [8].

DESIGN AND DISTRIBUTION OF THE QUESTIONNAIRE

The questionnaire was printed on nine sheets of A4 paper and included an instruction and 45 numbered questions. It was prepared in such a way that the annoyed person could fill it out directly, or a family member or case officer could do it, e.g., via an interview. The cover letter recommended that the annoyed person did it personally. In all cases, name and address of the annoyed person were registered. Most of the questions were structured in a multiple-choice form. A few questions required text to be entered.

Instructions

The respondents were encouraged to add comments in the large margins of the sheets, if the multiple-choice possibilities did not offer the relevant answer. It was pointed out that they were allowed to abstain from answering some of the questions, and that it was legal to give more than one answer in a question if appropriate. For these reasons the percentages of answers in a multiple-choice list will not necessarily sum up to 100%.

Depending on the situation and the answers given, some of the 45 questions would be irrelevant for some people. For this reason the respondents were sometimes told to skip questions and go to a subsequent question, depending on the answers already given. Some people were obviously too eager in answering the questions and did not make the correct jumps. These were kindly asked to fill out a new questionnaire, unless the error could be rectified in the data processing without any risk of misinterpretation.

Distribution

Questionnaires were sent to civic and regional environmental administrations throughout the country, to the secretariat of "Infralydens Fjender" and to a number of acoustic consultants in Denmark. It was furthermore available in PDF-format from the internet homepage of the Department of Acoustics, Aalborg University. People were encouraged to copy and distribute it freely.

Because of the distribution form, it is not known how many copies that were actually distributed, and the responses cannot be used to estimate the number of annoyed persons, the geographical distribution of the problems, or any similar statistics. The responses must simply be taken as examples of cases where a person experiences some kind of trouble, which he or she believes is caused by low frequency noise or infrasound.

202 questionnaires were returned, most of these within the first months following the launch of the campaign in August 1998. 4 persons did not respond to a request of clarification in connection with incorrect jumps, thus leaving 198 responses for analysis.

RESULTS AND DISCUSSION

Nearly all questionnaires were filled out by the annoyed person and only a few by a family member or a case officer. The respondents were between 14 and 86 years of age with a mean of 55.7 years.

About two thirds of the respondents were female and one third were male. The only well established evidence of women having a better hearing than men, is at high frequencies, where the impairment of hearing with age differs between genders (ISO 7029 [9]). Even though the similarity of hearing between genders has not been fully confirmed at very low frequencies, the difference in number of respondents is more likely caused by social or psychosocial reasons.

Questionnaires were received from all over the country. Large and small cities as well as the countryside were represented. The density of responses was clearly higher in the region close to the secretariat of "Infralydens Fjender" than in other regions, since 31.3% of the responses were from that county, and the county covers only 6.8% of the population in Denmark. This might indicate more problems in this region, but more likely it demonstrates the society's success in using the press to make people aware of the survey (and of the problem).

Individuals' description of the sound

In the first question about the noise, the persons were asked to describe the sound in their own words, and eight blank lines were left for this purpose. Most of the respondents tried eagerly to give a detailed description of the sound. Naturally, there is a large variety in the answers but some expressions are frequently used, such as *"the sound...."*

"....is a deep humming/rumbling sound",
"....is constant and unpleasant",
"....creates a pressure in the ears",
"....affects the whole body",
"....sounds like coming from a large (idle running) engine of a truck, pump, ferry or aircraft",
"....is coming from somewhere far away, outdoor, and may be transmitted through the ground".

Many persons are apparently not able to localize the sound source directly. Therefore they make a number of speculations as to what the source may be. The impression of the source being far away and outside the house might be caused by lack of midrange and high frequencies. Then our common experience from sound transmission through walls and over long distances could create the illusion of a distant source, even if the sound is actually generated nearby.

Where and when?

In one question the persons were asked where they experience trouble from the sound. The responses in terms of statistical frequencies are shown in Figure 2. It is seen that nearly all of the persons indicate indoors in their home, either all over the home (81.8%) or at particular places (16.7%). Furthermore it is seen that troubles are experienced not only inside buildings, but also sometimes outside. Only few problems are seen at the job. Many people added margin comments on extra details, such as where in the home the sound is most intense, what their experience is at other places etc.

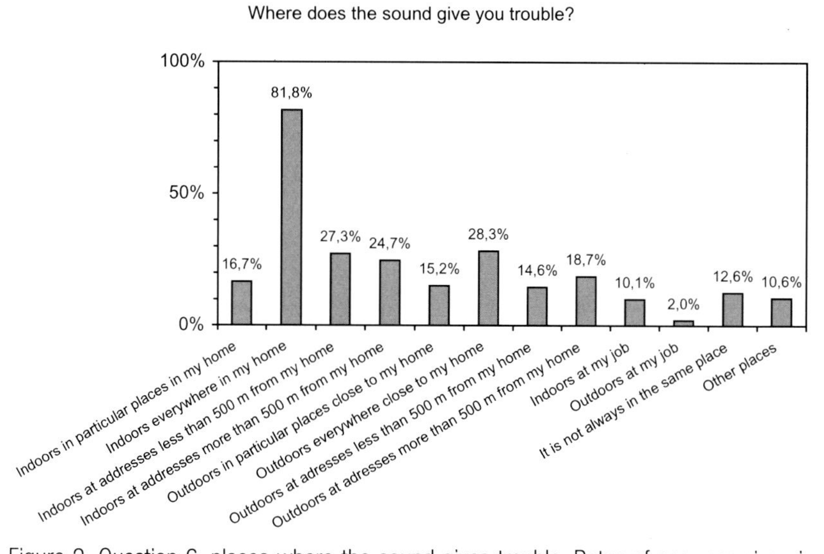

Figure 2. Question 6, places where the sound gives trouble. Rates of answers given in percentage of all respondents.

In another question the persons were asked which time of the day the trouble occurs. The answers were almost equally distributed between day, evening and night, however with a small preponderance in the nighttime (22:00-7:00). A vast majority marked two or three of the three given intervals.

Is there a sensory perception?

As mentioned, it has often been argued that some of the complainants might not actually *hear a sound*, but rather feel some general unpleasantness and put the blame on sound, only because of rumors about strange effects of infrasound and low frequency noise. In one question the persons were asked, whether they perceive the sound directly with their senses. In order not to bias the persons toward reporting of a false sensory perception, the wording of the question and the possible answers were carefully selected in order to make it perfectly 'legal' and not in any way doubtful to admit that the sound was not directly perceived.

The results of this question are given in Figure 3. It is seen that nearly all persons (92.9%) report that they hear a sound with their ears. Some persons (16.2%) report of a sensation in the ears but not as that of a sound. 98.0% answered one or both of the two first categories. Thus, nearly all respondents have a sensory perception related to the ears. Many have a sensation of vibrations, either in their body (43.9%) or of objects around them (28.8%).

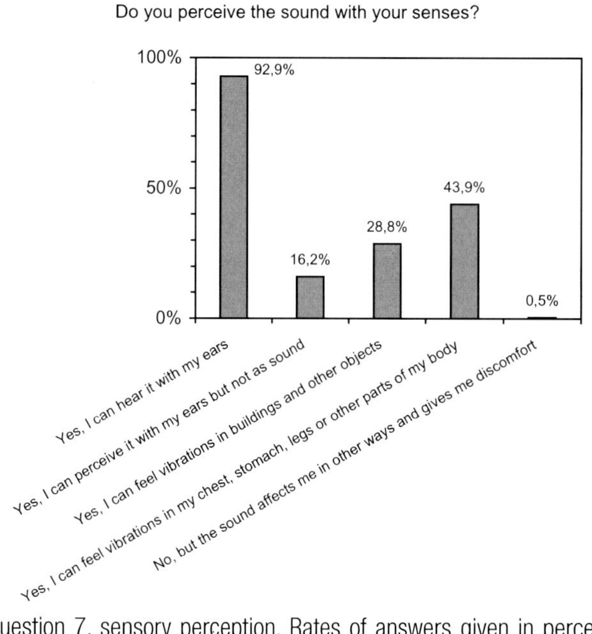

Figure 3. Question 7, sensory perception. Rates of answers given in percentage of all respondents.

Only 0.5% (a single person) did not report of a direct sensory perception. This person reported insomnia and headache, and as a reason for blaming infrasound or low frequency sound, the person reported that he or she had heard or read that it might be the reason.

In one question the persons were asked how long time they have to be in the sound before the trouble starts. Results from this question are given in Figure 4. Obviously, the trouble starts very soon for most of the persons, as 62.6% indicate "immediately" and 24.2% state "within a few minutes" (a few persons reported both of these answers, 83.3% answered at least one of them). The immediate occurrence of the trouble corresponds well with the fact that many of the troubles are connected directly to the sensory perception.

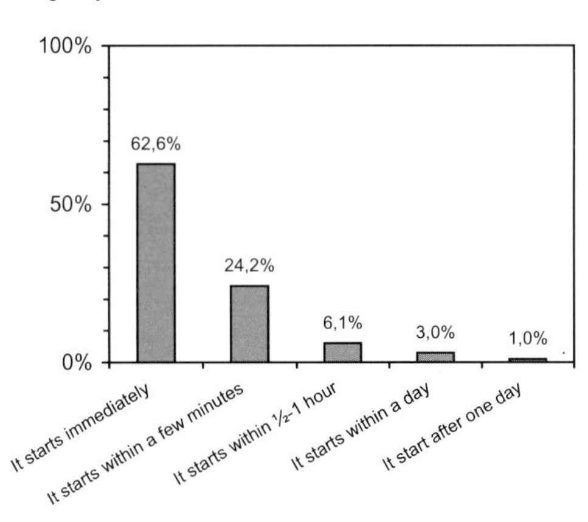

Figure 4. Question 28, time before trouble starts. Rates of answers given in percentage of all respondents.

Do other people hear the sound?

The persons with a reported sensory perception (i.e. all persons except one) were asked whether other people are able to perceive the sound as well. The results from this question are shown in Figure 5. A group of 38.1% reported that he or she is the only person who can hear it, while 28.9% indicated that a few persons can hear it. Only 14.2% indicated that the sound is audible to everybody.

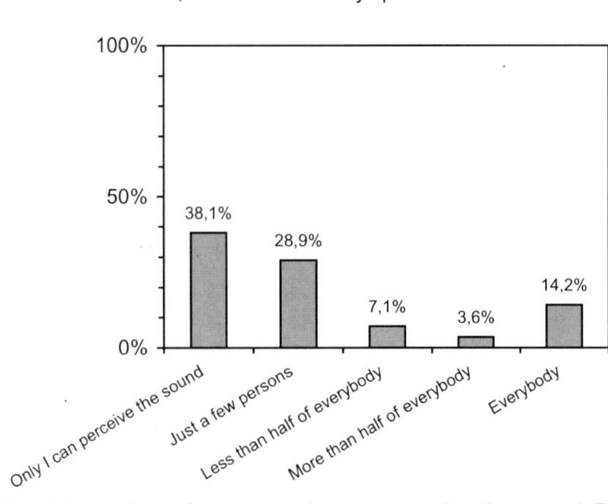

Figure 5. Question 14, number of persons who can perceive the sound. Rates of answers given in percentage of respondents with claimed sensory perception.

Some persons added extra information about exactly who can hear the sound, or mentioned that he or she lives alone and does not have visitors very often. In such cases there may be a bias in the answers, since more persons than indicated might be able to hear the sound, if only other people were being exposed to it.

In another question the persons were asked, whether other people had mentioned the sound without being made aware of it. This had happened in 34.5% of the cases.

Type of effects

The persons with a sensory perception were asked which kinds of trouble that are related to the sound. The question was split up into troubles directly related to the perception, and secondary effects, i.e., other kinds of trouble, which they believe are induced by the noise.

The answers from the question on troubles that are directly related to the perception are seen in Figure 6. A majority of the persons reported on problems like being disturbed when falling asleep or when reading, frequently paying attention to or being irritated by the sound, and being awakened from sleep. 76.1% consider the mere presence of the sound as a torment to them. An example from the "Others" category is *"pressure in the ears"*.

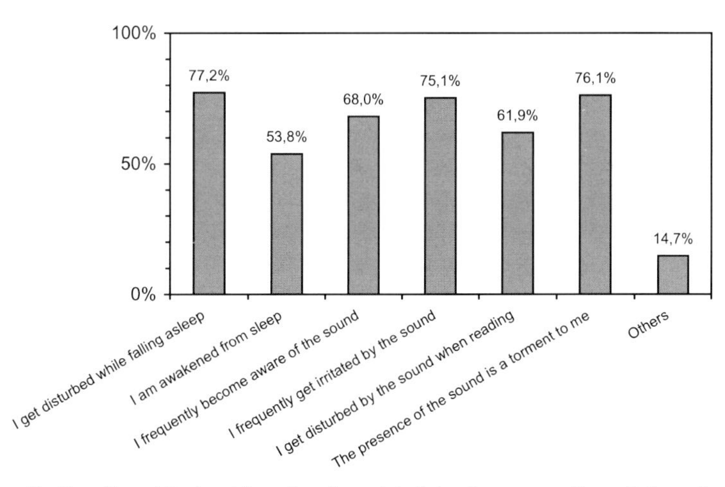

Figure 6. Question 16, troubles directly related to the perception. Rates given in percentage of respondents with a claimed sensory perception.

The answers concerning secondary effects are seen in Figure 7. The highest rates (close to 70%) occur for insomnia and lack of concentration, problems that are nearly directly related to the perception, and which were more or less reported already in response to the question on this. As examples of truly 'secondary' effects, many reported dizziness, headaches and palpitation. Examples from the "Others" category are stress, aggression, restlessness, nausea, fatigue, increased tension in muscles, and weak nerves.

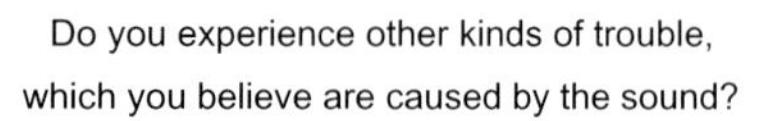

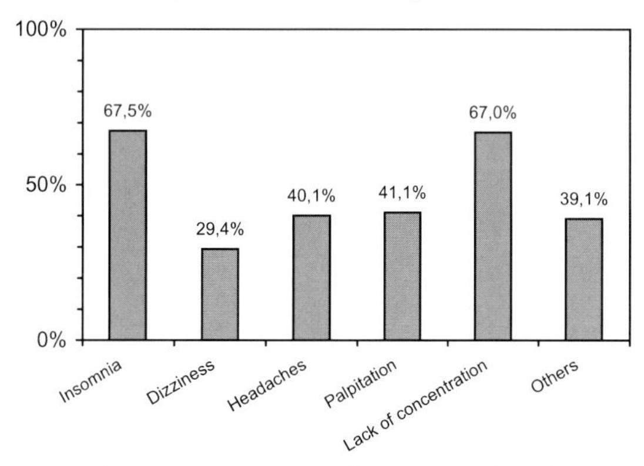

Figure 7. Question 17, troubles that are not directly related to the sensory perception. Rates of answers given in percentage of respondents with a claimed sensory perception.

Those persons who indicated secondary effects were asked, why they believe that infrasound or low frequency noise is responsible. 72.9% of them relate the secondary effects to the sound because it appears at the same place. Quite a few (36.2%) indicate that they have heard or read that the trouble they experience may be induced by sound.

Attempts to improve the situation

In one question the persons were asked what they have done in order to solve or relieve the trouble. Many have tried to use earplugs at night (62.9%) or during the day (34.0%), most often without any effect. 8.1% have moved to another house, and 48.7% consider doing it. 40.6% have consulted their general practitioner or a specialist, and 17.3% take medicine.

Complaints to authorities

64.6% of the responding persons have complained to the authorities about the noise. In 14.8% of these cases the complaint was rejected immediately. In 60.1% an official person has visited the complainant or an address in the neighborhood in order to evaluate the situation.

Noise measurements have been made in 48.4% of the cases in which an official complaint was filed, vibration measurements in 15.6%.

Typically, measurements did not reveal anything that was expected to give rise to problems (or be audible), and existing limits were usually not exceeded. (This refers to the explanations given by the annoyed persons; the authors have not had the opportunity to study the original measurement reports). Measurement difficulties are frequently reported, e.g., because of background noise or insufficient equipment. Some of the persons have expressed their distrust in the measurements and the limits.

Only 7.8% of those, who have complained to the authorities, indicate that their problem has been solved or partly solved. However, in an investigation like the present, there will be a natural bias toward a low number of persons for whom the problems have been solved, since these persons will be less motivated for filling out a questionnaire than those who still have a problem.

Are the troubles caused by a physical sound?

As mentioned, when measurements are made, only very low levels of low frequency and infrasonic noise are usually seen. The levels suggest that the sound would be inaudible or at least so soft that no complaints could be expected. It is a fact, though, that our knowledge of low frequency hearing is based on a few investigations with a limited number of subjects, and it cannot be excluded that there are individuals with a much better hearing at these frequencies, or an otherwise deviating hearing function, e.g., an unusually steep rise of loudness above the threshold. Møller and Pedersen [6] gathered a few cases of especially sensitive persons from the literature. If such extraordinary individual sensitivity is the reason for the troubles, it may not justify a lowering of the general limits, but a better understanding might lead to tools and solutions that could solve or relieve the trouble in specific cases. It is characteristic for many cases that the annoyed person, or even an alleged 'noise polluter', is willing to pay for a solution, if he or she only knew what to do.

The crucial question is, whether the trouble is induced by an external sound field or not, and if it is, which frequencies and levels are responsible. The authors have often been tempted to investigate in detail a few selected cases, e.g., with blind tests in the laboratory using recordings from the complainant's homes. However, we have refrained from doing this, since we imagine that there may be a variety of reasons for the complaints, and there would be a high risk of making wrong conclusions from a very limited and insufficient investigation.

The authors give a high priority to a detailed examination of a larger number of cases, and as seen below, a such investigation will follow. We are well aware that the investigation might show that external sound is responsible only in few or even none of the cases. Even that would be a valuable result, though, since it would pave the way for a constructive search for other possible reasons for the complaints. The uncertainty which is still connected to the matter has irrational consequences, e.g., power plants and factories being accused of 'polluting' entire regions with noise, worries about effects of sound based on pure speculation, worries that house prices will go down in 'polluted' areas etc. There are even examples of local authorities who have abstained from investigating straightforward cases of noise complaints – with rather loud noise that would be annoying for everybody – by referring to the difficulty in handling of low frequency noise problems!

SUMMARY AND CONCLUSION

The 198 respondents experience troubles mainly in and around their homes. Their verbal reports often describe *"a deep and humming or rumbling sound, like coming from a distant idling engine of a truck or pump"*. Nearly all respondents report of a sensory perception from the sound. In general they perceive it with their ears, but many have also a perception of vibrations, either in their body or of external objects. The sound disturbs and irritates during most activities, and many consider its mere presence as a torment to them. Many of the respondents report on troubles that are not directly related to the perception of the sound, e.g., insomnia, headache and palpitation, but which they associate with the sound, because they occur at the same place as the sound. In a majority of the cases, only a single or few persons can hear the sound, but there are also examples where it is claimed to be audible to everybody.

There are respondents from all over the country, however with a preponderance in the area where "Infralydens Fjender" has been particularly active. There are more women than men among the respondents. Many of the respondents have complained to the authorities, but most often this has not led to a solution. Typically, measurements have shown that existing limits are not exceeded. Sometimes authorities have rejected cases immediately without any investigation. The study is most likely biased toward having unsolved cases, since people with solved problems are less motivated for submitting a questionnaire.

Because of the simple distribution form of the questionnaire, the result of the investigation cannot be used to estimate the extent of infrasound and low frequency noise problems in the country, but the cases must be regarded as examples only. However, it can safely be concluded that there are more than just a few people in a small region, who experience various kinds of trouble, which they believe are caused by infrasound or low frequency sound.

The investigation has not proven that the troubles are due to external sound, but the fact that most of the respondents report that they perceive the sound with their senses motivates a further investigation, and it facilitates the design of blind tests.

FUTURE INVESTIGATION

On this background a continuation of the project was planned and initiated. 22 cases have been randomly selected for a detailed investigation. Sound measurements and calibrated recordings have been made at the places of the claimed exposure. Each recording will be played back in the laboratory to the actual complainant, using a pattern of blind tests to see whether the sound can be heard and recognized. Also, playback of filtered recordings is planned in order to encircle the frequencies responsible for the troubles. The playback will take place in a newly updated laboratory at Aalborg University [10], thus taking advantage of exposure facilities, which cover both the infrasonic and the low frequency range. Furthermore, all complainants will undergo a general medical check and detailed audiological and vestibular examinations, including examinations at low and infrasonic frequencies. The investigation matches well an investigation, which was planned in 1995 by a National Board of Health group of general physicians, epidemiologists, audiologists and engineers, but which was never carried out. Results are expected in 2006.

REFERENCES

[1] G. A. Brecher, "Die untere Hör- und Tongrenze", *Pflügers Arch. ges. Physiol.*, Vol. 280, 380, 1934.

[2] G. von Békésy, "Über die Hörschwelle und Fühlgrenze langsamer sinusförmiger Luftdruckschwankungen", *Ann. Physik*, Vol. 26, pp. 554-566, 1936.

[3] E. G. Wever and C. W. Bray, "The Perception of Low Tones and the Resonance-Volley Theory", *J. Psych.*, Vol. 3, 101-114, 1936.

[4] "*Orientering fra Miljøstyrelsen – Lavfrekvent støj, infralyd og vibrationer i eksternt miljø*", (in Danish), Orientering fra Miljøstyrelsen (Information from the Environmental Protection Agency), Nr. 9, 1997.

[5] ISO 389-7:1996, *Acoustics – Reference zero for the calibration of audiometric equipment – Part 7: Reference threshold of hearing under free-field and diffuse-field listening conditions*, International Organization for Standardization, Geneva, 1996.

[6] H. Møller and C. S. Pedersen, "Hearing at low and infrasonic frequencies", *Noise and Health*, Vol. 6, No. 24, pp. 37-57, 2004.

[7] Henrik Møller and Morten Lydolf, "Complaints of infrasound and low-frequency noise studied with questionnaires", *Proceedings of 9th*

International Meeting on Low Frequency Noise and Vibration, Aalborg, 17-19 May 2000, pp. 129-138.

[8] Henrik Møller and Morten Lydolf, *"En spørgeskemaundersøgelse af klager over infralyd og lavfrekvent støj – endelig rapport"* (in Danish), Department of Acoustics, Aalborg University, 134 pages, ISBN 87-90834-30-5, December 2002.

[9] ISO 7029:2000, *"Acoustics – Statistical distribution of hearing thresholds as a function of age"*, International Organization for Standardization, Geneva, 2000.

[10] Arturo Orozco Santillan and Morten Lydolf, "Low frequency test chamber with loudspeaker arrays for human exposure to simulated free-field conditions", *Proceedings of 10th International Meeting on Low Frequency Noise and Vibration and its Control*, 11-

Assessments of low frequency noise complaints among the local Environmental Health Authorities and a follow-up study 14 years later

Johanna Bengtsson, Kerstin Persson Waye
Department of Environmental Medicine, Göteborg University, Medicinaregutan 16, 41390 Göteborg, SWEDEN.
johanna.bengtsson@envmed.gu.se. kerstin.persson@envmed.gu.se

ABSTRACT
Interviews among a selection of 37 of the 289 Swedish local Environmental Health Authorities (EHA) were undertaken in order to assess the occurrence of complaints on low frequency noise. The study also aimed to evaluate whether the specific guidelines on low frequency noise, adopted in 1996, were used and how they performed when assessing low frequency noise. The results showed that most complaints of low frequency noise were due to noise from fan- and ventilation installations, amplified music, compressors and laundryrooms. According to 46% of the EHA, complaints due to low frequency noise had increased during the last two years, while the same percentage reported no change. When assessing low frequency noise, 62% of the EHA reported that the specific guidelines on low frequency noise based on third octave band analysis performed better or much better compared to the previous A-weighted guideline, and only one EHA thought it performed worse.

1. INTRODUCTION
In estimating the importance of low frequency noise as an environmental health problem a fundamental question is: How large is the problem, or more specifically how many people are negatively affected by low frequency noise? For transportation noise the number of exposed people is rather easy to assess based on standardised calculations that take into account e.g. the number of vehicles, their average speed, distance from the road to the dwellings. For low frequency noise it is more difficult to estimate the number of people exposed as no records exists on the number of low frequency sources that people may be exposed to, and also some sources may be generating a low frequency noise due to inaccurate installations. One way to estimate the effect is to use indirect estimates such as the prevalence of complaints to the authorities that are responsible for nuisance due to noise in the general environment.

Complaints directed to the local Health Authorities have been studied previously. Tempest [1989] carried out a questionnaire survey among 50% of the local Health Authorities in the United Kingdom. Complaints were prompted by a variety of sources, the most common being factories (35%), music (13%), traffic and other vehicles (11%) and commercial premises (9%). No estimation of the proportion of low frequency noise complaints in relation to other noise complaints was made. A questionnaire study among all local Environmental Health Authorities (EHA) in Sweden, gave the result that the three low frequency noise sources: fan- and ventilation noise, heavy vehicles and heat pumps comprised 71% of the total number of complaints of noise during a six month period in 1985 [Persson and Rylander 1988]. In that study, it was found that 32% of the EHAs reported that complaints on low frequency noise had increased during 1985 as compared to 1984, while 9% reported a decrease and 58% no change. The main reasons given for the increase were heat pumps, followed by fan- and ventilation installations and road traffic.

When evaluating the importance of complaints one has to bear in mind that although complaints are an important tool for the Environmental Health Authorities to prioritise the resources, the reasons for a complaint depend on several factors related to the authority, the individual and the noise source. Hence the number of complaints does not automatically reflect how many are actually disturbed, but can be seen as an indicator of disturbance.

As the studies referred to above were undertaken more than 10 years ago it would be of interest to investigate the situation today. Furthermore, in 1996, Sweden adopted specific guidelines on low frequency noise in the general environment [SOSFS 1996:7/E]. In short, the guidelines state that low frequency noise should be assessed by third octave band measurements in the frequency range of 31.5 to 200 Hz and the sound pressure levels given in Table I should not be exceeded in any third octave band. The measurement procedure is given in SP-Report 1996:10. For intermittent sounds such as amplified music, the guideline also states that an A-weighted level of 25 dB may be used. It was thus also of interest to investigate whether the specific guideline on low frequency noise was used in assessing low frequency noise and whether it had led to an impaired or improved situation.

2. AIMS

The aim of this study was to evaluate the occurrence of complaints related to low frequency noise directed to the EHAs and to see if the introduction of the specific Swedish guidelines on low frequency noise had influenced the EHA's assessment of low frequency noise. A secondary aim was to compare the results with a similar study carried out 14 years ago.

Table I. Guideline on low frequency noise (SOSFS 1996:7/E)

Third octave band (Hz)	Equivalent sound pressure levels (dB)
31.5	55
40	49
50	43
63	42
80	41.5
100	38
125	36
160	34
200	32

3. METHODS

In Sweden the EHAs are the authorities responsible for the environmental and health aspects including noise disturbance in the municipalities. The contact with the public is mainly complaint orientated. A complaint is a well-defined and recognized expression and it is registered at the EHA. Complaints are also one of the more important tools for the EHA to prioritise resources. We therefore considered the EHA to be the most reliable source of information as regards noise complaints in the general environment.

Among the 289 EHAs in Sweden, 41 EHAs were randomly selected. Of those, 37 (90.2%) participated. The selection of the EHAs was based on the number of inhabitants and geographical distribution of the municipalities. The selected sample comprised all eleven EHAs in municipalities with more than 100,000 inhabitants, ten EHAs in municipalities with 51,000 to 99,000 inhabitants and 20 EHAs in municipalities with less than 50,000 inhabitants. The number of inhabitants of the participating municipalities ranged from 6,000 to 740,000, with a median value of 57,500.

A questionnaire was distributed by mail together with an introductory letter describing the purpose of the investigation. It was stated that we wanted to ask about the prevalence of complaints on noise and low frequency noise in particular. A low frequency noise was defined as a noise with dominant sound pressure levels in the frequency range of 20 to 200 Hz and to further clarify, some examples of sources of low frequency noise, such as some ventilation systems, compressors, pumps, attenuated electronic music, heavy artillery shooting and explosions were given. Finally we asked to be put in contact with the officer that was most knowledgeable in these matters. Approximately one week later, telephone interviews were carried out with one environmental health officer per municipality.

The questionnaire comprised 35 questions divided into different sections: administrative matters, registered complaints on noise and low frequency noise, assessment of and actions undertaken against low frequency noise. Section I comprised questions on how the authorities perceived their competence on noise in general, the perceived possibility of preventing noise, routines to follow up complaints and legal processes. Section II comprised questions on the total number of complaints directed to the EHA during the last year (1999), complaints on noise and complaints on low frequency noise. In sections III and IV, specific questions were posted on the most common sources that caused the complaints on low frequency noise, how the EHA handled complaints due to low frequency noise, if they performed third octave band analysis and had access to relevant equipment. Finally, in section V the EHA was asked whether the number of complaints due to low frequency noise had increased, decreased or been the same during 1996–1997 as compared to 1998–1999.

4. STATISTICAL TREATMENT

Data were analysed for all municipalities and divided into three groups based on the number of inhabitants. Median values and percentiles are given for data with skewed distributions, while mean values and 95% confidence intervals of proportions are given for data showing normal distributions.

5. RESULTS

The results from the interviews showed that the total number of complaints directed to the EHAs during 1999 varied from 23 to 1,040, with a median value of 140. The total number of complaints on noise was 1578 and varied from 0 to 330, with a median value of 19. The number of complaints on noise in general comprised 15% (95% CI; 11.3%–18.1%) of the total number of complaints.

The number of complaints on low frequency noise directed to the EHAs varied from 0 to 215, with a median value of 7.

The number of complaints on low frequency noise per EHA is shown in Figure 1.

The proportions of low frequency noise complaints in relation to complaints on noise in general varied from 0 (8 EHA) to 100% (2 EHA). The mean value of the proportions was 35% (95% CI; 18.5 % – 51.5%). Overall the total sum of low frequency noise complaints in relation to the total sum of complaints on noise comprised 44%.

Another factor of interest was the incidence rate, i.e. the number of complaints per inhabitant. The incidence rate was calculated as complaints per 10,000 inhabitants to avoid very small numbers. For all municipalities the median value of incidence rate for complaints on low frequency noise per 10,000 inhabitants was 1.1 (25th perc – 75th perc; 0.25–2.4). The corresponding rate for noise in general was 3.3 (25th perc – 75th perc; 2.5–6.6).

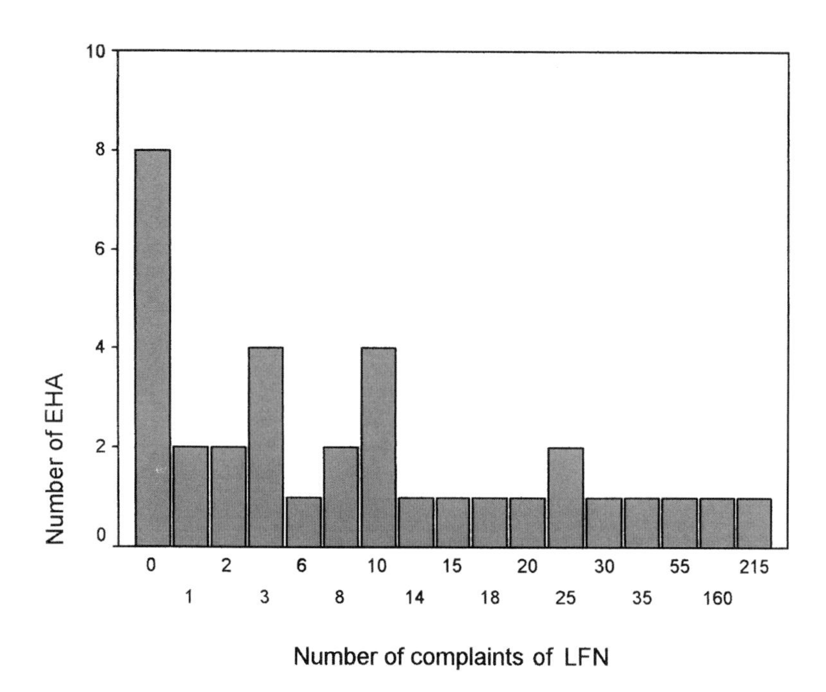

Figure 1. Number of complaints of low frequency noise per EHA

Table II gives the incidence rate for municipalities divided into the three categories: number of inhabitants less than 50,000, between 50,000–99,000 and more than 99,000 inhabitants.

Table II. The incidence rate related to municipality size

Municipality size	Median value of incidence rate per 10,000 inhabitants	25th percentile	75th percentile
<50,000	0.78	0	1.6
50,000–99,000	0.93	0.38	3.8
>99,000	1.6	0.86	3.0

The table indicates a somewhat higher incidence rate for the largest municipalities, however the differences between municipality categories were not statistically significant (Chi^2 =2.709, df=2, p=0.258, Kruskal Wallis test).

EHAs with more than 5 complaints on low frequency noise (18 EHAs altogether) were more closely analysed to see which noise sources was responsible for the largest proportion of complaints.

The relative proportion of the different noise sources emitting low frequency noise is shown in Figure 2.

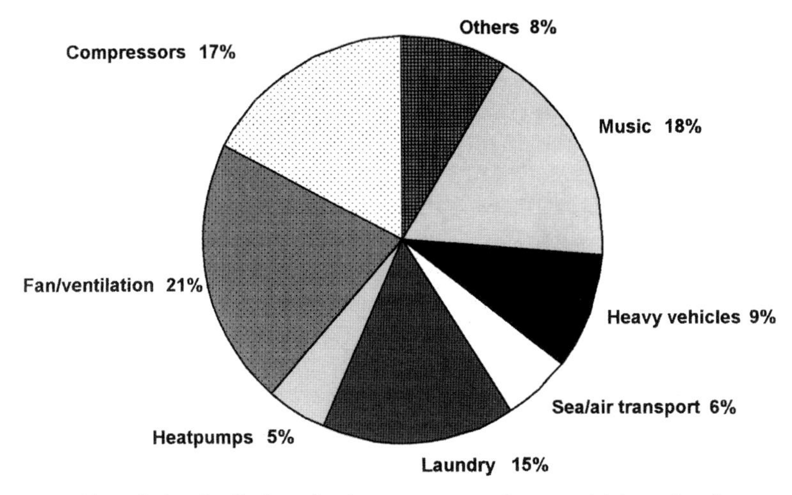

Figure 2. The relative distribution of noise sources causing complaints on low frequency noise

The analyses showed that complaints regarding fan- and ventilation installations comprised the largest part (21%), followed by amplified music (18%), compressors (17%), and laundryrooms (16%). Less frequent were complaints of noise from heavy vehicles (9%), sea transport and aircraft (6%) and heat pumps (5%). The occurrence of complaints on low frequency noise due to installations placed inside a building was similar to complaints due to installations placed outside a building. The sector named "others" includes different sources such as quarries, heavy artillery shooting, trains and road traffic.

To assess low frequency noise, most EHA used a combination of measuring third octave band sound pressure levels, dBC and the dBC-dBA difference, as well as listening to the noise. One third of the EHAs (30%) reported that they did not have enough access to relevant equipment. On the question of how the EHAs considered the specific guidelines [SOSFS 1996:7/E], involving third octave band analysis, performed when assessing low frequency noise complaints as compared to the previous guidelines in A-weighted noise levels, 62% reported that it performed better or much better, 35% did not know and only one EHA considered it to perform less well (Figure 3).

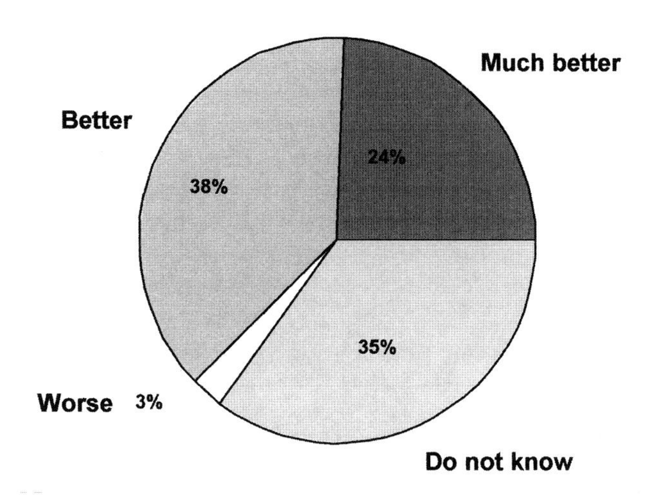

Figure 3. The proportion of EHA reporting on how the specific guideline on low frequency noise (SOSFS 1996:7/E) performed relative to the earlier used A-weighted guidelines

Finally, 46% of the EHAs estimated that complaints on low frequency noise had increased, 3% that they had decreased and 46% that the number of complaints had been stable during the last two years. Reasons given for an increased number of complaints were an increasing number of restaurants with music, fan- and ventilation installations and also, an increased awareness among the inhabitants and among the EHAs regarding the problems with low frequency noise.

6. DISCUSSION

6.1 Methods

The investigated sample of EHAs comprised 13% of the EHAs in Sweden. With this rather small proportion it is possible that the results are not representative of all EHAs. The sample in this study was chosen to be representative of both population size and geographical distribution and, as the main aim was to get a deeper knowledge, an investigation of all EHAs was not possible. In this study and the previous [Persson and Rylander 1988] there were also no clear relationships between size of municipality and number of complaints on noise or low frequency noise per inhabitant, and it can thus be concluded that there are no strong indications that the results would have been different if the number of EHAs had been greater.

The investigation of the occurrence and type of complaints in this study, was based on a comprehensive questionnaire sent out in advance and a subsequent interview. This was done as some of the questions required the EHA to go through their records of registered complaints beforehand. We have however not been able to check the validity of the answers, i. e. how well the answers were in agreement with the actual registered complaints in each authority. This could have been done by going through the registered number of complaints in a randomly chosen sample, but was unfortunately not within the scope of this study. Furthermore there were no indications during the interviews that the numbers given would be incorrect. If present, this type of error would probably apply to both complaints in general, complaints of noise and low frequency noise, and it would hence not affect the relative distribution of these complaints. It would however affect the certainty of the numbers given and also the information on which sources caused complaints. The error is estimated to be rather similar between this study and the previous.

A third factor that should be considered is how well the data on complaints reflect the occurrence of noise disturbance in the general environment. As mentioned in the introduction, the reasons for a complaint depend on several factors such as the accessibility of the authority, the individuals' earlier experience, education level and attitude to the noise source, the presence of local opinion and possibilities (perceived and otherwise) of influencing the noise source [Miller 1978]. It has also previously been found, based on experience in England and USA, that only a small number of people who are annoyed will actually formalise their distress into a complaint [Miller 1978, Daryl 1978, U.S. EPA 1974]. Taken together it can be concluded that if a complaint is present it can be taken as an indicator of the presence of annoyance, while no indication is given of the extent of annoyance. There are no indications that the factors influencing the reasons for whether a complaint is put forward or not in Sweden have changed considerably between 1985 and today.

6.2 Results

The results indicate that the proportion of low frequency noise complaints in relation to noise in general was lower in this investigation compared to the study carried out in 1985 [Persson and Rylander 1988]. The proportion of low frequency complaints of the total complaints on noise was in this study 44% compared to 71% found in the previous study. As this study comprised a selection of EHAs, it may be more relevant to compare the mean value of the relationship between the total number of complaints and low frequency noise for each EHA. The mean value was 35%.

There are several possible reasons for the different results between this study and the one carried out in 1985. One reason could be that the number of complaints *on noise in general* has increased. In the previous study, the total number of complaints on noise to 284 EHAs registered during a six months period, was 1,026 and of those 728 were categorised as low frequency noise complaints. Assuming that the 6 month period was representative for the whole year this totals 2,052 complaints on noise in general and 1,456 complaints on low frequency noise during a year. In this study the total number of complaints on noise in general was 1,578 from a sample of 37 EHAs and of those 692 were categorised as low frequency. In the previous study, the median values of complaints on noise in general were 4, and for low frequency noise 2. In this study the median values of complaints on noise in general were 17 and for low frequency noise 7. Although, the comparison is hampered by the different sample selection, and a somewhat different method, the data indicate that the different results may be due to a relatively greater number of complaints on *noise in general* being directed to the EHA today. Another factor that would give the same results would be if noise sources in the previous study were incorrectly classified as low frequency. This is not unlikely as the knowledge of low frequency noise in 1985 was less widespread. Furthermore, the classification in 1985 was in most cases based mainly on the C-weighted levels and possible octave band measurements, which made classification less precise. Since then, the knowledge of low frequency noise among the EHAs has increased to a great extent, mainly due to the introduction of specific guidelines on low frequency noise.

In this study the dominant sources of low frequency noise complaints were fan- and ventilation installations, followed by music, compressors and laundry-rooms. In the previous study, fan- and ventilation installations were also the largest source of complaints followed by heavy vehicles and heat pumps. The relative proportion of complaints between the sources asked for in the previous study and this study is strikingly similar. Fan- and ventilation noise complaints are/were reported about twice as often as heavy vehicles and heavy vehicles are/were reported about twice as often as heat pumps.

In the study by Tempest [1989], factory noise was the dominant source followed by music, traffic and other vehicles and commercial premises. Music was not specifically asked for in the study carried out in 1985, but from the results in the present study it can be concluded that it is an important source for low frequency noise complaints today.

As regarding changes over time compared to a previous reference period, the results from the previous study showed that 32% reported an increase, 9% a decrease and 58% an unchanged number. This study found that 46% reported an increase, 3% a decrease and 46% an unchanged number. These results indicate that low frequency complaints are an important issue that still seems to grow.

In the previous study ten EHAs stated in an open question that present regulations for controlling low frequency noise were not adequate. The data from the present study show that nearly two third of the EHAs stated that the specific guideline on low frequency noise based on third octave band analysis was much better or better compared to the previous A-weighted guidelines when assessing low frequency noise. This is worth observing as the specific guidelines on low frequency noise are more time-consuming, require a higher degree of competence and more sophisticated equipment. It is also worth noting that 35% did not know if the specific guidelines were better or worse and most of those did not have access to equipment that could measure third octave band sound pressure levels. The lack of relevant equipment may thus be an impediment for appropriate and equal assessments of complaints between municipalities.

7. CONCLUSIONS

The interviews among a selection of the EHAs in Sweden indicates that the proportion of low frequency noise comprised 35% of the total complaints on noise. The study also shows that complaints on low frequency noise are still an important and probably growing issue.

Most of the EHAs preferred the specific guidelines on low frequency noise based on third octave band analysis to previous A-weighted guidelines when assessing low frequency noise, even though it is more time-consuming and requires more competence and advanced equipment.

REFERENCES

1. Daryl N M. Basic subjective response to noise. In: *Handbook of Noise Assessment* (N M Daryl ed.), New York, Van Nostrand Reinhold. Pp 3--38, 1978.

2. Miller J D. Effects of noise on people. In: *Handbook of Perception* (E C Carterette and P F Morton, eds.) Academic press. Vol IV, pp 609–640, 1978.

3. U. S. Environmental Protection Agency, *EPA report No 550/9-74004*. Information on levels of environmental noise requisite to protect public health and welfare with an adequate margin of safety. 1974.

4. Persson K and Rylander R. Disturbance from low frequency noise in the environment: A study among the local environmental health authorities in Sweden. *J. Sound. Vib.* 1988, 12: 339–345.

5. SOSFS 1996:7/E. Indoor Noise and High Sound-Levels. *General Guidelines issued by the Swedish national Board of Health and Welfare*, 1996.

6. SP-REPORT 1996:10. *Measurement of low frequency sound in rooms.* Swedish National Testing and Research Institute, Borås, Sweden.

7. Tempest W. A survey of low frequency noise complaints received by local authorities in the United Kingdom. *J Low Freq Noise Vibr.* 1989, 8:45--49.

Low Frequency Noise Annoyance: The Behavioural Challenge

Stephen Benton and Orna Abramsom-Yehuda
Human Factors Research Group, Department of Psychology, University of Westminster, U.K.

INTRODUCTION

The common theme to which many Low Frequency Noise (LFN) researchers subscribe to is that here is a phenomenon that is consistently under-rated in terms of its status as an environmental pollutant (Benton & Leventhall, 1994, Persson-Waye 1995, Bengtsson et al, 2000). The difficulties surrounding the development of an effective and systematic approach to the quantification of LFN incidence and associated impact have centred upon source detection, identification, location and annoyance loading. The quantification of each and all of these aspects is often complicated by the combination of significant 'individual differences' in sensitivity to LFN and the relatively low sound pressure levels (SPLs) which can be associated with disturbance, annoyance and stress (Persson-Waye et al, 2000). Whilst research has sought to clarify and delineate problems within each of these areas, the approach has been one that tends towards an in depth analysis within the boundaries of each particular area. In practice, Environmental Health Officers (EHOs) actively engaged in resolving noise complaints are faced with composite problems produced by the interaction of all three areas, the expression of which is firmly embedded within the behavioural.

The restricted development of an overall effective and systematic LFN 'complaint-handling' methodology has meant that EHOs are usually reliant upon existing and dB(A)) driven protocols, when initiating case assessments. The capability of EHOs', and related health care agencies, to provide effective assessment, practical advice/solutions and resolution are integral parts of, and confounding elements within complainants' experience of exposure to LFN. Failed resolution of the situation through the application of inappropriate measurement produces a number of behavioural and emotional consequences, which actively undermine assessment results, core relationships (e.g. between the complainant and the EHO) and which act to exacerbate the impact of the physical and psychological features in order to create second order stress effects.

Our paper attempts to gauge the scale and type of burden that LFN complaints place upon the relevant agencies. Is it the case that LFN complainants remain within the system longer than most other categories of complainant? Is the system response to LFN complainants likely to lead to and heighten stress symptoms?

BACKGROUND

Most assessments of LFN, as an environmental pollutant, have necessarily centred on comparisons made against other noise impact criteria. Such assessments of impact will be guided by reference to a number of established criteria and categories of subjectivity that include, speech intelligibility (ISO, 1975) annoyance, sleep deprivation and performance degradation (Cooper, and Quick, 1999). Each of these categories has seen the development of empirically based protocols which, under well defined conditions of exposure, have led to the production of criteria designed to protect health and the quality of life as captured within increasingly internationalised standards. This trend is in response to the rapid growth in technological developments (ranging from transport to home used items, from industrial to an individual scale e.g. air conditioning) and associated transport infrastructures with the consequent environmental impact of noise. Clear-cut procedures of assessment and weighting are available for a number of key 'noise impact' categories including *annoyance*. The widespread application of standards and measurement techniques are an indication of the extent to which subjective and physical attributes of an 'impact' have been reliably correlated. However, as proponents of separate or discrete weighting networks for LFN are likely to note, this reliability has not been extended to include the effective handling of LFN complaints, with principal difficulties associated with exposure to low SPL of LFN.

Empirical findings have provided a raft of guidelines, objective measurement and procedural bodies of evidence that provide the EHO with ways in which to quantify the physical noise signature permitting the clarification of complainants experiences. Effective quantification enables all parties to the complaint to identify the source, or likely source, to agree upon existent and perceived levels of the noise and also on what would constitute effective steps to be taken towards a solution. This level of understanding and co-operation forms an essential part of the puzzle for the complainant, and it should serve to validate, in explicit terms, their individual experience. The increased access to and sharing in, professional and expert explanations of the physical parameters contribute to regaining a sense of control over their environment. Before the complainant is able to regain control they will need to be able to utilise concepts that are able to convey their personal experience in a manner that creates a coherent description of both the 'behaviour' of the noise and their behaviour to it. One of the consequent problems in the application of measurement and assessment practices that are frequently inappropriate to instances of LFN is that cases tend to 'stick' in the EHO system and while the relative numbers may be low the time taken to affect a durable solution is disproportionately high.

In general, the expertise provided by the EHO, both in content and style, acts as neutral ground from which a common language of representation and explanation should operate. From this the complainants are able to seek and achieve a degree of consensus and support for their situation, symptoms and anxieties.

Complainants report that one of the most debilitating aspects of noise is its 'intrusiveness'. They lose control over the quality of sound in their personal environment. Where noise is involved it seems that one person's choice of sound, or activity is another person's noise, and loss of personal space. The need for EHO's to address the psychological impact of noise is an integral element in initiating steps that will build an effective intervention and resolution. The subjective problems associated with the physical impact of the stimulus occupy one level of psychology; this can be assessed in relative terms of interference, loudness and pitch (intrusiveness) and to some degree annoyance. However, there also exists a secondary subjective impact and this originates from the methods of assessment themselves, a form of supra stimulus impact, not of the noise but rather from the behavioural context generated under conditions of exposure to the noise and the assessment protocols which come to define the experience for the complainant.

It is in this area that LFN is particularly likely to be problematic for the EHO. The area of incongruence between the complainant's experience and the EHO's findings is stark, a situation rich in conflicting perceptions and judgments raising serious challenges for resolution as the criteria upon which these judgments are based are polarised and frequently take on the property of being partisan. Many times initial assessments may at best downplay the noise as a problem or at worse measurements taken fail to substantiate even the existence of a noise. All of which is compounded by the fact that, usually by the time the EHO is called in individuals would already be experiencing unwanted subjective effects.

OBJECTIVE AND SUBJECTIVE PARAMETERS:

This paper is concerned with the extent to which assessment procedures are able to respond to the contextual psychological aspects associated with noise complaints. Clearly the influence of contextual matters will vary in respect to how well the assessment procedures are calibrated to the noise signature. For LFN cases, where a widespread and standardised measurement protocol has yet to be established, there is a high probability that a mismatch between objective measurements and contextual psychological aspects will characterise the situation. It is suggested that it is important to provide a 360-degree approach to the problem, one that includes both the physical, psychological and the contextual. By the time the EHO arrives the noise has already registered within the contextual, shaping the perceived priorities and influencing the quality and type of communication from the complainant. Communication, under these conditions of individuals sliding into coping failure, will tend to be focused and inflexible. The feature here being that the complainant has developed a way of making sense of their experience within the terms of reference accessible to them and consistent with the overall behavioural context (e.g. their understanding of how and why the noise behaves the way it does). The rules and justification for a complainant's personalised context may be difficult to communicate and are often supported by perception of implicit relationships rather than the explicit and objective ones. This form of implicit knowledge (experience that is difficult to communicate and demonstrate to others because it has been thoroughly internalised) is characteristic of expert knowledge in that it has tacit validity, which is often difficult to connect to objective points of reference. While tacit knowledge is frequently valid and leads to coherent personal judgments they are notoriously prone to confirmation bias (Kaufman, 1990). The individual tends to seek for confirming instances, in order to sustain an internally consistent evaluation of their experience, while disregarding or downgrading instances (findings) that run counter to their evaluations. In this way the tacit assessment criteria are sustained. This is not to suggest that dissociation from reality has occurred. Far from it, it is just as likely that the experience has been interpreted in a manner that would be common to others if they had shared the same route to acquiring it! A summary of the complainant's personal and behavioural context is shown in Figure 1.

Many complainants report that EHO's undervalue the distress that they are experiencing and this often becomes a confounding aspect in subsequent exchanges. The context of experience for the complainant fails to be translated into accessible concepts likely to be achieved through the application of objective assessments. The common ground sought by the EHO, from which to 'make sense' of the complaint, fails to hold as the evidence does not support the context within which the complaint is observed. The problem to be solved may already include the behaviour of the complainant. From the complainants point of view, as is often noted (Guest, 2002) it is soon the failure of the EHO's behaviour to measure up to the 'expectations' for solution and support that undermine the relationship and subsequent exchange of information.

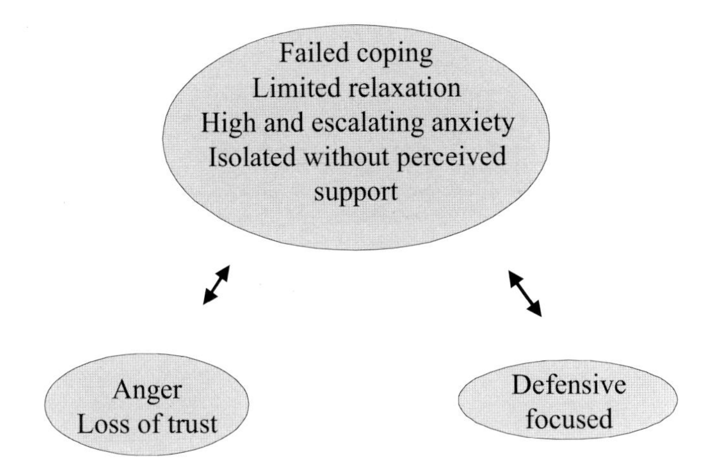

Figure 1. Behavioural Context: Barriers to Connecting with the Problem and the Person

TACIT KNOWLEDGE AND THE INTERVENTION PROCESS:

How to improve the quality of information to be exchanged between the complainant and the EHO? If this process could be enhanced, then the development of neutral (if not objective) ground and a commonly accessible frame of reference (a combination of the physical, psychological and contextual) could ensue. Broadening the framework of the problem to include the dynamics underlying the failed resolution of perceptions would allow for the development of improved 360 assessment protocols for LFN complaints. This is particularly the case as the current route towards resolution is dominated by reliance upon inappropriate and single measure assessments.

Where the physical attributes available are often confounded by some assessment practices, the non-physical and contextual attributes may offer a useful set of secondary indicators available for inclusion in the overall appraisal of the complaint. In order to up grade the level of information available to both parties it could be supportive of resolution to develop protocols that inhibit the interpersonal disablers, which commonly act to distort communication between EHO's and LFN sufferers. Perhaps lessons could be learnt from the field of negotiation and conflict resolution where parties are commonly derailed by persistent and antagonistic perceptions of what counts and what doesn't. Analysis, subjective symptoms combined with the behavioural context may prompt alternative interpretations and generate intervention options

CONCLUSION:

The behavioural context associated with LFN exposure in the environment offers a rich source of secondary information and may provide routes towards enhanced practice for the EHO. The structured collection of such information would also provide a valuable source for 'common ground' between the complainant and the EHO, an important element in developing a sustainable and practical base for improved coping, reduced isolation and improved communication for the complainant.

REFERENCES

Benton, S. and Leventhall, H.G. (1994). The Role of 'Background Stressors' in the Formation of Annoyance and Stress Responses. *Journal of Low Frequency Noise and Vibration*; 13.

Bengtsson, J., Persson Waye, K., Kjellberg, A., and Benton, S. (2000) Proceedings of the 9th International Meeting. *Low Frequency Noise and Vibration.* ISBN no. 87-90834-06-2.

Cooper, C. and Quick, J. (1999) Stress and Strain. Oxford: Health Press.

ISO (1974) Assessment of Noise with Respect to its Effects on the Intelligibility of Speech. ISO Technical Report TR 3352-1974.

Guest, H. (2002): Inadequate standards currently applied by local authorities to determine statutory nuisance from LF and infrasound. *10th International Meeting Low Frequency Noise and Vibration and its Control.* York UK (Editor: H G Leventhall), 61-68.

Kaufman, B.E. (1990) A New Theory of Satificing. *Journal of Behavioural Economics.* Spring 35-51.

Persson-Waye, K. (1995). Doctoral thesis., Goteborgs University, ISBN 91-628-1516-4. *On the Effects of Environmental Low Frequency Noise.*

Persson-Waye, K., Rylander, R., Bengtsson, J., Clow, A., Hucklebridge, F., and Evans, P. (2000). *Proceedings of the 9th International Meeting. Low Frequency Noise and Vibration.* ISBN no. 87-90834-06-2.

Comparison of objective methods for assessment of annoyance of low frequency noise with the results of a laboratory listening test

Torben Poulsen

Ørsted•DTU, Acoustic Technology, Building 352, Technical University of Denmark, DK-2800 Lyngby, Denmark

ABSTRACT

Subjective assessments made by test persons were compared to results from a number of objective measurement and calculation methods for the assessment of low frequency noise. Eighteen young persons with normal hearing listened to eight environmental low frequency noises and evaluated the annoyance of the noises. The noises were stationary noise with and without tones, intermittent noise, music, traffic noise and impulsive low frequency noise. The noises were presented twice in a random order at L_{Aeq} levels of 20 dB, 27.5 dB and 35 dB. The assessment methods were those used in Sweden, Germany, The Netherlands, Poland and Denmark. It was found that the Danish assessment method gave the best relation to the subjective assessments made by the test persons. An important property of this method is that it includes a 5 dB penalty for noises having an impulsive character.

1. INTRODUCTION

Different measurement and calculation methods for the assessment of annoyance due to low frequency noise have been proposed during recent years. The noise limits or criteria values used in the various assessment methods differ. As a consequence of this ambiguity the Danish Environmental Protection Agency has asked for an investigation, where the subjectively assessed annoyance due to a number of real life examples of low frequency noise was compared to the predicted annoyance using different assessment methods. Such an investigation could indicate the best suitable method for assessment of low frequency noise or it could indicate a need for an adjustment or a revision of the method presently used in Denmark. The Danish assessment method was published in Information No. 9/1997 from the Danish Environmental Protection Agency, "Low frequency noise, infrasound and vibration in the environment" [1]. Results from the project have been presented at the "Low Frequency Noise and Vibration" conferences, [2], [3], [4].

In [1] a general description is given of the generation and transmission of low frequency noise and of the properties of hearing in the low frequency and infrasound region. Recommended measurement and assessment methods for annoyance are described and recommended limit values are stated for environmental infrasound and low frequency noise. Contrary to usual the measurement and assessment procedures for environmental noise, such as road traffic noise or industrial noise, measurement of environmental infrasound or low frequency noise should be made indoors in dwellings. Sound in the frequency range below 20 Hz is defined as infrasound. The G-weighting function standardised in ISO 7196 [5], relates closely to the shape of the hearing threshold in the infrasound region. The loudness and annoyance due to infrasound increase very quickly with increasing level. The hearing threshold for single tones is usually about 95 dB(G), and tones with a 20 dB higher level are expected to be sensed as very loud. It can be assumed that infrasound below the hearing threshold is not annoying.

There is an obvious need for an investigation where the subjective annoyance due to typical examples of low frequency noise are compared to different objective measures of the level of the same noises. The present investigation was restricted to low frequency noise. No infrasound was included.

2. LISTENING TESTS

The listening tests were made in a standardised listening room [6] of dimension 7.52 × 4.75 × 2.76 m. Eight different noise examples were used, presented at three different levels. All presentations were made twice and the sequence of the presentations was randomised. Prior to the listening tests, the test persons were trained using four noise examples. After each presentation the test person gave evaluations of the noises on a paper form.

2.1 Noises

The noises are listed in Table I The traffic noise should serve as a reference noise, because there is a well-described relation between the level of road traffic noise and the annoyance of this type of noise. The other noises had all strong low frequency content.

Noise 1 was from a densely trafficked six-lane highway, having broadband characteristics and is almost continuous. Since it was filtered to simulate an indoor measurement, the tonal character of the engine noise from passing heavy vehicles is clearly audible and tire noise is also obvious. Noise 2 consists of a series of very deep, rumbling single blows from a drop forge. The noises 3, 4, 5, and 6 each have one tonal component. Noise 7 has three tones but two of them are at a low level. Noise 8 has a characteristic rhythmical pulsating sound due to the drums.

Table I. Description of the noises used in the listening tests

No.	Name	Description	Tones, characteristics
1	Traffic	Road traffic noise from a highway	None – broadband, continuous
2	Drop forge	Isolated blows from a drop forge	None – deep, impulsive sound transmitted through the ground
3	Gas turbine	Gas motor in a power-and-heat plant	25 Hz, continuous
4	Fast ferry	High speed ferry; pulsating tonal noise	57 Hz, pass-by
5	Steel factory	Distant noise from a steel rolling plant	62 Hz, continuous
6	Generator	Generator	75 Hz, continuous
7	Cooling	Cooling compressor	(48 Hz, 95 Hz) 98 Hz, continuous
8	Discotheque	Music, transmitted through a building	None, fluctuating, loud drums

The duration of all the noise presentations was 2 minutes. The noises were either recorded indoors or filtered to simulate indoor noise. They were recorded on DAT tape and transferred to the hard disk of a PC where they were edited digitally. The noises were presented to the test persons at A-weighted nominal levels of 20 dB, 27.5 dB, and 35 dB. In the listening room the sounds were measured at the listening position and subsequently analysed to obtain the objective levels of the noises. The noises were played directly from the PC via a D/A converter to a crossover filter and via four separate amplifiers to two broadband loudspeakers (KEF 105) and two subwoofers (Amadeus Sub). A detailed description of the set-up can be found in [7].

The noises were filtered in order to compensate for resonant modes in the listening room. The listening position in the room was chosen so that only one resonant mode (at 45.5 Hz) influenced the sound level. This mode increased the sound pressure level by about 18 dB at the listening position and therefore a notch filter with 18 dB attenuation at 45.5 Hz (corner frequencies at 40.5 Hz and 50.5 Hz) was used in all presentations.

An 'outdoor-to-indoor' filter was used for the noises that were recorded outdoors. The filter represents the reduction index of ordinary building materials and construction principles [8], [9] and [10] and was defined in the range 16 Hz–4000 Hz. From a subjective evaluation, the noises sounded 'natural' in the listening room. All the noises had a pronounced low frequency characteristic.

2.2 Test Subjects

Eighteen young persons (9 males and 9 females) with normal hearing were chosen for the listening tests. The age of the test persons was between 19 and 25 years. Pure tone audiometry was carried out in the frequency range 125 Hz to 8000 Hz with a Madsen Midimate 602 audiometer, equipped with Sennheiser HDA 200 earphones. The calibration of the audiometer was made using the values from [11] which are practically identical to ISO 389-8 [12]. Hearing threshold levels at or below 15 dB HL were accepted in the frequency range 125 Hz to 4000 Hz, and a hearing threshold level at 20 dB at a single frequency (including 8 kHz) was also accepted. The average hearing threshold of the listeners shows a slight decrease (less than 10 dB) at 6 kHz and 8 kHz.

In addition to the conventional audiometry, the hearing threshold in the low frequency range was determined. The tests were made using pure tones at 31 Hz, 50 Hz, 80 Hz, and 125 Hz with a Two Alternative Forced Choice method [13]. These hearing threshold measurements showed results that were less than 10 dB from the standard hearing thresholds given in ISO 389-7 [14].

2.3 Subject's Task

Test persons were given a written introduction to the tests, and they could ask about the procedure throughout the tests. A full training session was made prior to the listening tests. Information about the sound examples was given after all the tests were finalized.

The tests persons answered four questions after each presentation:

- 'How loud is the sound?' (on a scale labelled "not audible" in one end and "very loud" in the other end)
- 'How annoying do you find the sound if it was heard in your home during the day and the evening?' (on a scale labelled "not annoying" in one end and "very annoying" in the other)
- 'How annoying do you find the sound if it was heard in your home during the night?' (on a scale labelled "not annoying" in one end and "very annoying" in the other)
- 'Is the noise annoying?' (answer yes or no).

The response was made by making a mark on a horizontal line. All the lines were 10 cm long, and the response was measured in cm with a ruler and thus all data are given as figures between 0 and 10.

3. RESULTS OF THE LISTENING TEST

Table II shows the average subjective evaluation made by the listeners of the annoyance during night. The loudness question and the annoyance at day/evening question gave similar results.

Table II. Subjective assessment by the reference group of the annoyance from the noise examples if the noise was heard at night. Annoyance rating is given on a scale from 0 (not annoying) to 10 (very annoying)

Nominal presentation level	20 dB	27.5 dB	35 dB
Noise example	Subjective annoyance night	Subjective annoyance night	Subjective annoyance night
Traffic noise	1.6	3.4	5.2
Drop forge	4.3	5.9	6.9
Gas turbine	0.9	2.5	5.2
Fast ferry	0.9	3.2	5.4
Steel factory	1.0	2.7	4.9
Generator	1.7	3.2	5.0
Cooling compressor	2.7	4.4	6.0
Discotheque	3.0	5.4	6.7

It can be seen from Table II that the subjectively assessed annoyance increases when the same type of noise is played at a higher level. It can also be seen that the different types of noise are not assessed equally annoying. The noises from the drop forge, the discotheque and the cooling compressor are evaluated as more annoying than the other noises. It can also be seen that the traffic noise is just as annoying as many of the low frequency noises.

A statistical analysis of the data was performed although the data were not perfectly normally distributed. It was found that the noise, the nominal level, the measured dB(A) level and the low-frequency level ($L_{pA,LF}$), are all significant factors in the evaluations from the test persons. The repetition number (round 1 or round 2 with the same noise presentation) is not a significant factor, which shows the absence of a training effect.

4. ASSESSMENT METHODS FOR LOW FREQUENCY NOISE

A number of different methods have been suggested for the assessment of low frequency noise. The Danish method [1] has already been mentioned. The other methods used in the present investigation are the standardised German method [15], the Swedish method [16], a recent Polish method [17], and two different methods from the Netherlands [18] and [19]. All these methods are used to assess the annoyance due to low frequency noise, based on the indoor noise level. The methods give different guidelines or criteria for the allowed noise level and the administrative procedures used in the individual countries to enforce the criteria for low frequency noise are very different.

Table III gives an overview of the main features of the various methods.

Table III. Overview of the objective methods

Assessment method	Type of measurement	Criterion	Impulse correct
Danish	1/3 oct. bands, 10 Hz – 160 Hz, A-weighting	Limits	Yes
German A- level	1/3 oct. bands, 10 Hz – 80 Hz, A-weighting of levels above hearing threshold	Limit re. hearing threshold	No
German tonal	1/3 oct. bands, 10 Hz – 80 Hz, level of tone(s)	Limits	No
Swedish	1/3 oct. bands, 31.5 Hz – 200 Hz	Curve	No
Polish	1/3 oct. bands, 10 Hz – 250 Hz	Curve	No
Dutch proposal	1/3 oct bands, 10 Hz – 200 Hz	Curve	No
C-level	1/3 oct bands, 10 Hz – 160 Hz	(Limit)	(Yes)

4.1 Danish Method

The Danish method [1] gives recommended limit values for low frequency noise and infrasound. The noise is measured in several positions indoors, and is analysed in 1/3-octave bands. The nominal A-weighting corrections are added to the spectrum values, and the weighted spectrum is summed to form the A-weighted level of the noise in the frequency range 10 Hz–160 Hz. The resulting level is called $L_{pA,LF}$. A direct measurement of the A-weighted level, $L_{pA,LF}$, is not possible since the minimum limit of the tolerance for the A-weighting filter is undefined (i.e. minus infinity) below 20 Hz.

In the Danish method a table of recommended limit values is used for the assessment of the noise. In dwellings the A-weighted equivalent level (averaged over 10 minutes) shall not exceed 20 dB $L_{pA,LF}$ in the evening and the night (18–07) or 25 dB $L_{pA,LF}$ in the day period (07–18). In offices etc the $L_{pA,LF}$ level shall not exceed 30 dB, and in other rooms in business premises the limit is 35 dB. If the noise has an impulsive character, the limits are reduced by 5 dB.

4.2 German Method

In the German method [15] low frequency noise is defined as noise where the C-weighted noise level is at least 20 dB higher than the A-weighted level, based on either equivalent levels or maximum levels.

If the noise is evaluated as 'low frequency', a 1/3-octave frequency analysis is made. The method considers the frequency range 10 Hz–80 Hz, but in special situations the 8 Hz and / or the 100 Hz band can be included. The method applies to rooms in dwellings where people stay or rest. In an Annex to the method a range of limits or criteria values are given for the day period (06–22) and for the night period (22–06).

In the German method, a distinction is made between tonal noise and noise without tones. If the level in a particular 1/3-octave band is 5 dB or more above the level in the two neighbouring bands, the noise is said to be tonal.

For tonal noise, the level of the frequency band with the tone is compared to the hearing threshold (L_{HS}) in the same band. It is then found how much the tone is above the threshold. The levels in the other frequency bands are not taken into account. The limit value for the equivalent level of the tone in the day period is: 5 dB in the 8 Hz–63 Hz bands, 10 dB in the 80 Hz band, and 15 dB in the 100 Hz band. The same assessment method applies to the maximum level of the noise; here the limit values in the same three frequency ranges are 10, 15, and 20 dB. In the night period all the limits are reduced by 5 dB, and thus the limits for the equivalent level of the tones are 0 dB, 5 dB, and 10 dB.

If the noise is not tonal, the limit for the A-weighted equivalent level (10 Hz–80 Hz) is 35 dB during daytime and 25 dB during the night. The A-weighted level is calculated by adding the A-weighting corrections to only those levels that are above the hearing threshold. As opposed to the Danish method, the contributions from levels below the threshold are disregarded. The corresponding limits for the maximum levels are 45 dB and 35 dB.

4.3 Swedish Method

The recommendations from the Swedish National Board of Health and Welfare [16] give guidance on an assessment as to whether noise under different conditions may have health effects. The recommendation comprises a criterion curve of recommended maximum levels of low frequency noise in rooms used for living. The curve covers the frequency range 31.5 Hz–200 Hz and applies to the equivalent level of the noise. A measurement method is specified and is described in a report from the Swedish Testing Institute [20]. If the noise level exceeds the criteria curve in any 1/3-octave band, the health and environmental authorities may characterise the noise as a nuisance.

4.4 Polish Method

The Polish method applies a threshold curve. This is defined in the frequency range 10 Hz–250 Hz, and corresponds to 1/3-octave levels each giving an A-weighted level of 10 dB (i.e. 10 dB above the inverse A-weighting correction). The criterion curve is called L_{A10} [17].

The noise is considered annoying if both of these conditions are met:

- The spectrum of the noise exceeds the criterion curve L_{A10} in one or more 1/3-octave bands
- The spectrum of the noise exceeds the spectrum of the background noise

It is mentioned in [17] that usually the background noise is somewhat higher than the criterion curve at the highest frequencies, above 100 Hz.

4.5 Dutch Proposed Method

The proposed method [18] is intended for use in connection with the granting of environmental permission to industries and businesses. The method uses a criterion curve defined in the frequency range 10 Hz–200 Hz. In the upper part of the frequency range the criterion curve agrees well with the Swedish criterion curve. At the lowest frequencies, where the Swedish curve is not defined, it corresponds to the hearing threshold as specified in the German method. It is expected that annoying low frequency noise will occur if the criterion curve is exceeded in one or more 1/3-octave bands.

Dutch Criterion for Audibility

This method is described in [19]. It is intended for use in cases where people complain about low frequency noise in order to decide if audible low frequency noise occurs. The aim of the method is not to verify whether the noise is annoying or not. The method employs a hearing threshold based on the best 10% of a non-selected population aged 50–60 years. The threshold curve is used in the frequency range 20 Hz–100 Hz.

4.6 C-weighted Sound Pressure Level

In the German method the difference between the C-weighted and the A-weighted sound pressure level is used to determine if low frequency noise is present. Similar rules of thumb have regularly been mentioned in the literature. The C-weighted sound pressure level has been included in the analysis because it has been claimed that the C-weighting should give a better description of low frequency noise than the A-weighting.

5. CRITERION CURVES

The criterion curves from the different assessment methods are shown in Figure 1.

The 'Dutch proposal', 'Swedish', and 'Polish' are criterion curves directly aimed at assessing if the noise is annoying. The two first curves differ only in the frequency range 50 Hz–80 Hz, where the Swedish curve is clearly lower than the Dutch proposal. In the entire frequency range the Polish curve is lower than the two other curves. Here it must be remembered that the background noise is also part of the Polish criterion, which will often have a relieving influence on the criterion curve at high frequencies, but this part of the method is not considered here.

The 'German' curve is a hearing threshold curve and is used as a criterion for tones in the noise. It permits tones to exceed the curve by 5 dB during daytime, and a higher exceedence is allowed at higher frequencies. The curve 'Dutch audibility' is used in cases with complaints in order to decide whether there is audible noise in the relevant frequency range. The curve is not used to determine if the noise is annoying. It can be seen that the German and the Dutch threshold curves are almost identical, and that they almost coincide with the curve 'Dutch proposal' below 40 Hz.

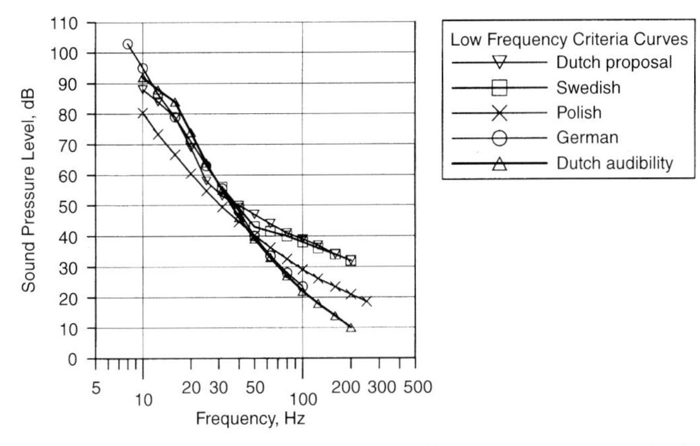

Figure 1. Comparison of criteria curves from the different assessment methods.

6. COMPARISON OF THE SUBJECTIVE EVALUATION OF ANNOYANCE WITH THE RESULT OF OBJECTIVE MEASURES

The test persons evaluated the annoyance in two situations: if the noise was heard in the day and the evening, and if the noise was heard at night. Figure 2 shows that there is a very close relation between these two assessments. Generally the annoyance at night is slightly larger than the annoyance in the day/evening. The relation between the pair of

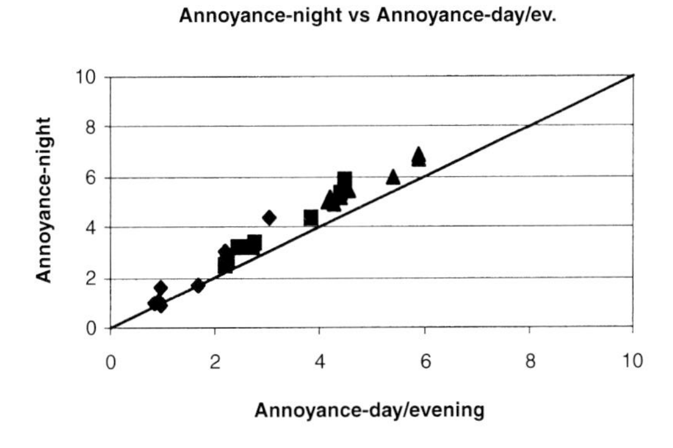

Figure 2. Assessment of the annoyance in the day and evening period and the annoyance at night for the same noise examples.

evaluations can be described by the correlation coefficient, which is as high as 0.9885. The following comparisons are therefore made only for the subjective assessment of the annoyance in the night period. The various methods, criteria and curves are used in relation to the 1/3 octave spectrum of the noises. In every analysis the objective metric is chosen as the x-parameter, and the subjective annoyance evaluation (which is the same in all cases) is the y-values.

6.1 Danish Method

Table IV shows the result of the use of the Danish method for the various noises. The second column shows the excess of the Danish limit. For the night period the limit value is $L_{pA,LF}$ = 20 dB, but since the drop forge as well as the discotheque are considered as impulsive noises, the limit for these noises is 15 dB. The third column gives the average assessment made by the test persons.

Table IV. Subjective evaluation of the various noise examples shown together with the objective 'assessment' by use of the Danish method ($L_{pA,LF}$)

	Nominal level					
	20 dB		27,5 dB		35 dB	
Noise	**Excess**	**Subjective annoyance**	**Excess**	**Subjective annoyance**	**Excess**	**Subjective annoyance**
Traffic noise	-0,3	1,6	7,0	3,4	14,5	5,2
Drop forge	6,9	4,3	14,2	5,9	21,5	6,9
Gas motor	-0,2	0,9	7,3	2,5	14,8	5,2
Fast ferry	0,1	0,9	7,5	3,2	15,0	5,4
Steel factory	-1,0	1,0	5,6	2,7	13,1	4,9
Generator	-1,3	1,7	8,6	3,2	16,1	5,0
Cooling compressor	-0,4	2,7	6,5	4,4	14,0	6,0
Discotheque	3,7	3,0	10,7	5,4	18,1	6,7

The data from table IV are shown in Figure 3.

Figure 3 illustrates the subjective evaluation as a function of the excess of the Danish criteria. It is seen that a straight line (not shown in the figure) can represent the group of points. This line is found by linear regression (least squares method, function LINEST of an Excel spreadsheet). The regression line has the formula:

$$y = 1.61 + 0.26 * x$$

where 'y' represents the average subjective evaluation made by the test subjects and 'x' represents the excess over the limit in dB.

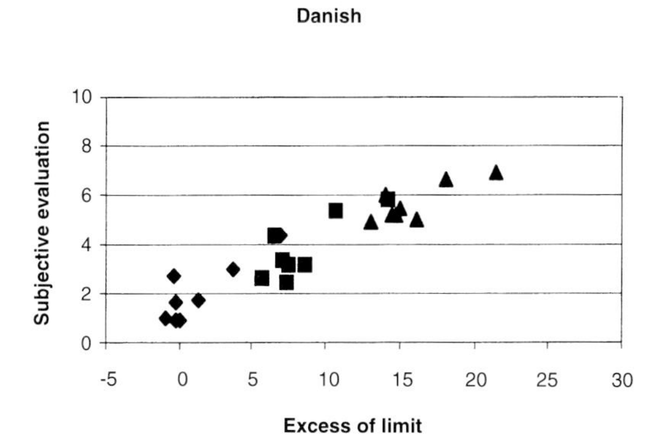

Figure 3. Relation between the Danish assessment method and the subjective evaluation. Diamonds: low presentation level; squares: intermediate presentation level; triangles: high presentation level.

The regression line does not explain how well it represents the group of points. For this purpose we can use the residuals, which are the vertical distances between the points and the line. We also use the average of all y-values and the y-values determined by the regression method for each x-value. The determination coefficient or 'degree of explanation' (r2) is defined as:

$$r2 = SSe / (SSe + SSr)$$

where SSe is the residual sum of squares, and SSr is the regression sum of squares. In practice r2 is calculated by the Excel function LINEST.

If r2 equals 1.00 there is a perfect linear relationship between the points. If r2 is close to zero, the regression line cannot be used to explain the relation between x and y. In other words, the r2 value indicates how well the points can be described by a straight line. The closer the value is to 1, the better the description.

The relation between the x- and the y-values can also be described by the correlation coefficient, r. This is calculated as the ratio between the covariance of x- and y-values, and the product of the x-variance and the y-variance:

$$\rho = \text{covariance}(x, y) / (\sigma x * \sigma y)$$

The covariance is calculated as the deviation between the x-value and the x-average, multiplied by the deviation between the y-value and the y-average. The x-variance, sx, is the deviation between the x-value and the x-average squared. Similarly the y-variance, sy, is the deviation between the y-value and the y-average squared. The correlation coefficient is calculated by use of the Excel function CORREL. It explains the degree of relation between the x and the y values and gives a coarse indication of the shape of the swarm of points. If r is close to 1 (or -1) the shape of the group of points has the shape of a 'cigar' around the regression line. If r is close to zero, the points lie in a diffuse cloud and there is no obvious relation between x and y.

There is an important difference between the degree of explanation and the correlation coefficient. The degree of explanation, r2, assumes a functional relationship between y and x (e.g. the subjective evaluation is 'caused by' the noise). The correlation coefficient does not assume such causality and can be calculated from any two datasets.

The values calculated for the Danish method are summarized in Table V.

Table V. Data from linear regression and from correlation analysis for the Danish method

Slope	**Intersection (x = 0)**	**Degree of explanation, r2**	**Correlation coefficient, ρ**
0.26	1.61	0.88	0.94

6.2 German A-level

Figure 4 shows the A-weighted levels of the noise examples, calculated according to the German standard DIN 45 680. The level is calculated as the sum of the A-weighted levels of those 1/3-octave bands that exceed the hearing threshold. All the noise examples are used in this calculation, including those examples where the noise contains tones.

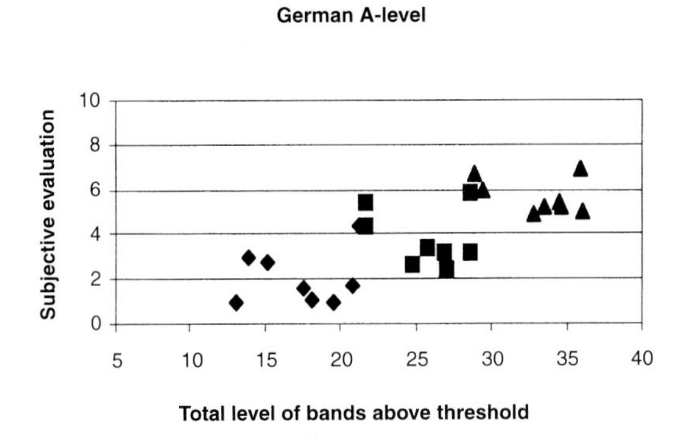

Figure 4. Illustration of the relation between the German assessment method using the A-weighted level and the subjective evaluation.

It is seen that the points fall in two groups (two lines); the upper points are the noise examples from drop forge, discotheque, and compressor. The German method (in the present interpretation) obviously cannot give a sufficient assessment of impulsive noise (drop forge and discotheque). The degree of explanation, r2, is 0.54, see table VI

Table VI. Overview of the results from regression analysis of the relation between the subjective evaluations and the different assessment methods

Assessment method	Slope	Intersection (x = 0)	Degree of explanation, r2	Correlation coefficient, ρ
Danish	0.26	1.61	0.88	0.94
German A-level	0.19	-0.98	0.54	0.73
German tonal	0.16	1.58	0.52	0.72
Swedish	0.21	2.10	0.57	0.76
Polish	0.20	1.00	0.50	0.71
Dutch proposal	0.17	2.67	0.40	0.64
C-level	0.15	-1.82	0.44	0.66

6.3 German Tonal Method

Figure 5 shows the relation between the subjective evaluation and the tone level above the hearing threshold for those noises that contain tones (according to the German assessment method). In Figure 5 only one point per noise is shown. This point corresponds to the tone with the greatest excess above the threshold.

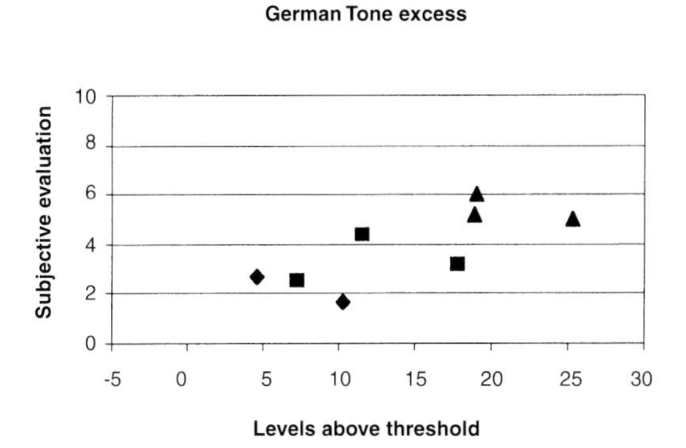

Figure 5. Illustration of the relation between the German assessment method for tonal noise and the subjective evaluation.

The Gas Turbine has two tones above threshold at the highest presentation level (triangle) and one at the intermediate level (square). There is no tone above threshold at the lowest presentation level.

The Generator has one tone above threshold at the highest level (triangle), at the intermediate level (square) and at the lowest level (diamond). The tone is at 80 Hz and thus the level has been reduced by 5 dB according to the German assessment method.

The Cooling Compressor has two tones (50 Hz and 100 Hz) above threshold at the highest presentation level (triangle) and at the intermediate level (squares). As one of the tones is at 100 Hz the level for this tone must be reduced by 10 dB according to the German assessment method. This makes the other tone (at 50 Hz) the greatest. There is only one tone above threshold at the lowest presentation level (diamond).

6.4 Swedish Method

The Swedish criterion curve must not be exceeded in any 1/3-octave band. Figure 6 shows the subjective assessment as a function of the greatest excess. The degree of explanation, r2, is 0.57.

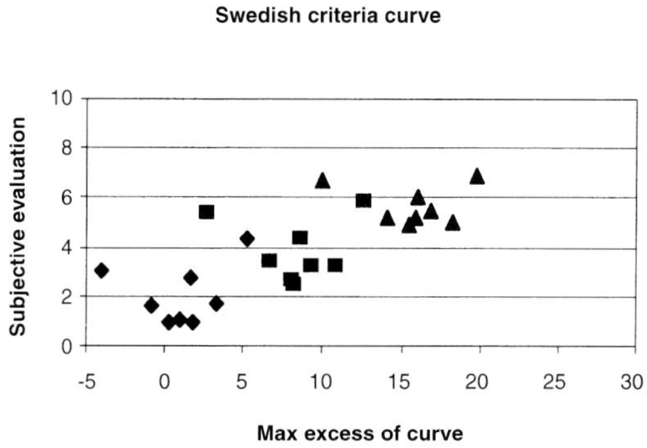

Figure 6. Illustration of the relation between the Swedish assessment method and the subjective evaluation.

It may be seen from Figure 6 that three points fall 'to the left' of the rest of the points. These three points are from the discotheque. Obviously this type of noise should have been assessed about 10 dB 'higher' for the points to fit into the rest of the points and the relative low values of r2 and r are mainly caused by these points. Removal of the discotheque points increases the degree of explanation to 0.81.

6.5 Polish Method

Figure 7 shows the excess over the Polish criterion curve, which is a curve of 1/3-octave band levels each of which corresponds to an A-weighted level of 10 dB. The other part of the Polish method, which deals with the excess of the background noise level, has not been considered.

The noises fall in two rather distinct groups, where the discotheque, the drop forge and the cooling compressor are 'to the left' of the remaining points. They should have been assessed at least 5 dB 'higher' by the method to align with the other points. If the groups of points from each nominal level are looked upon separately (diamonds, squares and triangles), it appears that the points in each group shows a tendency to a downwards slope to the right; that is, the subjective evaluation decreases as the max excess increases. The degree of explanation, r2, is 0.50 for the polish method.

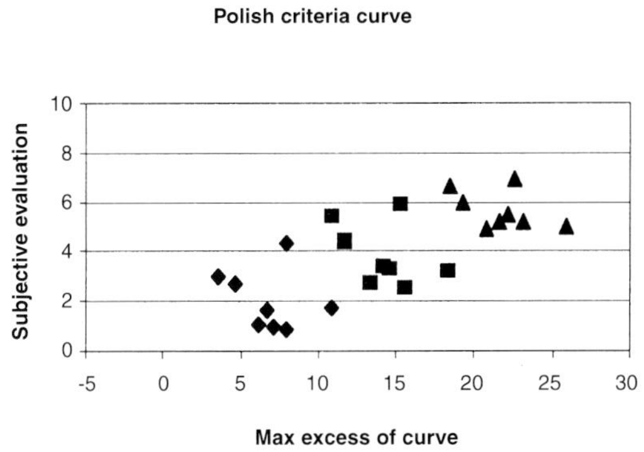

Figure 7. Illustration of the relation between the Polish assessment method and the subjective evaluation.

6.6 Dutch Proposal

The proposed assessment method for use with environmental approval of industries also employs a criterion curve. Figure 8 shows that the points are fairly spread, and no clear picture can be seen. It appears that the points from the same nominal level slope 'the wrong way' as was seen with the Polish method. As an example (filled squares) it can be seen

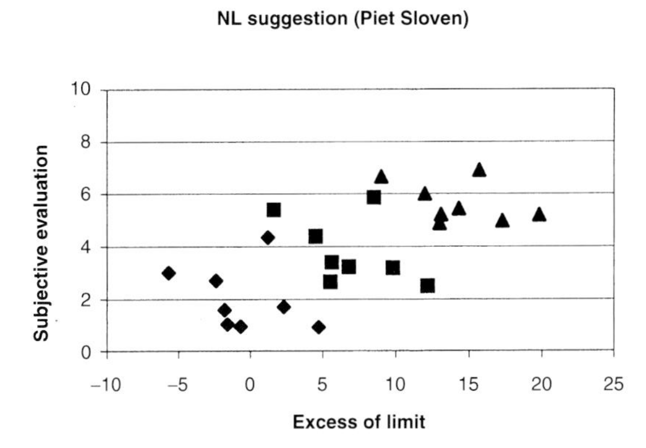

Figure 8. Illustration of the relation between the Dutch proposed assessment method by Sloven and the subjective evaluation.

that when the excess increases from 1 dB to 12 dB, the subjective evaluation decreases from 5.5 to 2.5. The degree of explanation is 0.40.

6.7 C-weighted Level

Figure 9 shows the relation between the C-weighted level of the noises and the subjective evaluation. It can be seen that the spread of the points is very large. Only the frequency range 10 Hz to 160 Hz is included in the calculation of the C-weighted level.

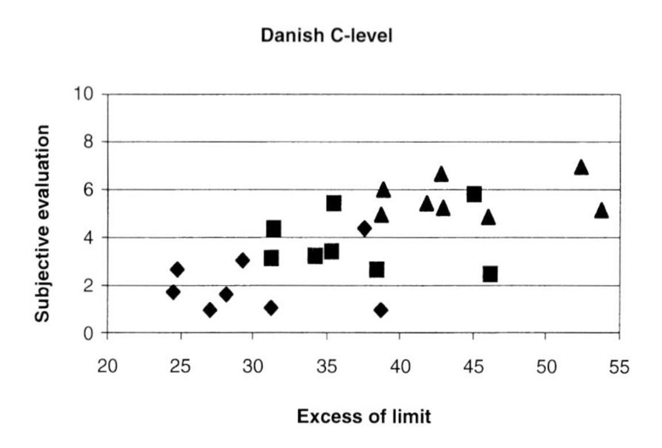

Figure 9. Illustration of the relation between the C-weighted sound pressure level and the subjective evaluation.

Overview of the Results

The results from the above analysis are summarized in Table VI.

The assessment method with far the best correlation between the subjective and the objective assessment is the Danish method, using the A-weighted level in the frequency range from 10 Hz to 160 Hz. The second best method is either the Swedish method, based on a criterion curve, or the German method using the A-weighted level.

7. DISCUSSION

The present investigation has been performed as a typical laboratory experiment in contrast to a field investigation. The advantage of a laboratory experiment is that it is possible to control almost all the experimental conditions (noises, levels, duration, presentation sequence, test subjects, etc). The disadvantage is that the presentations may not be realistic enough.

The response sheet contained all the questions on a single sheet. It has been argued that the different questions should have been on separate sheets in order to avoid biassing effect from one question to the next. Also the wording of the questions might be revised even though no test subjects reported any difficulties.

The number of subjects could be increased in order to improve the accuracy of the results but for a group of 18 it is believed that an increase of the number of test subjects would not change the general results dramatically.

The noises constitute a reasonably broad selection of low frequency sounds. The noises were selected to represent typical low frequency noises known to produce community claims. In retrospect it would have been an improvement to include more noises with an impulsive character in order to better 'test' the impulse penalty in the Danish method. All noises had clearly a low frequency character partly because of an outdoor-to-indoor filtering of the noises recorded out of doors. Traffic noise was included in order to serve as a reference noise, but due to the outdoor-to-indoor filtering the traffic noise was converted into another low frequency noise.

The criteria and evaluation methods used in this investigation are all based on some kind of measurement of the noise level. There is a clear connection between the noise level and the experienced annoyance and thus it makes sense to use such criteria and evaluation methods.

In the statistical analysis it was seen that the data deviated somewhat from a normal distribution partly caused by the saturation effect from the fixed endpoints of the scale. Despite this deviation in the distribution of the data the statistical analysis showed the expected effects and thus no attempt was made to correct for the saturation effect in the data.

The maximum excess over a criterion curve or the excess over a noise limit was used as input data for the analysis because this is the way the criteria curves and the noise limits are used in practise. In relation to a noise limits this is straightforward because the level of the noise is calculated according to some rule and compared to the limit.

For the criteria curves, on the other hand, the procedure may constitute a problem as only a single frequency band of the noise is used in the comparison and not the whole spectrum. Only the band where the maximum excess occurs is taken into account and excess at other frequency bands is neglected. From Figure 1 it is seen that the different criteria curves differ somewhat above 40 Hz and this means that the various criteria curves will give very different results if the excess occur in this frequency range. It also means that the excess decision will be very dependent on the inherent measurement uncertainty in the measurement of the spectrum. The calculation of a level – based on a spectrum – is much less sensitive to measurement uncertainty as the uncertainties are 'averaged' in the calculation process.

The measurement uncertainty is inversely proportional to the bandwidth of the analysing filter and also inversely proportional to the duration of the measurement (i.e. the integration time). This means that a one-third-octave analysis of a low frequency noise must be extended over a long period of time in order to keep the uncertainty below a certain limit. It is common practice to require that the standard deviation of repeated measurements shall be less then 0,2 dB. This corresponds to an integration time (in seconds) greater than 471/B where B is the bandwidth in Hertz of the analysing filter. For the one-third-octave filter at 10 Hz this means an integration time of almost five minutes. At 40 Hz a one-minute integration time is necessary and at 1000 Hz two seconds are needed. The noise signal should be stable over this period of time but this is not always the case in practice.

Uncritical use of criteria curves may be misleading. Some of the curves (e.g. the German and the Dutch) are hearing threshold curves and can therefore only be used to predict whether a noise is audible or not. The excess cannot predict the annoyance of the noise. This will depend on the shape of the noise spectrum.

The use of weighting functions (such as G- and A-weighting) will not automatically give a loudness or annoyance measure. In the conventional audible frequency range it is well known that neither the loudness level contours nor the A-weighting can predict the loudness of complex sounds. It is believed that loudness is a major component of annoyance. Loudness is related to the level and the spectrum of the noise. Annoyance is therefore also dependent on level and spectrum but annoyance is also influenced by (or dependent on) many other factors and these factors cannot be described by physical measurements of the noise.

8. CONCLUSION

A laboratory investigation of the annoyance of low frequency noises has been performed and the subjective evaluations were compared to the noise limits and criteria curves for low frequency noise used in the European countries. Eighteen normal hearing test subjects listened to eight different noises and evaluated the loudness, the annoyance during the day/evening and the annoyance at night. All noises had considerable low frequency content.

The results show that the Danish measuring method describes the subjectively experienced annoyance better than the measuring methods used in other countries. This result relies on the 5 dB impulse noise penalty included in the Danish method. The decision about whether or not a 5 dB penalty shall be applied to a specific noise is based on a purely subjective judgment and therefore the Danish method could be improved at this point. The Swedish method is almost as good as the Danish method if the (impulsive) discotheque sound is omitted from the analysis. The Swedish method is based on a specified criterion curve (in contrast to the Danish noise level calculation) and as such more sensitive to random measurement uncertainties.

An almost perfect correlation was found between the annoyance at day/evening and the annoyance at night. The annoyance at night is slightly lager than the annoyance at day/evening. The difference in the annoyance ratings between day and night corresponds to a level change of about 5 dB.

REFERENCES

1. D-EPA, *Low frequency noise, infrasound and vibration in the environment (In Danish). Information no. 9.* 1997, Danish Environmental Protection Agency.

2. Mortensen, F.R. and Poulsen, T. *Annoyance of low frequency noise and traffic noise Proc., 9th Intl. Meeting on Low Frequency Noise and Vibration*. 2000. Aalborg, Denmark.

3. Poulsen, T. *Annoyance of Low Frequency Noise (LFN) evaluated by LFN-sufferers and non-sufferers. in Joint Baltic-Nordic Acoustical Meeting, August 2002*. 2002. Lyngby, Denmark.

4. Poulsen, T. *Laboratory Determination of Annoyance of Low Frequency Noise. in Proc. 10th Int. Meeting on Low Frequency Noise and Vibration, September 2002*. 2002. York, UK.

5. ISO, *ISO 7196 Acoustics - Frequency weighting characteristic for infrasound measurements*. 1993, International Organization for Standardization: Geneva, Switzerland.

6. IEC, IEC *268-13. Sound system equipment. Part 13: Listening tests on loudspeakers*. 1985.

7. Mortensen, F.R., *Subjective evaluation of noise from neighbours with focus on low frequencies. Main project*. 1999, Department of Acoustic Technology, Technical University of Denmark, DK2800 Lyngby.

8. DSB, *Noise project: Isolation against noise. Technical solutions. (In Danish)*. 1987.

9. Wyle, *Preliminary evaluation of low frequency noise and vibration. Reduction retrofit concepts for wood frame structures*. 1983, Wyle Research Report WR 83-26.

10. D-EPA, *Assessment of low frequency noise from ferries (In Danish)*. 1997, Danish Environmental Protection Agency.

11. Han, L.A. and Poulsen, T., *Equivalent threshold sound pressure levels for*

Sennheiser HDA 200 earphone and Etymotic Research ER-2 insert earphone in the Frequency range 125 Hz to 16 kHz. Scandinavian Audiology, 1998. **27**(2): p. 105-112.

12. ISO, *ISO 389-8 Acoustics - Reference zero for the calibration of audiometric equipment - Part 8: Reference equivalent threshold sound pressure levels for pure tones and circumaural earphones.* 2001, International Organization for Standardization: Geneva, Switzerland.

13. Buus, S., Florentine, M., and Poulsen, T., *Temporal integration of loudness, loudness discrimination and the form of the loudness function.* Journal of the Acoustical Society of America, 1997. **101**(2): p. 669-680.

14. ISO, *ISO 389-7 Acoustics-Reference zero for the calibration of audiometric equipment- Part 7: Reference threshold of hearing under free-field and diffuse-field listening conditions.* 1996, International Organization for Standardization: Geneva, Switzerland.

15. DIN, *DIN 45680: Messungen and Bewertung tieffrequenter Geräuschimmissionen in der Nachtbarschaft - Beiblatt 1: Hinweise zur Beurteilung bei gewerblichen Anlagen.* 1997, Deutsche Norm DIN.

16. SOSFS, *Indoor Noise and High Sound Levels. General guidelines issued by the Swedish national board of health and welfare,* in SOSFS *1996:* 7/E. 1996, Socialstyrelsen, Sweden.

17. Mirowska, M. *Evaluation of Low Frequency Noise in Dwellings. New Polish Recommendation. in Proc., 9th Intl. Meeting on Low Frequency Noise and Vibration. 2000.* Aalborg.

18. Sloven, P. *Structured approach of lfn-complaints in the Rotterdam region. in Proc., 9th Int. Meeting on Low Frequency Noise and Vibration. 2000.* Aalborg.

19. Berg, G.P. and Passchier-Vermeer, W. *Assessment of low frequency noise complaints.in Inter-Noise 99.* 1999.

20. SP-INFO, *Recommendation for measurement of sound levels in rooms with low frequencies (in Swedish).* 1996, Statens Provningsanstalt.

Blast densification method: Sound propagation and estimation of psychological and physical effects

Hiroyuki Imaizumi, Yasumori Takahashi, Motoharu Jinguuji, and Sunao Kunimatsu
Geo-Technology and Environmental Assessment Research Group, Institute for Geo-Resources and Environment, National Institute of Advanced Industrial Science and Technology, 16-1 Onogawa, Tsukuba, Ibaraki 305-8569, Japan
E-mail: hiroyuki.imaizumi@aist.go.jp

ABSTRACT
Experimental results on the propagation of blasting sounds from underground explosions for preventing liquefaction of ground are described. Psychological and physical effects of blasting sounds are estimated by comparison with previous studies on the effects and evaluation of continuous low-frequency sounds. The blasting sounds generated by underground explosions were impulsive low-frequency sounds, which predominantly involved frequencies of several Hz. Frequencies above 100 Hz tended to attenuate to a greater degree during propagation from the blasting area. Sound pressure levels at frequencies above 20 Hz exceeded the hearing threshold level as well as the threshold for rattling of building fittings for low-frequency sounds even at a point about 400 m distant from the blasting area. To predict the attenuation of the blasting sounds and estimate the appropriate distance for reducing the possibility of complaint, results of numerical calculations by the parabolic equation method together with meteorological data are shown.

1. INTRODUCTION
When water-saturated sand layers, which are located below the ground-water level, are forced into strong motion by an earthquake, the layers frequently behave like a liquid under specific conditions, an effect that is called liquefaction. Liquefaction gives rise to the settlement, flotation, and demolition of underground structures such as the foundation piles of buildings and industrial facilities, and of gas and water pipelines; thus, it has a significant influence on daily life. Several measures for preventing these phenomena have been developed, and around three quarters of them are classified as densification methods.[1] The blasting densification method (BDM) is one such method: Explosives are detonated underground and their impulsive energy increases the density of the weak structure of the sand layers. This method has advantages including cost and duration of the operation. On the other hand, the generation of

impulsive low-frequency sounds and vibrations due to underground blasting is sometimes a disadvantage as compared to other methods from an environmental point of view. For this reason, an engineering assessment of BDM is being carried out in Japan.[2,3]

Blasting operations have been used worldwide, especially in mining and engineering work, because mining by blasting is much more efficient than other methods. The noise that is generated during blasting operations generally consists of impulsive sounds of short duration with high sound pressure. Complaints due to the perception of the low-frequency sounds or the rattling of building fittings have sometimes occurred even at long-ranges of propagation, since the blasting sounds predominantly involve low-frequency components.[4] With respect to the noise generated by underground explosions, the generation mechanisms, frequency characteristics, and influences upon dwellings have been investigated.[5] Furthermore, the relationship between the amount of explosive used in engineering works and the peak sound pressure level has been reported from a practical point of view.[6] Comprehensive studies on sound and vibration due to blasting have also been conducted.[7] In addition, the effects of atmospheric conditions and ground surfaces on the propagation of low-frequency sounds have been examined, and the attenuation of low-frequency sounds has been predicted and compared to the results of numerical calculations.[8,9] The influence of continuous low-frequency sound/infrasound has been evaluated individually from both psychological and physical points of view.[10,11] G-weighting[12] is defined in ISO 7196 as the frequency characteristics for infrasound of 1 to 20 Hz and reflects the perception and response of people to it.

At present, there are no international regulations concerning low-frequency noise, although some recommendations and guidelines have been decided individually in some countries.[13,14] No criteria for low-frequency noises have been decided in Japan. However, a technical committee in INCE/Japan proposed a measuring method and basic specifications of the measuring equipment for low-frequency noise/infrasound in 1991.[15] Each local government has been conducting procedures according to this method.

There are several points in common between characteristics of the blasting sounds generated by the BDM and those in the previous studies described above. However, there exist some differences as follows:

1) Explosives are placed in the weak sand layers instead of hard ground such as bedrock,
2) The BDM is a new application of the blasting operation in Japan,
3) If the BDM is to be applied close to residential areas, more attention to the surrounding circumstances has to be paid than is usual under blasting operations.

In this study, we describe the characteristics of the blasting sounds generated by underground explosions in a field experiment for the BDM and examine their psychological and physical effects, by comparison with previous studies of the effects and evaluation of low-frequency sounds. In addition, numerical calculations using the parabolic equation method are used to predict noise propagation and to estimate the appropriate distance from residential areas for reducing the possibility of complaints.

2. EXPERIMENTAL WORK

2.1 Experimental Site and Specifications of the Blasting

The field experiment was carried out on flat, reclaimed ground near Tokachi Port in Hokkaido, Japan (Figure 1). The reclaimed ground consisted mainly of sand, and its area was about 4,200 m^2. Figure 2 is a plan view of the blasting area showing the arrangement of the boreholes. In the area, 127 boreholes were excavated in the reclaimed ground, and 844 kg of explosives were placed in the boreholes. The blasting area was divided into 4 areas: A, B, C, and D. The numbers of boreholes were 54, 48, and 25 in areas A, B, and C, respectively. Figure 3 shows cross-sections of the boreholes that illustrate the locations of the explosives. In each borehole, two explosive charges were set up at different depths below the ground surface. The total weight of explosives per borehole was 7 kg. In order to fix the explosives in the boreholes, small stones were installed between the explosives as stemming material. The depths of explosives in areas A and B were the same, while the 2 explosives were set up slightly deeper in area C. Figure 4 shows the flow of the blasting in the field experiment. The explosives in the boreholes of areas A and B were ignited according to a constant time sequence with 0.1 s delay (dotted lines in Figure 2). After completing the explosions in areas A and B, ignition in

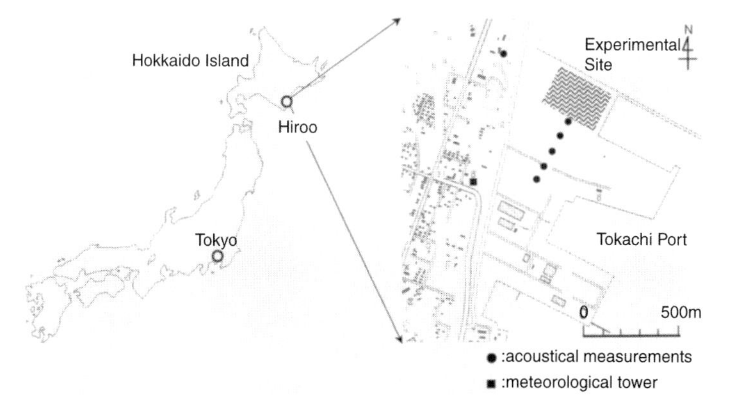

Figure 1. Location of the field measurement for BDM and a detail map of the experimental site near Tokachi Port.

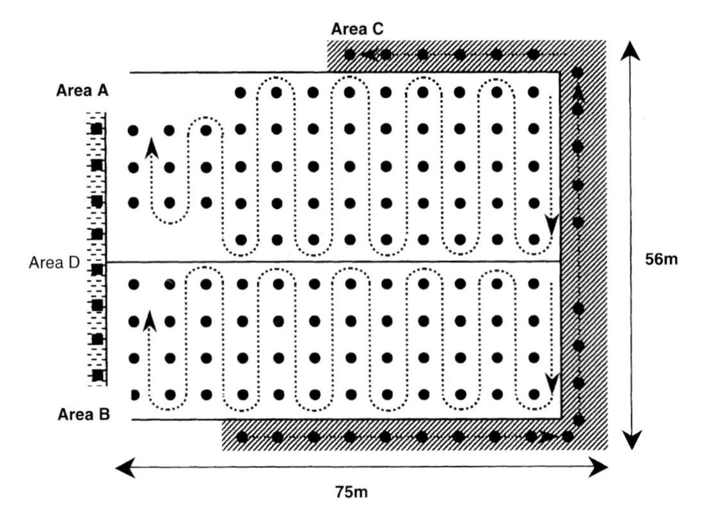

Figure 2. Plan view of the blasting area. Filled circles (●) and squares (■) are positions of the boreholes and another series of explosives, respectively.

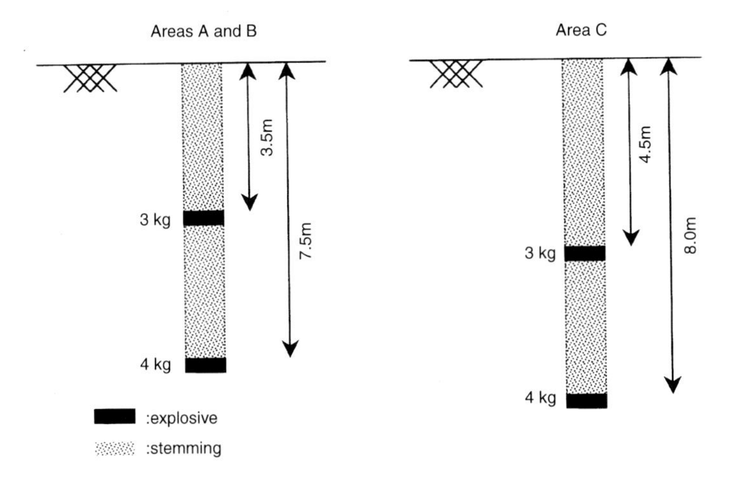

Figure 3. Charge patterns of explosives in the boreholes.

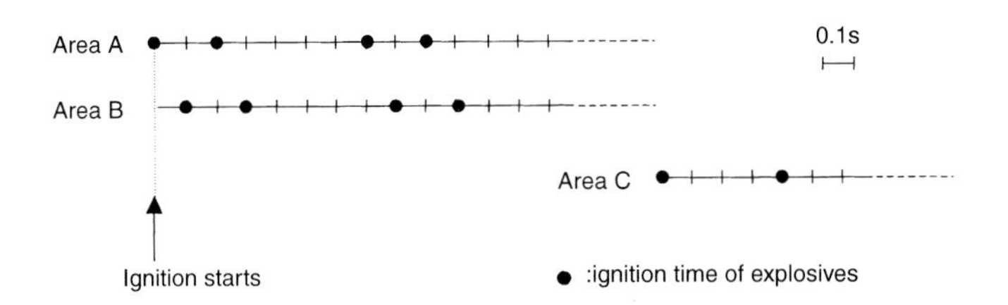

Figure 4. Time sequence of the blasting in areas A, B, and C.

area C was started. In areas A and B, the time delay of ignition between the 2 explosives in each borehole was 0.2 s, while that in area C was 0.5 s. The time intervals of the ignition between boreholes in areas A, B, and C were 0.7, 0.7, and 0.5 s, respectively. Total duration for completing the ignition of all the explosives was 62.3 s. Area D contained another series of explosives, and was ignited after the completion of area C.

2.2 Measuring Apparatus

As shown in Figure 1, blasting sounds generated by underground explosions during the BDM experiment were observed at 6 measuring points by low-frequency sound level meters (RION, type NA-17). The frequency characteristic of the low-frequency sound level meters was flat between 1 to 1 kHz, and the response was down by 1 dB/octave below 1 Hz and by 2 dB/octave over 1 kHz. The 5 measuring points were located southwest of the blasting area, and the points were 30, 100, 200, 300, and 382 m distant from the blasting area, respectively. The blasting sounds measured at these points were recorded on a DAT recorder. The DAT recorder had a flat frequency response from DC to 5 kHz. Another low-frequency sound level meter was set up at a measuring point about 300 m northwest of the blasting area. There was a 25 m difference in elevation between the measuring point and the experimental site. In the experiment, the frequency-weighting "F" and the time-weighting "S" were applied to the data collected from all the low-frequency sound level meters. Recorded blasting sounds were analyzed in 1/3-octave bands by a real-time frequency analyzer (Brüel & Kjær, type 2133).

Meteorological data during the experiment were recorded at a meteorological tower located southwest of the blasting area. An ultrasonic anemometer and a thermometer were set up on the top. These meteorological sensors were 40 m different in elevation relative to the surface of the blasting area. Wind speed, wind direction, and atmospheric temperature were observed every 5 s, and variations in the meteorological data were recorded on a chart.

3. RESULTS AND DISCUSSION

3.1 Propagation of Blasting Sounds

Figure 5 shows sound pressure-time waveforms for 5 s after ignition of the explosives measured at points 30, 100, and 300 m distant from the blasting area. At the 30-m measuring point, impulsive increases of sound pressure are clearly observed after ignition of the explosives underground. It is suggested that those are primarily generated by rapid motion of the ground surfaces around the boreholes by the underground explosions. As the propagation distance increases, the sound pressures from the blasting sounds decrease rapidly, while the waveforms taken at the further measuring points are relatively similar to that at 30 m, and some sound-pressure peaks still remain. From observation of the time waveforms, high-frequency components in the blasting sounds that propagated to the 300-m point were obviously attenuated more than others. By comparing the waveforms synchronized in time as in Figure 5, sound waves that include relatively high-frequency components are observed in the initial part of the waveforms, especially for the more distant receivers. These may have been generated by excitation of the air induced by blasting vibrations traveling faster than the blasting sounds.

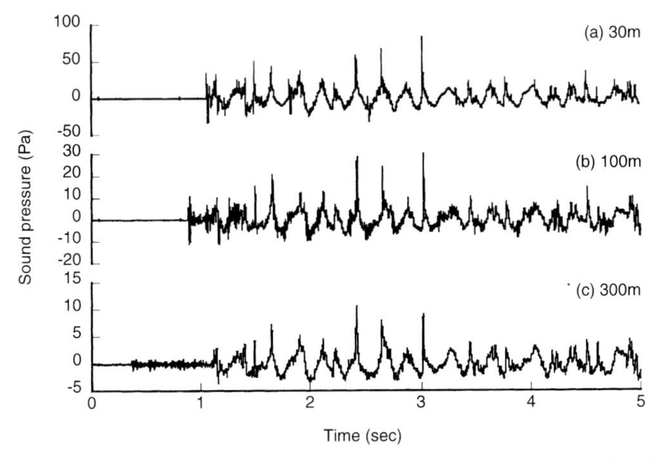

Figure 5. Comparison of sound pressure-time waveforms measured at points (a) 30 m, (b) 100 m, and (c) 300 m distant from the blasting area.

Results of frequency analyses in 1/3-octave bands of the blasting sounds are shown in Figure 6. Among all the measuring points, the primary frequency components are below 10 Hz and the frequency band of 4 Hz is predominant. The peak frequency probably corresponds to the time delay of the ignition between the upper and lower explosives in the boreholes. On the other hand, frequencies higher than 100 Hz decrease remarkably with increasing propagation distance. Moreover, the sound pressure levels that were observed at the measuring point 300 m northwest are relatively similar to those observed 200 m southwest.

Figure 7 plots the excess attenuation at each measuring point relative to the sound pressure level measured at 30 m. The tendencies of the excess attenuation are similar to each other, except at the point 382 m southwest. Excess attenuation at frequencies below 25 Hz are around -5 dB up through the 300-m southwest point, and the reflected waves at these frequencies seem not to be strongly attenuated and be changed in phase on the ground surface. On the other hand, the sound pressure levels above 100 Hz decrease gradually, as the propagation distance increases. The components at frequencies below 10 Hz attenuate remarkably at the

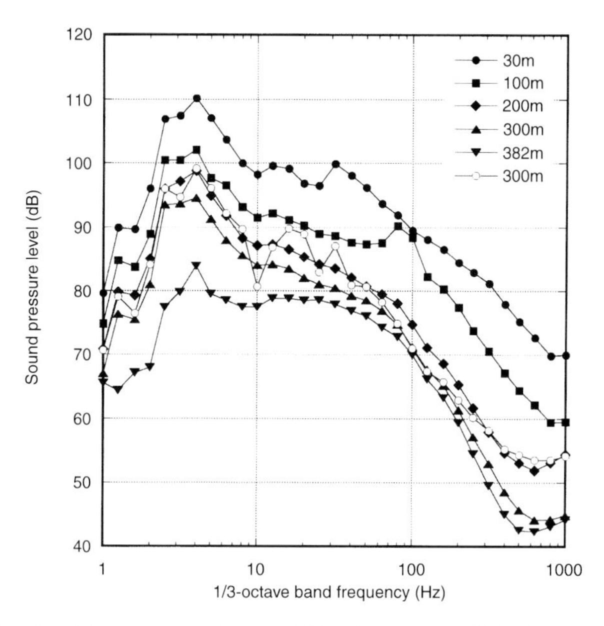

Figure 6. Results of frequency analyses in 1/3-octave bands of blasting sounds.

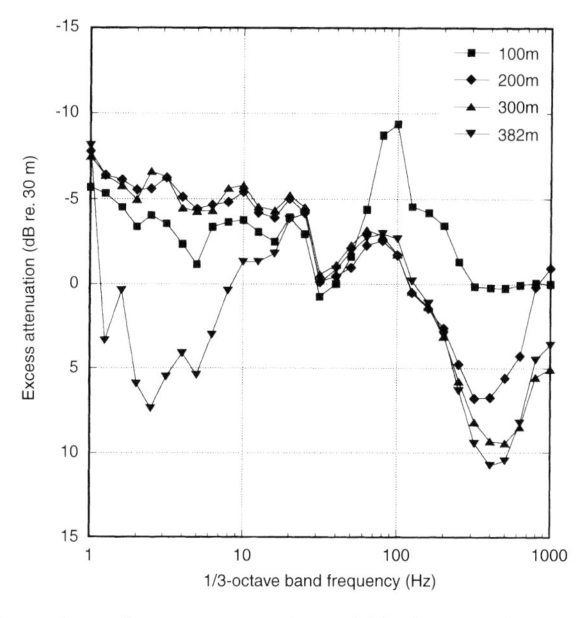

Figure 7. Comparison of excess attenuations of blasting sounds measured by the southwest receivers.

382-m measuring point. This may be caused by interference with reflected waves on several piles of rocks about 20 m behind the measuring point.

3.2 Comparison with Previous Investigations into the Evaluation of Low-Frequency Noises

It is well known that both direct human perception and the rattling of building fittings as a secondary effect have stimulated complaints against low-frequency sounds. We compared previous studies on the influence of low-frequency sounds to the blasting sounds created during the BDM, although the estimation method for impulsive sounds such as those due to blasting is still under consideration. Figure 8 shows both the thresholds of human sensation (curve A)[10] and the rattling of Japanese building fittings subjected to incident low-frequency sine waves (curve B).[11] The effects of low-frequency sounds are divided into 4 domains of frequency and sound pressure by the curves A and B.[7] Each is summarized as follows:

I. There may be no complaints at all.
II. There is no rattling, but some people sense sounds similar to a driven motor.
III. Sometimes the sound of the rattling is detected and complaints are made about a weird feeling.
IV. Due to both human perception and the rattling, strong complaints are frequently made.

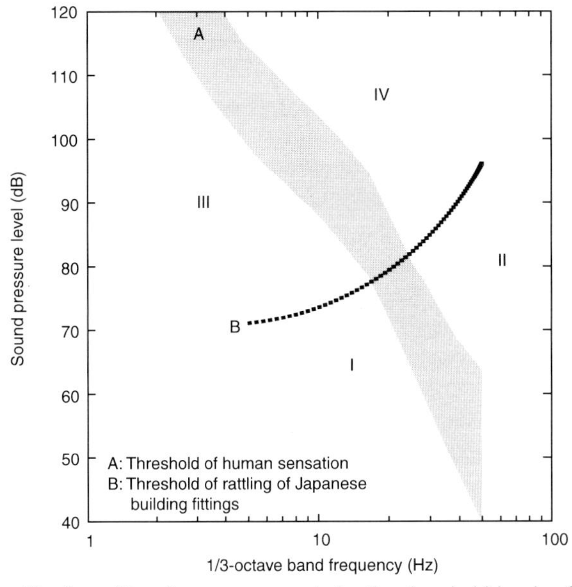

Figure 8. Classification of low frequency sounds by the threshold levels of both human sensation and rattling of Japanese building fittings (see text for explanation).

Figure 9 compares the frequency analyses shown in Figure 6 to both thresholds in Figure 8. In addition, the sound pressure levels at which 50% of people have oppressive or oscillatory feelings for low frequency sounds are plotted by curve C.[16] At the measuring point 30 m distant from the blasting area, sound pressure levels at frequencies above 3 Hz are larger than the threshold level of sensation (curve A), and those below 50 Hz are also greater than the threshold for rattling (curve B). The frequency range from 3 to 50 Hz corresponds to area IV, in which complaints will occur frequently. Moreover, the sound pressure levels of frequency components above 12.5 Hz are larger than those at which half of the people have oppressive or oscillatory feelings (curve C). While the sound pressure levels of the blasting sounds decrease as the propagation distance increases up to 300 m, the levels at frequencies below 25 Hz and above 20 Hz are still greater than the thresholds of rattling and human sensation. These sound pressure levels are classified into areas III and II, respectively. Moreover, the levels at around 40 Hz are slightly larger than the sound pressure levels for the threshold of oppressive or oscillatory feelings. Even at the 382-m measuring point, the predominant frequencies of around 4 Hz are more than the threshold for rattling, and the levels at frequencies above 20 Hz are larger than the sensation threshold level. From these results, complaints of perception of low-frequency sounds and the rattling of Japanese building fittings may occur even up to 400 m from the blasting area.

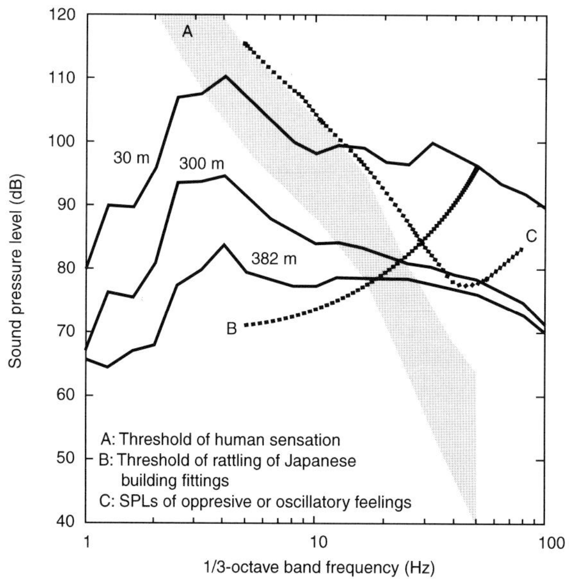

Figure 9. Comparison between 1/3-octave band frequency analyses of blasting sounds and threshold levels of both human sensation and rattling of Japanese building fittings.

Figure 10 shows the distance attenuation of G-weighted sound pressure levels. The G-weighted levels were numerically calculated by a computer program, in which signal processing of the sound level meter is digitally simulated.[17] In the program, frequency and time weightings were chosen appropriately, and the weighted sound pressure level can be obtained from the time waveforms of blasting sounds measured by the frequency-weighting "F". The G-weighted sound pressure level was around 112 dB close to the blasting area and remained about 96 dB even at 300 m. Up to the 300-m measuring point, the G-weighted sound pressure levels decreased by an almost constant gradient with increasing propagation distance. The large decrease at the 382-m point is due to the remarkable attenuation at the frequency range of infrasound as shown in Figure 7. On the other hand, the G-weighted level measured at 300 m northwestward is about 10 dB higher than that at the same distance southwestward.

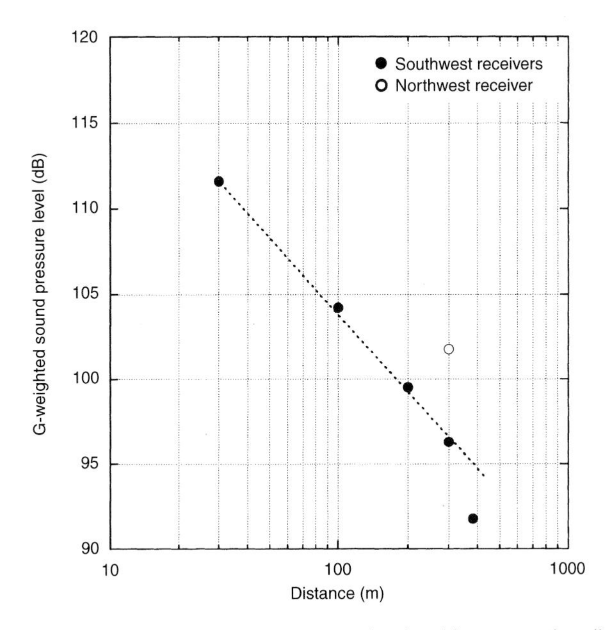

Figure 10. Change of G-weighted sound pressure levels with propagation distance.

From actual field investigations in Japan,[18] disturbance of sleep seems to appear at a G-weighted sound pressure level of around 100 dB. In addition, other field investigations[19] have reported that complaints about low-frequency sounds fluctuating largely with impulsive components occurred at G-weighted sound pressure levels between 84 to 100 dB. Thus, in the present study, the G-weighted sound pressure levels of 90 to

95 dB that were recorded around 400 m from the blasting area may result in both sleep disturbance and complaints about the blasting sounds.

3.3 Numerical Calculation of Attenuation of Blasting Sounds Using the Parabolic Equation Method

The parabolic equation (PE) method[20,21] is a very effective and powerful technique for solving various wave equations and can be applied to give numerical solutions for predicting sound propagation in a realistic outdoor environment. The advantage of the PE method is that it is able to treat complicated environments in a relatively simple way by assuming that the sound wave always propagates outward from a sound source without any backscattering. Therefore, the PE method can approximate a complicated boundary value problem with a simpler initial value problem, and the boundary value problem can result in solvable differential equations.

If we apply the axial symmetric approximation to the three-dimensional Helmholtz equation, the following two-dimensional Helmholtz equation is obtained:

$$\left[\frac{\partial^2}{\partial x^2} + \left(\frac{\partial^2}{\partial z^2} + k^2\right)\right]\Psi = 0, \tag{1}$$

where we use x - z coordinates, and Ψ and k are the velocity potential and wave number, respectively. By replacing the z-dependent part of eq (1) by an operator Q, we obtain

$$\left(\frac{\partial}{\partial x} - i\sqrt{Q}\right)\left(\frac{\partial}{\partial x} + i\sqrt{Q}\right)\Psi = 0. \tag{2}$$

Considering only the single outgoing sound propagation term of eq (2), we can write

$$\left(\frac{\partial}{\partial x} - i\sqrt{Q}\right)\Psi = 0. \tag{3}$$

We choose a solution that is an outgoing wave with a slowly varying factor $\varphi(x, z)$:

$$\Psi(x, z) = \varphi(x, z)\exp(ik_0 x), \tag{4}$$

where k_0 is a reference wave number. Substituting eq (4) into (3), eq (3) becomes

$$\frac{\partial \varphi}{\partial x} = \mathrm{i}\left(\sqrt{Q} - k_0\right)\varphi. \tag{5}$$

For sufficiently short-range steps, the operator Q can consider to be independent of range. Then eq (5) can be rewritten as

$$\exp\left[-i\frac{\Delta x}{2}\left(\sqrt{Q}-k_0\right)\right]\varphi(x+\Delta x)=\exp\left[i\frac{\Delta x}{2}\left(\sqrt{Q}-k_0\right)\right]\varphi(x). \quad (6)$$

If the exponential operator in eq (6) is approximated by a linear expansion, it becomes

$$\left[1-\left(i\frac{\Delta x}{2}\right)\left(\sqrt{Q}-k_0\right)\right]\varphi(x+\Delta x)=\left[1+\left(\mathrm{i}\frac{\Delta x}{2}\right)\left(\sqrt{Q}-k_0\right)\right]\varphi(x).$$

Now, if the following approximation is applied for the operator $\sqrt{Q}$, we obtain the wide-range parabolic equation.

$$\sqrt{Q}\equiv k_0\sqrt{1+q}\approx k_0\frac{1+\frac{3}{4}q}{1+\frac{1}{4}q}$$

Figure 11 shows the fluctuations of atmospheric temperature, wind speed, and wind direction observed at a 40 m height from the ground surface before and after the field experiment. As shown in the figure, atmospheric conditions in this period were relatively calm. Wind speed in

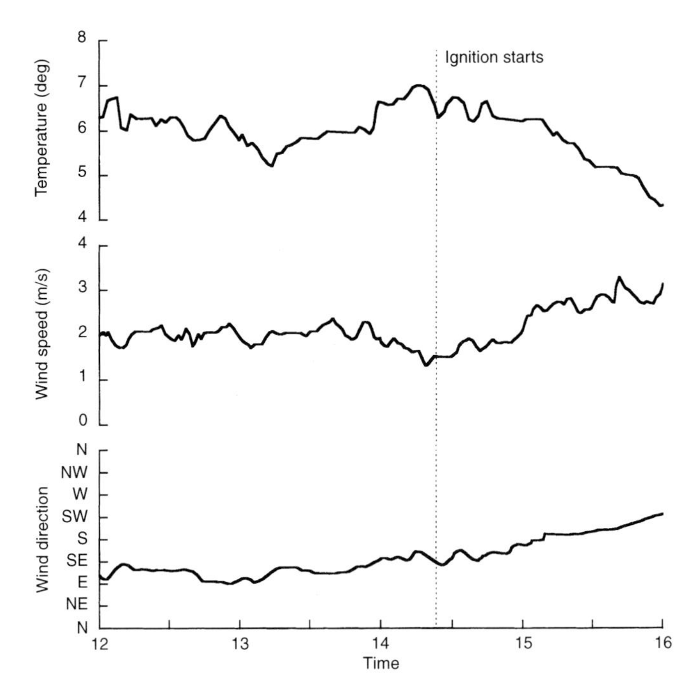

Figure 11. Variations of atmospheric temperature, wind speed, and wind direction measured at a meteorological tower before and after the field measurement. A dotted line shows the time when ignition started.

the experiment was almost constant at around 1.5 m/s, and the direction of wind changed slightly from southeast to east. Generally, the speed of sound, $c(z)$, varies vertically from the ground surface and is approximated by[21]

$$c(z) = \begin{cases} c_0 + a \ln(z/z_0), & z > z_0, \\ c_0, & z \leq z_0. \end{cases}$$

c_0 is the speed of sound calculated from the atmospheric temperature on the ground surface, and z_0 is a parameter related to the roughness of the surface.[22] z_0 was assumed to be 0.01, because the ground surface of the experimental site was flat and smooth. Moreover, the refraction parameter a was obtained from fitting the calculated speed of sound from meteorological data with the above equations. Figure 12 is an estimation of the sound speed profile above the ground surface during the field experiment, for which the refraction parameter a is equal to -0.13. The speed of sound decreases slightly with altitude.

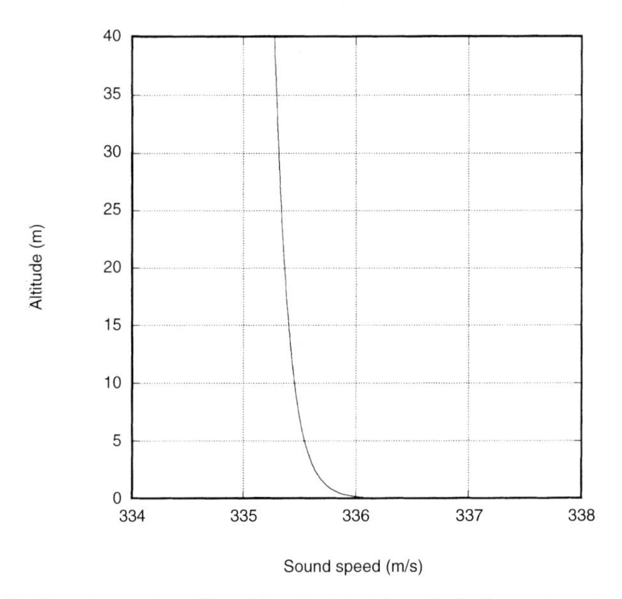

Figure 12. Sound-speed profile, c(z), calculated from both the meteorological data and logarithmic approximation.

In calculations by the PE method, the model proposed by Wilson[23] was applied to obtain the ground impedance, and the effective flow resistivity was assumed to be 800 kPa sm^{-2}. As described in the previous section, the explosives were located over a wide area, and therefore may not be treated as a point source. In order to compare the results to numerical calculations, the sound pressure levels that had been obtained from a single 4-kg shot of explosives in series with the blasting in area D were used.

Relative sound pressure levels for frequency bands at 4, 8, 16, 40, and 160 Hz calculated by the PE method using the estimated sound speed profile in the field experiment are shown in Figure 13. The filled circles indicate results of the measurements. The predicted and the measured levels agree fairly well within a 400-m propagation range from the blasting area. A difference of less than 5 dB was seen at a frequency of 4 Hz at the 382-m point. This was caused by reflected sounds from several piles of rocks located just behind the measuring point and by the PE method that cannot handle backscattering of sounds principally. Based on the predictions shown in Figure 13, no frequency band attenuates significantly under the condition of weak upward refraction, even after long-range

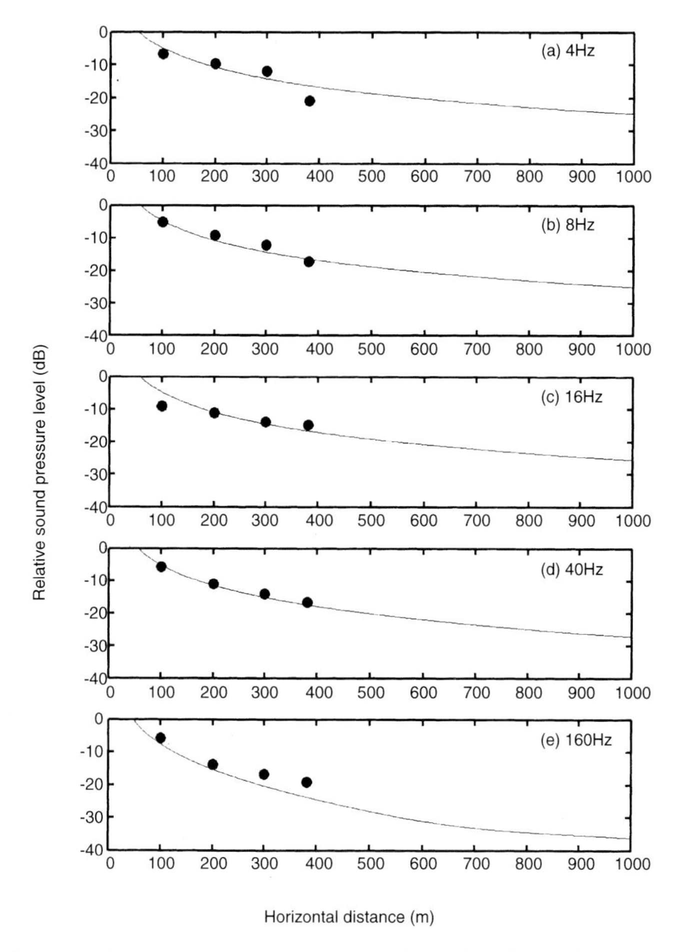

Figure 13. Comparisons of relative sound pressure levels in between the measurements (●) and calculations (solid lines) by the PE method for frequencies of (a) 4 Hz, (b) 8 Hz, (c) 16 Hz, (d) 40 Hz, and (e) 160 Hz.

propagation up to 1 km. The predicted attenuation of the blasting sounds in the field experiment at predominant frequencies such as 4 and 8 Hz is around 25 dB at 1 km away, so the possibility of complaints about the blasting sound will likely still remain since the sound pressure levels decrease by only several dB as compared to the level observed at 400 m.

As a case study we estimated the relative sound pressure levels under larger sound speed profiles above the ground surface than occurred in the field measurement. Relative sound pressure levels at the 5 frequency bands for the refraction parameters of a = 1.0 and -1.0 are plotted in Figure 14. Differences between the attenuations calculated under both upward and downward refractions at frequencies of 4 and 8 Hz are

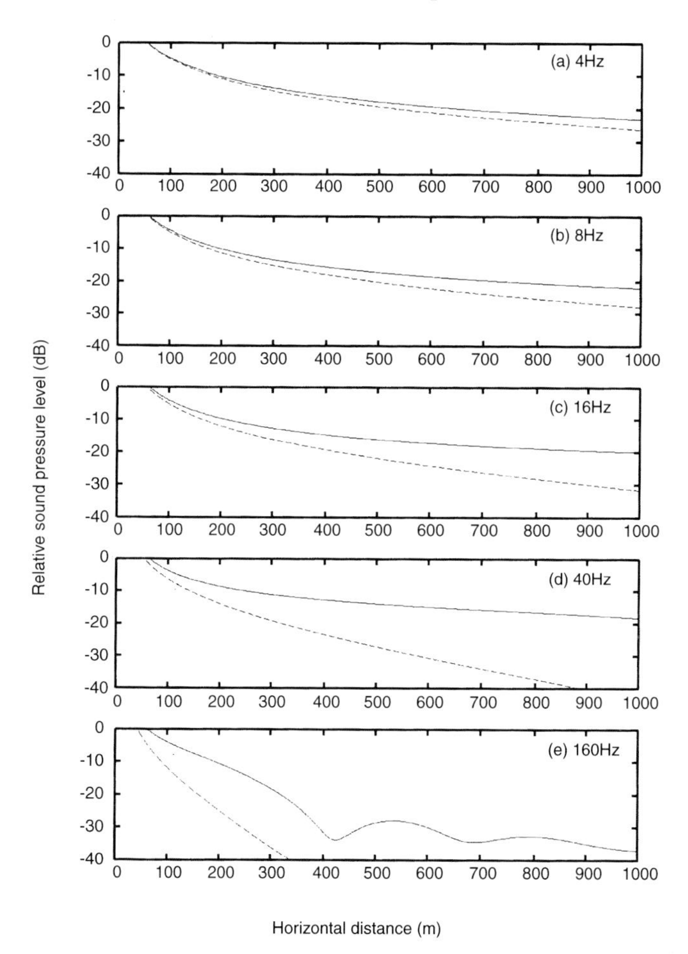

Figure 14. Relative sound pressure levels estimated by larger sound speed profiles with refraction parameters a = 1.0 (downward refraction: solid lines) and a = -1.0 (upward refraction: broken lines). Other features are the same as in Figure 13.

small and are less than 10 dB even 1 km distant from the blasting area. In addition, as compared to the relative levels of the field measurement, the differences from the cases of the larger sound speed profiles are within several dB. However, the sound pressure levels at higher frequencies propagated in the upward refraction drop rapidly with increasing propagation distance, while the levels under downward propagation even increase gradually except at 160 Hz.

Figure 15 compares the predicted sound pressure levels at 1 km under different meteorological conditions to several effects of the low-frequency sound. For the estimated condition in the period of the field experiment, sound pressure levels at frequencies of 5 and 6.3 Hz exceed the threshold of the rattling of building fittings and those at frequencies over 25 Hz may be close to the threshold of human sensation. Even with stronger upward and downward refractions, frequencies at 5 and 6.3 Hz were estimated to be over the threshold of the rattling. In addition, the levels above 40 Hz to under severe downwind conditions seem to exceed the hearing threshold level. On the other hand, most of the frequency bands except at 5 and 6.3 Hz in the severe upwind condition, were classified into area I, in which there may be no complaints at all. Consequently, the complaints for the rattling of building fittings may still

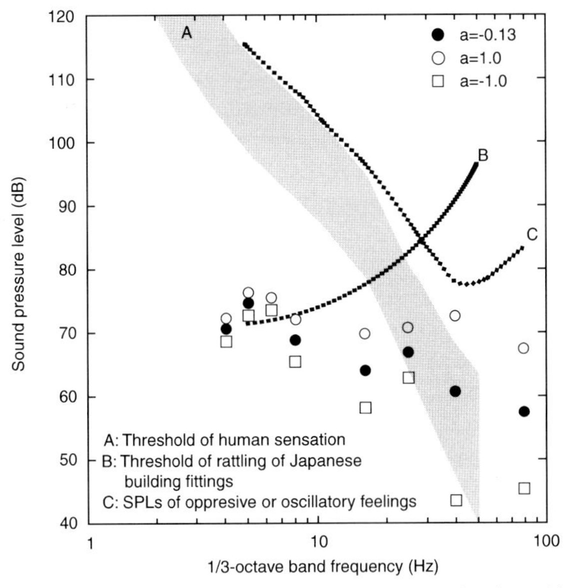

Figure 15. Comparisons between the predicted sound pressure levels at 1 km away from the blasting area under different meteorological conditions and several effects of low-frequency sounds. Filled circles shows the levels calculated from the meteorological condition in period of the field experiment (refraction parameter $a = -0.13$). Open circles and squares show the levels calculated by $a = 1.0$ and $a = -1.0$, respectively.

occur even at 1 km distant from the blasting area under these meteorological conditions. Therefore, it is recommended based on these numerical calculations and comparisons that some measures near the sound source, such as control of the charge for the BDM, should be taken for preventing and/or reducing the physical and psychological effects of the blasting sounds.

4. CONCLUSIONS

Characteristics of blasting sounds generated by underground explosions in a field experiment of the BDM were investigated and compared to previous studies of the evaluation and effects of low-frequency sounds. In addition, the parabolic equation method was applied to noise prediction and estimation of the appropriate distance for reducing the possibility of complaints about the blasting sounds.

An impulsive increase of sound pressure following the detonation of the explosives was observed in sound pressure–time waveforms of the blast sounds at the nearest measuring point, while sharpness in the time waveforms decreased during distant propagation because of the attenuation of high-frequency components. The blasting sounds consisted predominantly of frequencies below 10 Hz. Attenuation of the predominant frequencies was small even after long-range propagation.

Sound pressure levels at the measuring point close to the blasting area exceeded thresholds of both human sensation and the rattling of building fittings. The levels at frequencies above 20 Hz were also over the level at which 50% of people have oppressive or oscillatory feelings. Even at the most distant measuring point, the sound pressure levels suggested the occurrence of rattling of building fittings and still exceeded the threshold of human sensation at relatively higher frequencies.

The G-weighted sound pressure level at the nearest measuring point was more than 110 dB and decreased at a constant gradient with increasing propagation distance. From comparison with field investigations of low-frequency sounds, the G-weighted levels observed in the present study may possibly cause sleep disturbance and the occurrence of complaints even at the most distant measuring point.

Numerical predictions using the parabolic equation method with observed meteorological data agreed fairly well with measurements up to a range of 400 m. The calculated attenuations at the predominant frequencies of the blasting sounds at 1 km away did not greatly increase under the same meteorological conditions as the field experiment. A case study showed that sound pressure levels at higher frequencies were influenced more by a larger sound speed gradient than those at the predominant frequencies of 4 and 8 Hz. And, as compared the results of numerical calculation to the effects of low-frequency sounds, the predominant frequencies may cause some complaints even at a receiver

1 km away. This indicates that a greater distance might be necessary to reduce the levels at predominant frequencies under the conditions of the field measurements. These results suggest the importance of implementing some measures close to the sound source, such as charge control of the blasting, in order to prevent physical or psychological effects of the blasting sound and to thereby reduce the possibility of complaints.

ACKNOWLEDGEMENTS

This work was supported by the Ministry of the Environment, Japan.

REFERENCES

1. T. Ishiguro and H. Shimizu, "Better countermeasures against liquefaction," J. Japan Society of Civil Engineering **83,** 17-19 (1998) (in Japanese).

2. S. Kunimatsu *et al.*, "Ground vibration characteristics by blast densification method," Proc. Annual Meeting of Japan Society of Civil Engineering, 656-657 (1999) (in Japanese).

3. H. Imaizumi *et al.*, "Characteristics of blast low frequency noise/infrasound by blast densification method and the comparisons with several evaluation methods for low frequency noise/infrasound," J. INCE/J **26,** 341-350 (2002) (in Japanese).

4. T. Isei *et al.*, "Long range propagation and effects of blast sound due to explosion of high explosive," Proc. Inter-Noise **98,** 1119-1122 (1998).

5. E. Kuroda *et al.*, "Air vibration from underground explosions," Kogyo Kayaku **46,** 229-236 (1985) (in Japanese with English abstract).

6. M. Shioda, "Low frequency noise by blast," J. INCE/J, **13,** 326-336 (1989) (in Japanese).

7. Committee of low frequency noise of INCE/JAPAN, Noise and vibration due to blasting (Sankai-do, Tokyo, 1996) (in Japanese).

8. L. R. Hole *et al.*, "Effects of strong sound velocity gradients on propagation of low-frequency impulse sound: Comparison of fast field program predictions and experimental data," J. Acoust. Soc. Am. **102,** 1443-1453 (1997).

9. J. S. Robertson, "Low-frequency sound propagation modeling over a locally reacting boundary with the parabolic equation," J. Acoust. Soc. Am. **98**, 1130-1137 (1995).

10. Y. Tokita, "On the evaluation of infra and low frequency sound," J. Acoust. Soc. Jpn. **41**, 806-812 (1985) (in Japanese).

11. Environment agency, Japan, Study on the effects of infra- and low-frequency sound to houses and buildings (Environment agency, Japan, 1978).

12. ISO 7196-1995, "Acoustics - Frequency-weighting characteristics for infrasound measurements." (1995).

13. M. Mirowska, "Evaluation of low-frequency noise in dwellings. New Polish recommendations," J. Low Freq. Noise Vib. and Active Control **20**, 67-74 (2001).

14. J. Jakobson, "Danish guidelines on environmental low frequency noise, infrasound and vibration," J. Low Freq. Noise Vib. and Active Control **20**, 141-148 (2001).

15. INCE/JAPAN Technical committee, "Measuring method of low-frequency noise and infrasound," Tech. Report **11**, 1-38 (1991) (in Japanese).

16. T. Nakamura *et al.*, "Fundamental study on sensation and evaluation of human beings for low frequency sounds," Environmental Science, 1-20 (1979) (in Japanese).

17. S. Kunimatsu *et al.*, "Multi-evaluation of blast sound propagation attenuation by using digital simulation for both frequency and time weighting," J. INCE/J **17**, 261-267 (1993) (in Japanese with English abstract).

18. H. Ochiai and A. Yokota, "Evaluation of low frequency sound by G-weighted sound pressure level," Proc. Annual meeting of INCE/J, 191-194 (1998) (in Japanese).

19. H. Ochiai *et al.*, "The present status of complaints for low frequency noise," Proc. Technical meeting of INCE/J, 5-12 (1998) (in Japanese).

20. M. West *et al.*, "A tutorial on the parabolic equation (PE) model used for long range sound propagation in the atmosphere," Appl. Acoust. **37,** 31-49 (1992).

21. K. E. Gilbert and M. J. White, "Application of the parabolic equation to sound propagation in a refracting atmosphere," J. Acoust. Soc. Am **85,** 630-637 (1989).

22. H. A. Panofsky and J. A. Dutton, Atmospheric Turbulence – Models and methods for engineering applications – (John Wiley & Sons, 1984), Chapter 6.

23. D. K. Wilson, "Simple, relaxational models for the acoustical properties of porous media," Appl. Acoust. **50,** 171-188 (1997).

Annoyance of low frequency noise (LFN) in the laboratory assessed by LFN-sufferers and non-sufferers

Torben Poulsen
Ørsted•DTU, Acoustic Technology, Building 352, Technical University of Denmark, DK-2800 Lyngby, Denmark
E-mail: dat@oersted.dtu.dk

ABSTRACT
In a series of listening tests, test subjects listened to eight different environmental low frequency noises to evaluate their loudness and annoyance. The noises were continuous noise with and without tones, intermittent noise, music, traffic noise and low frequency noises with an impulsive character. The noises were presented at L_{Aeq} levels of 20, 27.5 and 35 dB. The main group of test subjects (the reference group) comprised eighteen young persons with normal hearing. A special group of four subjects who had reported annoyance due to low frequency noise in their homes was also included. It was found that the special group generally assessed the annoyance of the noises much higher, especially the annoyance at night.

1. INTRODUCTION

Annoyance from low frequency noise and infrasound is a 'hot topic' on both the scientific and the political scene. Some people are very annoyed by this kind of noise and much debate has occurred about noise limits and especially about the measurement methods related to the limits. The measurement of low frequency noise is difficult because it can be hard to isolate the noise in question. In [1] a comparison is made between the results of various objective measuring methods and the subjective annoyance experienced in a laboratory test.

In the same investigation, [1], the subjectively assessed annoyance from a number of environmental low frequency noises was evaluated by two groups of test subjects. One group, the reference group, consisted of 18 young normal-hearing persons. The other group, the special group, consisted of four persons who suffer from low frequency noise in their homes. The present paper is about a comparison between the results from the reference group and the special group. The signals in the investigation were restricted to low frequency sounds. No infrasound signals were used.

Sound in the frequency range below 20 Hz is defined as infrasound. The G-weighting function standardised in ISO 7196 [2], has relates closely to the shape of the hearing threshold in the infrasound region. The loudness and annoyance due to infrasound increase very quickly with increasing level. The hearing threshold for single tones is usually about 95 dB(G), and tones with a 20 dB higher level are expected to be sensed as very loud. It can be assumed that infrasound below the hearing threshold is not annoying.

Whereas infrasound is a well-defined concept, low frequency noise is not. Low frequency noise may be defined to comprise the frequency range 10 Hz – 160 Hz [3], but other assessment methods may define other frequency ranges (usually within 8 Hz – 250 Hz). Some assessment methods use the spectrum of the noise (1/3-octave spectrum measured indoors), and compare this spectrum to a criterion curve. Other methods calculate a level and compare this to a limit.

Depending on the actual conditions, many types of noise can be regarded as low frequency noise. The firing rate of many diesel engines is usually below 100 Hz, so road traffic noise can be regarded as low frequency noise as well as (diesel) train noise or noise from ferries. Similar considerations apply to engines or compressors in industries or co-production plants. Burners can emit broadband low frequency flame roar. Low frequency noise can be noise or vibration from traffic or from industries, totally or partly transmitted through the ground as vibration and re-radiated from the floor or the walls in the dwelling. By this method of transmission, frequencies above approximately 20 Hz are attenuated. It is a general observation that indoor noise is perceived as more 'low-frequency-like' than the same noise heard out of doors.

2. LISTENING TEST

The listening tests were identical to those described in [1] where more detailed information may be found. Eight noises were used (see table I), presented at three different levels. All presentations were made twice and the sequence of the presentations was randomised. Prior to the listening tests, the subjects were trained using four noise examples. After each presentation the subject gave his/her evaluations of the noise on a paper form.

Noise No 1 has a broadband character and is almost continuous. Noise No 2 consists of a series of very deep, rumbling single blows from a drop forge. The noises 3, 4, 5, and 6 have a tonal component. Noise No 7 has three tones but two of them are at a low level. Noise No 8 has a characteristic rhythmical pulsating sound due to the drums.

The duration of each noise was 2 minutes. The noises were either recorded indoors or filtered to simulate indoor noise. The presentation levels were 20 dB, 27.5 dB, and 35 dB (A-weighted levels) measured at the listening position. The noises sounded 'natural' in the listening room and had a pronounced low frequency characteristic.

3. TEST SUBJECTS

Eighteen young persons (9 males and 9 females, age between 19 and 25 years) comprised the reference group for the listening tests. The reference group was the main group of listeners. In addition four persons, the special group, who have reported annoyance due to low frequency noise in their homes were included in the listening tests. The special group were all members of a society against low frequency noise in the home and they were selected independently by this society for the investigation. The special group consisted of two of each gender and the ages were between 41 and 57 years. No measurement or other kind of objective quantification was made of the noise in their homes.

Table I. Description of the noises used for the listening tests

No.	Name	Description	Tones, characteristics
1	Traffic	Road traffic noise from a highway	None – broadband, continuous
2	Drop forge	Isolated blows from a drop forge ransmitted through the ground	None – deep, impulsive sound
3	Gas turbine	Gas motor in a power-and-heat plant	25 Hz, continuous
4	Fast ferry	High speed ferry; pulsating tonal noise	57 Hz, pass-by
5	Steel factory	Distant noise from a steel rolling plant	62 Hz, continuous
6	Generator	Generator	75 Hz, continuous
7	Cooling	Cooling compressor	(48 Hz, 95 Hz) 98 Hz, continuous
8	Discotheque	Music, transmitted through a building	None, fluctuating, loud drums

Pure tone audiometry was carried out over the frequency range 125 Hz to 8000 Hz using a Madsen Midimate 602 audiometer, equipped with Sennheiser HDA 200 earphones. The calibration of the audiometer was according to the values from [4] which are practically identical to ISO 389-8 [5]. Hearing threshold levels at or below 15 dB HL were accepted in the frequency range 125 Hz to 4000 Hz, and a hearing threshold level of 20 dB at a single frequency (including 8 kHz) was also accepted. The average audiogram is seen in Figure 1.

The threshold of the reference group was within the chosen 15 dB limit whereas the special group showed a hearing loss at the frequencies above 2 kHz. Some of the persons in the special group had a considerable hearing loss, partly due to age. It is assumed that the high frequency hearing loss does not influence the subjective evaluations of the low frequency noises used in the present investigation.

The hearing threshold at 31 Hz, 50 Hz, 80 Hz, and 125 Hz was also determined: This test was made with a Two Alternative Forced Choice method described in [6]. The average hearing threshold is seen in Figure 2.

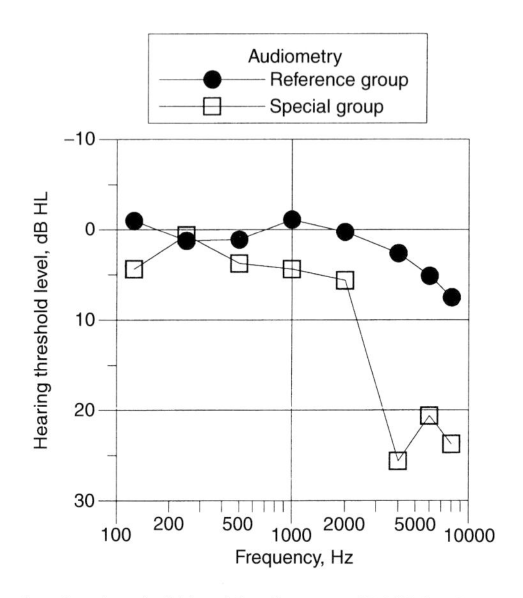

Figure 1. Average hearing threshold level (audiometry, dB HL) for the two subject groups.

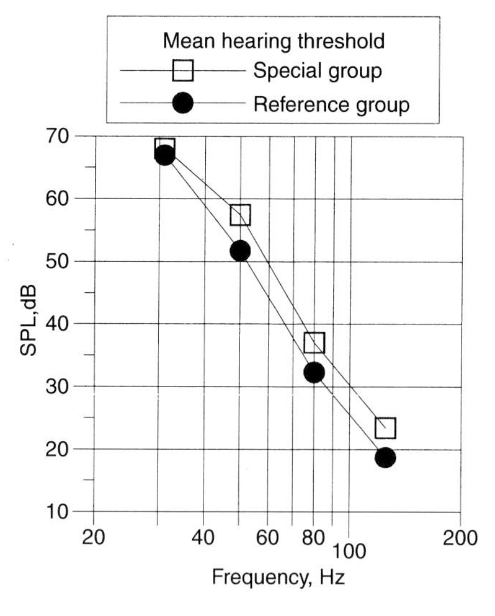

Figure 2. Average hearing threshold (dB SPL) at low frequencies for the two subject groups.

The low frequency hearing thresholds (31 Hz – 125 Hz) showed that the special group was about 5 dB less sensitive than the reference group. The threshold data show that the special group – who had reported annoyance due to low frequency noise in their homes – do not have higher (better) hearing sensitivity at the low frequencies. The hypothesis that the special group should be able to hear the low frequency sounds more easily than the reference group is not supported by these hearing threshold measurements.

3.1 Subject's Task

The test persons were given a written introduction to the tests. Information about the sounds was given after all the tests were finalized.

After each presentation the tests persons were asked to mark on a response line their answer to four questions:

- 'How loud is the sound?' (on a line labelled "not audible" in one end and "very loud" in the other end)
- 'How annoying do you find the sound if it was heard in your home during the day and the evening?' (on a line labelled "not annoying" in one end and "very annoying" in the other)
- 'How annoying do you find the sound if it was heard in your home during the night?' (on a line labelled "not annoying" in one end and "very annoying" in the other)
- 'Is the noise annoying?' (answer yes or no).

All response lines were 10 cm long, and the response was measured in cm with a ruler and thus given as a figure between 0 and 10.

4. RESULTS OF THE LISTENING TEST

As an example, Table II shows the average subjective evaluation – made by the reference group – of the annoyance during the night from the various sounds. Similar tables were made for the special group and for the annoyance during day/evening.

The subjectively assessed annoyance increases when the same noise is presented at a higher level. Different types of noise are not assessed equally annoying. The noises from the drop forge, the discotheque and the cooling compressor are evaluated as more annoying than the other types of noise.

A statistical analysis of the data from the reference group showed that the noise, the nominal level, the measured dB(A) level and the low-frequency level ($L_{pA,LF}$), are all significant factors. The repetition number (round 1 or round 2 with the same noise presentation) was not a significant factor, which shows the absence of a training effect.

A corresponding analysis was made with the data from the special group. Since this group has only four persons the data are very uncertain and highly dependent on random variations. The result of the analysis is showed that the noise level influenced the evaluations. The influence of noise type on the annoyance was just at the limit of being significant.

The statistical analysis was performed although the data were not perfectly normal distributed.

Table II. The average assessment of the night-annoyance made by the reference group. Annoyance ratings were given on a scale from not annoying (0) to very annoying (10)

Nominal presentation level	**20 dB**	**27.5 dB**	**35 dB**
Noise example	**Subjective annoyance Night**	**Subjective annoyance Night**	**Subjective annoyance Night**
Traffic noise	1.6	3.4	5.2
Drop forge	4.3	5.9	6.9
Gas turbine	0.9	2.5	5.2
Fast ferry	0.9	3.2	5.4
Steel factory	1.0	2.7	4.9
Generator	1.7	3.2	5.0
Cooling compressor	2.7	4.4	6.0
Discotheque	3.0	5.4	6.7

5. COMPARISON OF THE RESULTS FROM THE TWO SUBJECT GROUPS

Table III shows the subjective evaluation made by the special group for annoyance at night. Table III show the same data for the special group as Table II shows for the reference group. Table III shows that the subjectively assessed annoyance increases with increasing level (apart from the noise from the generator, which apparently is equally annoying at both a nominal level of 20 dB and at 27.5 dB). By comparing Table II and Table III it is seen that the special group assesses the noises much more annoying than the reference group does. The annoyance found by the special group at a nominal level of 20 dB corresponds almost to the annoyance reported by the reference group at a level of 35 dB.

An interesting result is that it is not the same noises that are evaluated as most annoying by the two groups. The reference group clearly found the drop forge, the discotheque, and the cooling compressor the most annoying. This rank holds at any of the three presentation levels. In contrast, the special group found the generator the most annoying (at the lowest presentation level) and the discotheque as one of the lesser annoying sounds.

The evaluations made by the two subject groups are compared in the following figures.

Table III. Results of the subjective evaluation of annoyance during the night time made by the special group. Annoyance ratings were given on a scale from not annoying (0) to very annoying (10)

Nominal presentation level	**20 dB**	**27,5 dB**	**35 dB**
Noise example	**Subjective annoyance at night**	**Subjective annoyance at night**	**Subjective annoyance at night**
Traffic noise	4.7	7.2	8.5
Drop forge	7.5	8.3	8.9
Gas motor	5.0	8.1	9.8
Fast ferry	6.6	8.8	9.3
Steel factory	5.9	8.2	9.3
Generator	8.4	8.3	9.0
Cooling compressor	7.4	8.5	9.1
Discotheque	6.0	7.9	8.6

Figure 3 shows that there is a good correlation between the assessments of loudness made by the two groups. The correlation coefficient is calculated to be 0.82. The special group generally finds the noise examples somewhat louder than the reference group does. The points are rather close to a line that would be offset from but parallel to the line indicated in Figure 3 (showing an assumed 1:1 relationship).

The relation between the assessments of day/evening annoyance of the two groups, Figure 4, is less clear. The correlation coefficient drops to 0.75 and especially the group of points from the highest nominal level (triangles) shows a considerable scatter. The special group finds the noises more annoying than the reference group does. On the average the

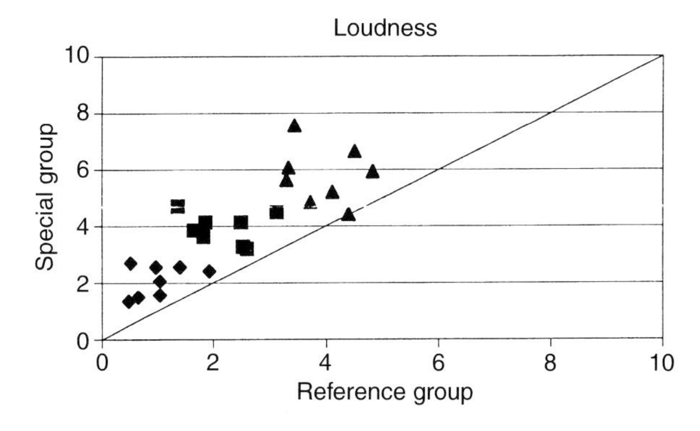

Figure 3. Assessments of loudness made by the reference group and the special subject group.

special group rate the annoyance at day/evening about 2 to 3 scale units higher that the reference group. An increase in the rating of 2 to 3 units corresponds roughly to an increase in level of about 10 dB.

For the assessment of annoyance at night, Figure 5, the picture is shifted. The special group finds the noises much more annoying at night than at day (or evening), and the difference between the assessments of the two groups increases considerably. Figure 5 shows a 'saturation' phenomenon, that is, one or more of the test persons in the special group uses the maximum indication of the annoyance scale. The correlation coefficient is 0.73. On the average the special group rate the annoyance at night about 4 to 5 scale units higher that the reference group. Such an increase in the rating of 4 to 5 units corresponds roughly to an increase in the level of about 17 dB.

The responses to the 'yes/no' question are shown in Figure 6. The figure show how many percent of the group that have marked the noise as annoying. The saturation is obvious as all (four) persons in the special group have marked several noise examples as annoying.

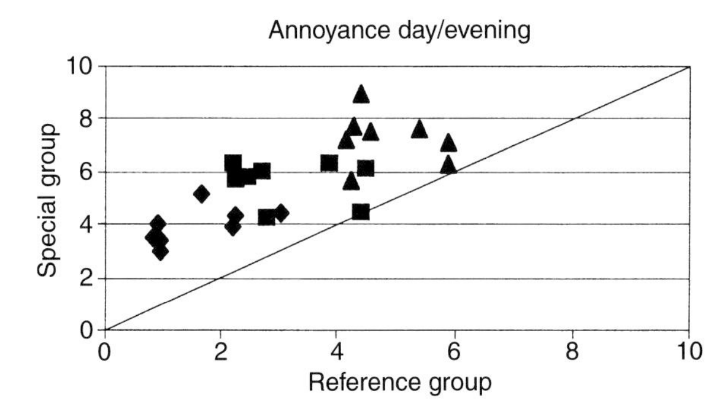

Figure 4. Assessments of annoyance during day/evening made by the reference group and by the special group.

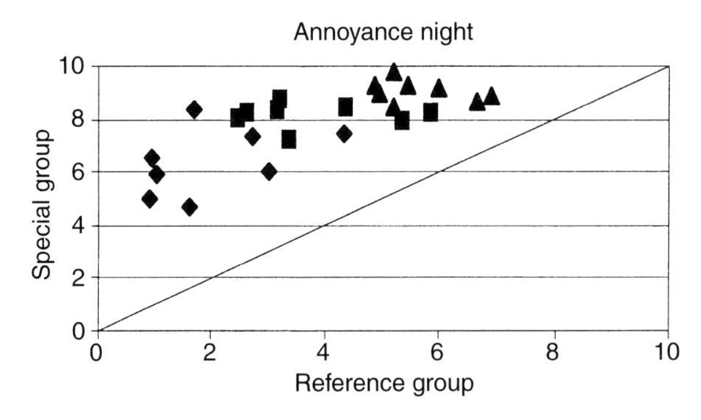

Figure 5. Assessments of annoyance at night made by the reference group and by the special group.

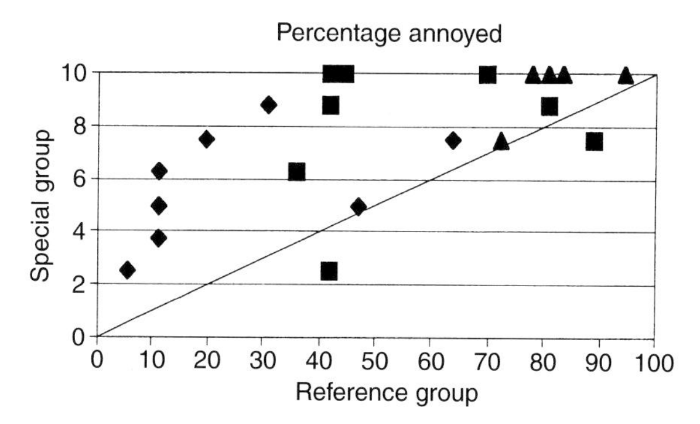

Figure 6. Percentage of yes-responses to the yes/no question made by the two subject groups.

5.1 Assessments Within the Special Group

The relation between loudness and annoyance (day/evening) made by the special group is almost linear and the correlation coefficient is as high as 0.96. The loudness ratings are less than the annoyance ratings and thus the noises are perceived more annoying than loud.

The relation between annoyance at day/evening and at night is illustrated in Figure 5. A non-linear relation due to saturation is clearly seen.

The saturation effect indicates the need (in this case) for a stronger assessment than 'very annoying'. The group of points from the middle level (filled squares) is evaluated 2–3 'units' more annoying when they occur at night than at daytime, but the points from the highest presentation level are only indicated 1–2 'units' more annoying. For the special group the annoyance generally increase by two 'units' from day to night corresponding roughly to a 10 dB change in the noise level. For the reference group the annoyance at night was generally rated about one 'unit' higher than at day – at all presentation levels. Such a one-unit

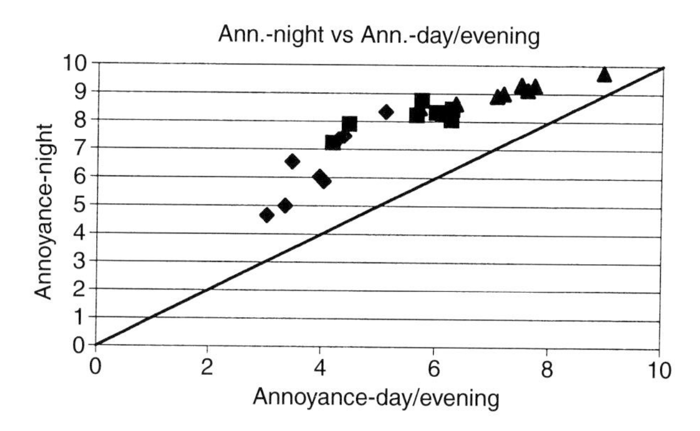

Figure 7. Assessment made by the special group of annoyance (night) and of annoyance (day/evening).

change in the rating corresponds approximately to a 5 dB change in the noise level and supports thus the 5 dB penalty in the noise limits at night.

6. SUBJECTIVE EVALUATION OF ANNOYANCE COMPARED WITH OBJECTIVE MEASURES

The special group's evaluation of annoyance in the night period was compared to a number of objective measures used in European countries. The details of the procedures are described in [1] and the analysis for the reference group is also given in that reference. Table IV shows the results of the analysis made for each of the objective assessment methods based on the annoyance ratings made by the special group.

Table IV. Summary of results of regression analysis of the relation between the assessments made by the special group and the different objective assessment methods

Assessment method	**Slope**	**Intersection (x = 0)**	**Degree of explanation, r2**	**Correlation coefficient, ρ**
Danish	0,16	6,52	0,60	0,78
German A-level	0,16	3,83	0,69	0,83
German tonal	0,05	7,99	0,39	0,54
Swedish	0,17	6,44	0,72	0,85
Polish	0,17	5,47	0,66	0,81
Sloven	0,15	6,84	0,59	0,77
C-level	0,09	4,40	0,31	0,55

It can be seen from Table IV that none of the assessment methods gives any particularly successful correlation to the subjective assessment made by the special group. Two examples are illustrated in Figure 8 and 9, the Swedish and the Danish method. The groups of points from the intermediate and the highest presentation level both line up reasonably well with a slightly sloping line in the upper part of the figures, while the group of points from the low presentation level appears very different in the two figures. In Figure 8 showing the Swedish method these points have a curved tail-like appearance, while they in Figure 9 showing the Danish method appear as a diffuse cloud.

The other assessment methods show results without any particular trend like it is seen with the Danish assessment method. Obviously there is no strong connection between the subjective assessment made by the special group and the objective results found by the objective measuring methods. It should be noted though that only four subjects are included in the special group. However, the results give rise to a number of questions about how low frequency noise in the environment is experienced and how it can be assessed.

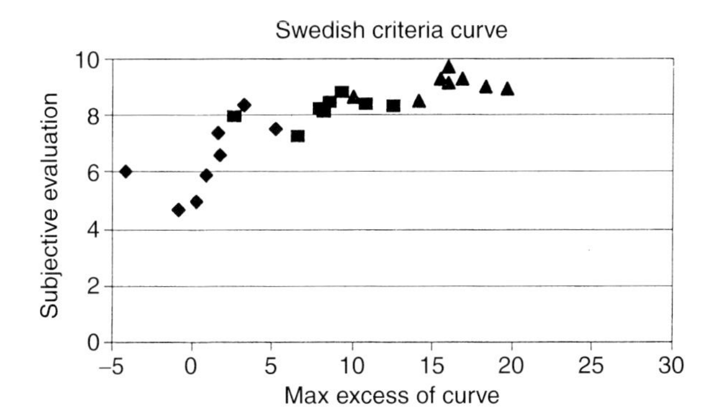

Figure 8. Illustration of the relation between the Swedish assessment method and the subjective evaluation made by the special group.

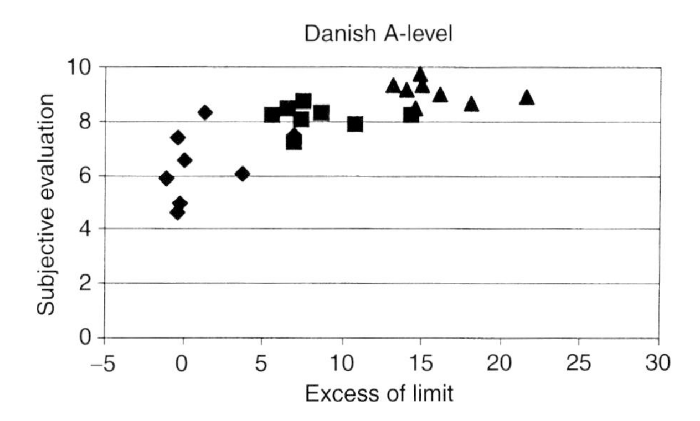

Figure 9. Illustration of the relation between the Danish assessment method and the subjective evaluation made by the special group.

6.1 Discussion of Results from the Special Group

The results show clearly that the special group made the annoyance evaluations differently from the reference group. The overall annoyance rating (averaged over all annoyance evaluations, presentation levels and noises) was 3.5 for the reference group and 6.7 for the special group. This difference can also be illustrated by ordering the noises from the most annoying to the least annoying. This is done in Table V.

The reference group has the Drop forge and the Discotheque on top of the list. These two noises have an 'impulsive' character and thus a 5 dB penalty is added to the calculated level in the Danish evaluation method. The Generator and the Traffic noise are in the middle of the list and the Gas turbine is evaluated as the least annoying.

For the special group the Generator is on top of the list whereas the Discotheque and the Traffic noise are evaluated as the least annoying. It is interesting that Traffic noise gives the lowest overall scaling for this group. The value 5.6 is well below the next one (Discotheque) at 6.2. The order of the noises could indicate that the special group put more attention to those noises, which resemble the typical low frequency noises that they complain about. On the other hand it has not been possible to demonstrate a biased, individual connection between the laboratory evaluations and the kind of noise they are bothered by at home.

Table V. Ordering of the noises from most annoying to least annoying for the two groups of test subjects

Order, reference gr.	**Average Scaling**	**Order, special gr.**	**Average Scaling**
Drop forge	5,1	Generator	7,3
Discotheque	4,6	Cooling compressor	7,2
Cooling compressor	4,1	Drop forge	7,0
Generator	3,1	Gas turbine	6,9
Traffic noise	3,0	Fast ferry	6,9
Fast ferry	2,9	Steel factory	6,8
Steel factory	2,7	Discotheque	6,2
Gas turbine	2,7	Traffic noise	5,6
Grand average	3,5	Grand average	6,7

7. DISCUSSION

A general discussion about the experimental method, noises and evaluation methods is given in [1] and is not repeated here. The discussion here will concentrate on the aspects related to the two subject groups.

The test subjects made their response by a mark on a 10 cm long horizontal line. The results from the special group of subjects showed in some cases a saturation effect that made the results more difficult to interpret. A way to alleviate the saturation effect could be to include an extension to the response line beyond the 'very annoying' mark. Another way of reducing the saturation effect could be to exchange the word 'very' with a stronger adjective like e.g. 'extremely'.

The number of subjects (18 in the reference group; 4 in the special group) could be increased in order to obtain more certainty in the results. For the reference group it is believed though that an increase of the number of test subjects would not change the general results. For the special group, an increase in the number of test subjects would certainly improve the validity of the group results. On the other hand it may be misleading to handle the persons with low frequency problems as a homogeneous group of subjects. The individual evaluations from this small group vary between test persons. The problems they experience are different and it might thus be more relevant to look at the results from these test persons individually. It has not been possible to demonstrate a connection between the laboratory evaluations and the kind of noise they experience at home.

The low frequency pure-tone hearing threshold was measured for both subject groups. The result showed that the special group was less sensitive to low frequency sounds than the reference group. Although this is an interesting result, it might be more informative to measure the loudness growth curve for the test subjects. It is believed that the loudness growth curve would be a much better predictor for annoyance than just the hearing threshold. Measurement of loudness growth is time consuming and is certainly not straightforward at low frequencies.

There is a good agreement between the annoyance evaluations from the reference group and the Danish measurement/calculation method [1]. The same good agreement is not found for the special group. This raises a question about the aim or objectives of a measuring method. Should such an evaluation method be made for the average person (the general population) or should a method be made with special emphasis on the persons who react more strongly to low frequency noise? The criteria and evaluation methods are all based on some kind of measurement of the noise level. For the reference group there is a clear connection between the noise level and the experienced annoyance and thus it makes sense to use such criteria and evaluation methods. For the special group this connection between level and annoyance is less clear and thus an evaluation method based on noise level measurements may be of little value for this subject group. There is a need for further research, with larger number of such test subjects, into the factors, both subjective and objective, that determine the low frequency annoyance. The results from the special group also underline the need of more awareness about the limitations of objective methods based on sound level measurements.

8. CONCLUSIONS

A laboratory investigation of the annoyance of low frequency noises was performed for two subject groups. A reference group consisted of eighteen normal hearing subjects. A special group of consisted of four persons who were known to experience problems with low frequency noises. The subjects listened to eight different noises and evaluated the loudness, the annoyance at day/evening and the annoyance at night.

The low frequency hearing threshold of the four special test subjects was not found to be better than the hearing threshold of the ordinary test subjects.

The annoyance evaluations made by the four special test subjects were clearly different from the evaluations made by the ordinary test subjects.

The ratings were systematically higher. Especially at night the annoyance was rated as close to maximum and thus not dependent on the level of the noise. The four special test subjects were not annoyed by the impulsive noises to the same degree as the ordinary test subjects were. The connection between level and annoyance is unclear in the results of the special group and thus an evaluation method based on noise level measurements may be of little value for this subject group.

REFERENCES

1. Poulsen, T., *Comparison of objective methods for assessment of annoyance of low frequency noise with the results of a laboratory listening test.* Journal of Low Frequency Noise, Vibration and Active Control, 2003. 22(2): p. 117-131.

2. ISO, *ISO 7196 Acoustics – Frequency weighting characteristic for infrasound measurements.* 1993, International Organization for Standardization: Geneva, Switzerland.

3. D-EPA, Low *frequency noise, infrasound and vibration in the environment (In Danish). Information no. 9.* 1997, Danish Environmental Protection Agency.

4. Han, L.A. and Poulsen, T., *Equivalent threshold sound pressure levels for Sennheiser HDA 200 earphone and Etymotic Research ER-2 insert earphone in the Frequency range 125 Hz to 16 kHz.* Scandinavian Audiology, 1998. **27**(2): p. 105-112.

5. *ISO, ISO 389-8 Acoustics – Reference zero for the calibration of audiometric equipment – Part 8: Reference equivalent threshold sound pressure levels for pure tones and circumaural earphones.* 2001, International Organization for Standardization: Geneva, Switzerland.

6. Buus, S., Florentine, M., and Poulsen, T., *Temporal integration of loudness, loudness discrimination and the form of the loudness function.* Journal of the Acoustical Society of America, 1997. **101**(2): p. 669-680.

Psychological analysis of complainants on noise/low frequency noise and the relation between psychological response and brain structure

Toshiya Kitamura*, Masaki Hasebe, and Shinji Yamada***
**University of Yamanashi, Takeda 4 Kofu 400-8511 Japan. Kitamura@ms.yamanashi.ac.jp*
***University of Hokkaido, Jusanjo-Nishi 8 Kitaku Sapporo 060-8628 Japan*

ABSTRACT
In Japan there are two kinds of low frequency noise (LFN) problems. One is LFN that can be heard directly in a house and causes discomfort, and the other is LFN that rattles windows or doors and causes annoyance. Authors met about 100 complainants on noise or low frequency noise. Hearing thresholds of some complainants were measured and Yatabe-Guilford personality inventories of some complainants were carried out. We observed many complainants and tried to analyze the complainants' mind by the psychoanalytical method. Their minds have three layers. The 1st layer is the basic desire for survival and good life etc. The 2nd layer is personality or consideration obtained by experience and the 3rd layer concerns movement, speech or action etc. Many complainants lack the 2nd layer on noise/low frequency noise and their movements and reactions on noise/LFN appear directly from the 1st layer of basic desire without consideration of the 2nd layer. The findings of the three layers are discussed in relation to present knowledge on how the different parts of the brain are organized.

1. EXAMPLES OF LOW FREQUENCY COMPLAINANTS

1.1 Case 1

Figure 1 shows the hearing threshold measured in our laboratory and the low frequency noise measured in the complainant's room (windows are closed). The sound source is fan exhaust noise from duct exits in a factory. The measured low frequency noise is nearly equal to the average threshold. Complainants are annoyed very much by low frequency noise and the family moved to another rental flat. In particular a female complainant fears of physiological affection by the LFN. The countermeasures were to fit silencers and fix the base of the fan rigidly. The tone of 50 Hz disappeared and the level decreased by 5 dB. LFN in the house is now very little over the level of ISO-389-7 and the existence of LFN is not apparent. The fear disappeared and they could live in their original house. The hearing threshold of this female complainant is almost the same to the average threshold measured in Yamanashi University. In this case psychological counseling by one of the authors helped the recovery.

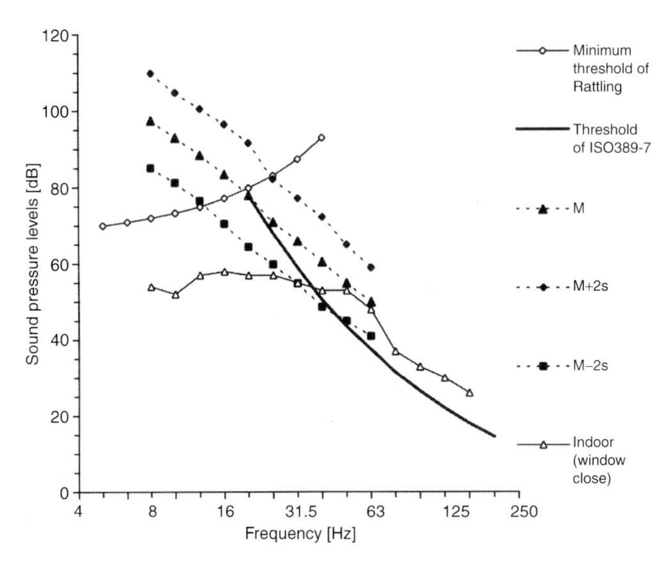

Figure 1. Hearing thresholds of LFN and the level in complainant's house.
M : Average threshold (measured in Yamanashi Univ.)
s : Standard deviation (measured in Yamanashi Univ.)

1.2 Case 2

Figure 2 shows the measured level at the complainant's house. The sound source is a textile factory and an exhaust duct. The factory worked until 10 p.m. and the level is a little over the average hearing threshold. The factory and the complainant's house are very close and audible noise can be heard at an open window. At 1 a.m. the factory did not work and the complainant (female) recognized that the factory did not operate. But at 3 a.m. she woke up and she felt that the factory was in operation. But the factory did not operate. She is very nervous and when she woke up in the night, she has some confusion and misunderstood that the factory operated. The owner of the factory said that in the night the factory does not operate. The human relationship between the owner and the complainant is very bad.

Many complainants are classified into three groups as shown in Table I. The first group includes ordinary low frequency complainants. They are annoyed indoors with window closed. When they open windows, the annoyance decreases. They hear music indoors for masking the LFN.

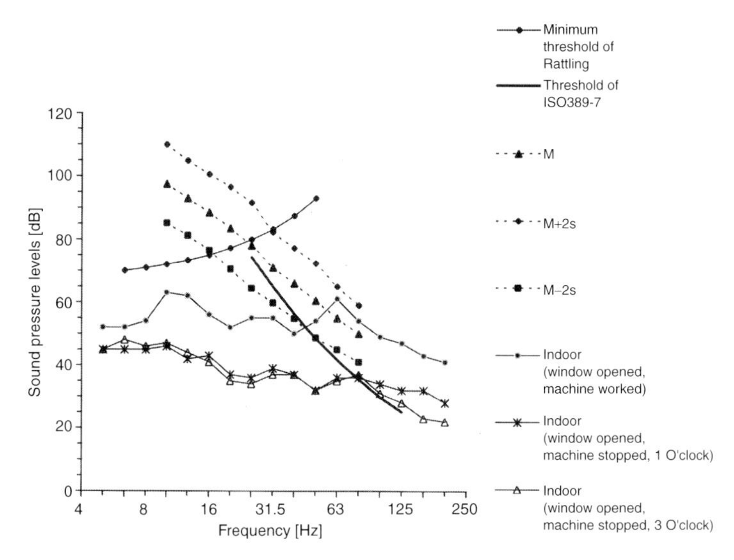

Figure 2. Hearing thresholds of LFN and the level in complainant's house.
M : Average threshold (measured in Yamanashi Univ.)
s : Standard deviation (measured in Yamanashi Univ.)

Table I. Classification of complainants

Ordinary LFN complainant	Annoyed, especially indoors Open window decreases annoyance
Complex complainant of LFN and noise	Window open: annoyed by audible noise Window closed: annoyed by LFN
Complainant with misunderstanding	Complains, but level is too low Can not recognize operation of the source Bad relationship with neighbours Sometimes complain of radio etc. Tinnitus

Another complainants complain of LFN and audible noise. When the windows are closed, they feel LFN and when the window is opened, they complain of audible noise.

LFN is difficult to understand scientifically and sometimes misunderstandings occurs. Some complaints have no correspondence with facility operation. Even if a machine stops, he/she complains. In many cases they have bad relationships with neighbours. Sometimes they have much hearing loss and tinnitus. They think that the noise in the ear comes from outside.

2. HEARING THRESHOLD OF COMPLAINANTS

Figure 3 shows the threshold of young listeners and the threshold of complainants measured in our laboratory. Sometimes it is said that the complainants are more sensitive than ordinary people and the thresholds of complainants are lower than that of ordinary people. But this figure shows that the thresholds are almost the same. The individual thresholds are included in the range of M±2s (M: average, s: standard deviation) of ordinary people. These complainants are not sensitive to LFN but they are sensitive to annoyance. When the level of LFN just exceeds the threshold, then they are easily annoyed.

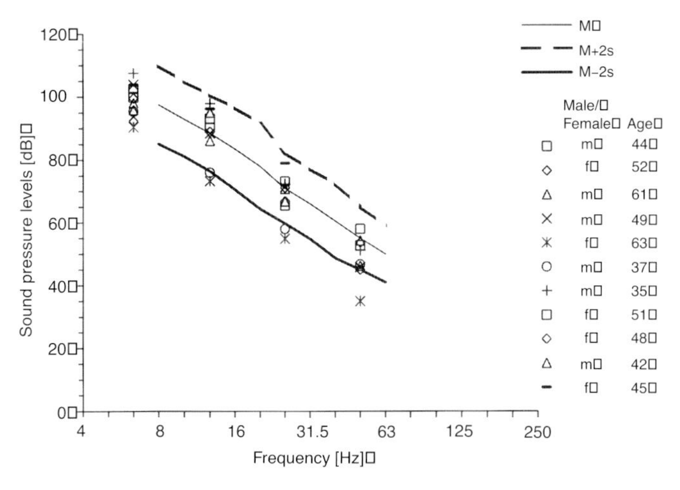

Figure 3. Threshold of complainants.
M : Average threshold (measured in Yamanashi Univ.)
s : Standard deviation (measured in Yamanashi Univ.)

3. THE YATABE-GUILFORD PERSONALITY INVENTORY(SEE APPENDIX I)

The Yatabe-Guilford Personality Inventory was used to assess 12 low frequency complainants. The result is shown in Table II. In the table the numbers show the number of complainants.

This result shows that the personality of "stable and active" appears a little more and "mediocrity" appears a little less. "Stable and positive"

Table II. Result of Yatabe-Guilford Personality Inventory of 12 low frequency complainants

A	0	B	1	C	0	D	1	E	0
A′	1	B′	1	C′	0	D′	4	E′	2
A″	0	AB	1	AC	0	AD	0	AE	1
Mediocrity		Unstable Active		Stable Passive		Stable Active		Unstable Passive	

is a good personality in normal life. There are not very clear personalities in these complainants but the number of the subjects is not sufficient. By the observation of one of the authors, complainants may be normal citizens in ordinary life.

4. PSYCHOLOGICAL ANALYSIS OF COMPLAINANTS

Complainants are annoyed and they have many problems on their minds. We meet these complainants and do psychological counseling for them. In counseling we analyze their mind by the psychoanalytical method. Much data on the psychoanalysis of complainants shows that normally human being have three layers in the mind.

1st layer: The human being has the desire for existence, appetite, sexual desire etc. in the base of mind. Commonly these desires are not clearly conscious. But in the base of mind these desires support human life and behaviour. The origin of these desires is the limbic system in the brain (1).

2nd layer: This layer is the main layer of the 'consideration' system and is derived from many experiences, in the learning and development of each human being. A human being can be conscious of this layer partially or vaguely. This gives rises to logical consideration of stimuli etc.

3rd layer: Speech, movement, behaviour which can be observed from outside.

We observed about 100 complainants and analyzed their minds from observation, conversation and measured sounds. They are all normal citizens but sometimes they lack the 2nd layer, especially with respect to low frequency noise or audible noise. Apart from the noise problems they have normal comprehension and they have the 2nd layer. But when they are annoyed by low frequency noise, they feel the annoyance directly at the 1st layer without the 2nd layer and they have no reasonable understanding of LFN problems.

5. THE STRUCTURE AND ROLE OF BRAIN

By the survey of literature (1) the structure and the role of brain in the human body are as stated in Table III.

The limbic system controls emotion, appetite, sexual behaviour and basic learning or memory. Emotion means anger, fear, unpleasantness, pleasantness, and award and aversive systems. Pleasantness and award systems means 'profitable' conditions for the living body and unpleasantness and aversive systems means 'non-profitable' conditions for the living body. The corpus amygdaloideum evaluates, whether the value of the phenomenon is 'profitable' or 'non-profitable' for the living body. Unpleasantness or annoyance occurs when the phenomenon or noise is harmful or 'non-profitable' to the living body.

Table III. Part of brain and role of brain

Part of brain	Role
Frontal association area	Comprehension, understanding, judgement, thinking, control
Limbic system	Control of emotion, affection, instinct, instinctive behaviour, pleasantness, unpleasantness
Hypothalamus	Centre of autonomic nervous system for respiration, circulating system, digestion, control of internal secretion
Corpus amygdaloideum	Biological value evaluation, judgement of value
Hippocampus	Memory, including memory of vision and audition Increase of pleasantness and unpleasantness by the memory of audition In the memory of emotion there is a connection between the hippocampus and the corpus amygdaloideum

The 1st layer of complainants corresponds to the limbic system.

The limbic system is an old brain and common with animals. It is a very important system for the survival existence of animals. Complainants may comprehend the biological evaluation of low frequency noise for the living body without consciousness.

The frontal association area controls the emotion or unpleasantness by means of knowledge, learning and memories. This frontal association area has many neural connections to the limbic system. The 2nd layer of complainants corresponds to this frontal association area.

The hypothalamus is the centre of the autonomic nervous system for respiration, circulation, digestion, and the control of internal secretions. When annoyance occurs, it affects respiration, heart rate and internal secretion (3).

6. TO SUPPORT COMPLAINANTS AND TO SOLVE PROBLEM

Complainants complain of low frequency noise, and psychological and physical discomfort. When we accept a complaint from a complainant, we make efforts to solve the problem by the following procedure, considering the psychological counseling methods.

1. When we receive complaints from a complainant by phone, we hear the complaints for one hour or a half. Basically at first we do not deny the contents of the complaints. We make an effort to get a good relationship with the complainant. Psychologically this good relationship means "rapport".
2. We analyze the contents of the complaints and classify them into acoustic, psychological and physical phenomena.
3. Sometimes complainants have fear of low frequency noise (LFN).

We try to decrease the fear. We explain simply that there are high levels of LFN in a car or in an airplane but we are not damaged directly.

4. We ask the complainant if there is measured data or not. If not, we ask the complainants to contact the city (local) authority and to measure the sound. If possible, we visit the complainant's house and measure the sound.
5. At the end of the conversation we say "You can contact us at any time" to keep the complainant's mind calm.
6. If there is a real LFN problem, we help the complainant to solve the problem. We recommend them to complain to the city authority or speak directly to the source without aggression. There are many methods to reach a settlement by the complainant, by the city authority, by mediation, by arbitration or by judge.
7. If the measured data is very small and he/she can not identify the on/off of an assumed source, we ask carefully about the time history of the complaint and the relationship with the neighbour etc. We try to assess the complainant's attitude and personality and make an effort to understand his/her mind totally. Normally there is a cause why he/she feels discomfort not by LFN but by another cause. In some cases there was formerly a sound problem but countermeasures were done and now the complainant cannot identify the on/off of the source. This situation is a psychological after-effect and normally there is a bad relationship with a neighbour.

   We discuss this situation gently with the complainant. But the complainant does not accept this easily. We wait for some months. The situation of the complainant changes in everyday life and there arises some chance to solve the problem. The role of counseling is to help and support the complainant to solve the problem by the complainant's own will.
8. If there is the possibility of tinnitus, we ask them to measure the hearing threshold. And we ask them whether they feel something in an anechoic room. Many people hear tinnitus easily in an anechoic room. And if the tinnitus sound resembles the sound in the house, we do a matching test, if possible.
9. Human beings have complex patterns of thinking and behaviour. We try to understand them totally, including LFN.

CONCLUSION

A complainant is a human being and human being has a control system to keep them stable. To solve low frequency or other noise problems it is important to think of the complainants totally psychologically and socially, and to help them not only with countermeasures but with counseling.

LITERATURE

(1) J. Senba, *Brain and living body control* (textbook of Broadcasting University), 1998 (in Japanese).

(2) Shinji Yamada et al., Difference of low frequency noise influence between sufferers and students, *Inter-noise* 85 in Munich, pp. 1001-1003, Sept. 1985.

(3) S. Yamada et al., Physiological effects of low frequency noise, *J. LFNV*, vol. 5, no. 1, pp. 14-25, 1986.

APPENDIX I

Yatabe-Guilford Personality Inventory

This personality inventory is basically developed by J.P. Guilford and improved by T. Yatabe, and is widely used in Japan.

This personality inventory has 200 questions and a subject selects one from three answers (yes, ?, no).

One example of a questionnaire
Are you joyful? yes ? no

This inventory classifies the personality into 5 categories. 5 categories are A (Average: Mediocrity), B (Blacklist: unstable and active), C (Calm: stable and passive), D (Director: stable and active), and E (Eccentric: unstable and passive).

Annoyance of low frequency tones and objective evaluation methods

Jishnu K. Subedi, Hiroki Yamaguchi*, Yasunao Matsumoto, Mitsutaka Ishihara
Department of Civil and Environmental Engineering, Saitama University, 255 Shimo-ohkubo, Saitama, 338-8570, Japan
**E-mail: hiroki@post.saitama-u.ac.jp*

ABSTRACT
Annoyance of low frequency pure and combined tones was measured in a laboratory experiment. Three low frequency tones at frequencies of 31.5, 50 and 80 Hz at four sound pressure levels, from about 6 dB to 24 dB above average hearing threshold, were selected as pure tones. The combined tones were combinations of two tones: the four levels of 31.5, 50 and 80 Hz tones and a constant level 40 Hz tone. The results showed that the rate of increase in annoyance of pure tones with increase in the sound pressure level was higher at lower frequencies, as reported in previous studies. The results for the combined tones showed that the increase in the annoyance of the combined tone compared to the annoyance of pure tone was dependent on the level difference of the two tones and their frequency separation. These results were compared with the evaluation obtained from different objective methods. The three methods were Moore's loudness model, the low frequency A-weighting and the total energy summation used as objective evaluation methods. Among the methods, the low frequency A-weighting gave the best correlation.

1. INTRODUCTION

The problem of low frequency noise in the environment is not new. There have been continuous efforts to understand the behaviour of these noises and their effects on human beings [reviewed in 1-3]. These studies have reported that the hearing threshold rises at lower frequencies, that the subjective responses, such as loudness and annoyance, increase rapidly at lower frequencies with the increase in the sound pressure level above the hearing threshold, and that the low frequency noises can have a variety of negative effects on humans. In order to reduce the negative effects from the low frequency noise in the environment, many countries have adopted different standards and criteria. For example, the limiting values of low frequency tonal noise in the daytime for different countries (cf. [4]) are shown in Fig. 1. As shown in the figure, most of them are near to the average threshold of hearing defined in ISO 389-7 [5]. This implies that there is a general agreement among these standards that low frequency sound just above audible level can produce negative effects on humans. The night-time values in some countries are lower than the values shown in the figure.

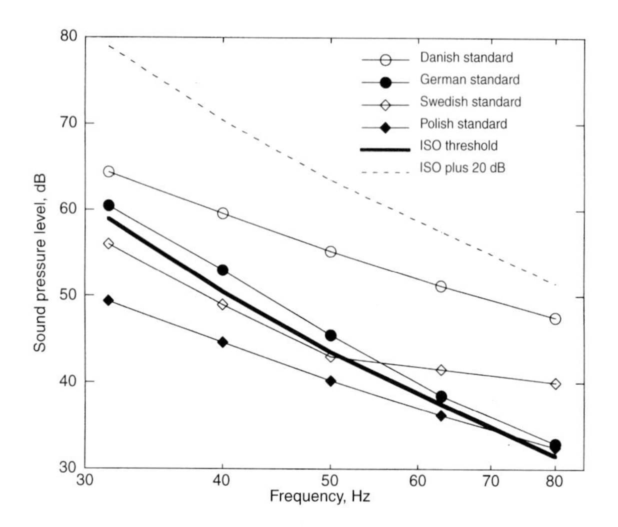

Figure 1. Low frequency noise limits in different countries (cf. [4]). Hearing threshold curve defined in ISO 389-7 and curve representing 20 dB plus ISO threshold are also shown.

Broner [1] has mentioned that the primary negative effect due to low frequency noise is annoyance, at least for lower sound pressure levels. There have been various studies to measure the annoyance of low frequency sounds [4, 6-8]. Moller [6] and Andresen and Moller [7] obtained equal annoyance contours of pure tones for frequencies at 4, 8, 16, 31.5 and 1000 Hz. They measured the annoyance of these pure tones starting at a sound pressure level of about 15 dB above average hearing threshold defined in ISO 389-7 [5] for 31.5 Hz and at similar levels for other frequencies, too. Their results showed that the rate of increase in annoyance with the increase of sound pressure level is frequency dependent, i.e., the rate increases with the decrease in the frequency. Inukai *et al.* [8] measured equal unpleasantness sound pressure levels of pure tones from 10 Hz to 500 Hz. Five different levels of unpleasantness from "not at all unpleasant" to "very unpleasant" were measured for the tonal sounds. The rate of increase in the unpleasantness level with the increase of sound pressure level showed a trend similar to that observed in the case of the annoyance level. The lowest level of unpleasantness in their measurement, i.e., "somewhat unpleasant", occurred at about 20 dB above the average hearing threshold. However, their results also showed that the acceptable limits in different environments are much lower than the lowest unpleasantness level.

The measurements of annoyance and unpleasantness of pure tones mentioned above showed that small changes in sound pressure level cause relatively large changes in annoyance at lower frequencies. However, the trend needs to be investigated at sound pressure levels just above the hearing threshold, a region where low frequency sound has caused problems in living environments.

Low frequency noises in the environment are usually complex sounds containing energies in a wide frequency range. The annoyance of these complex sounds cannot be predicted by superposition of the annoyance due to pure tones. Bryan [9] used nine different noises from real situations to assess annoyance, and hypothesized that the slope of their spectra rather than their absolute levels may be important in determining the annoyance. Goldstein and Kjellberg [10] measured the annoyance of band noise with different slopes of the frequency spectrum. Their results contradicted the hypothesis of Bryan, and they concluded that "the slope of the frequency spectrum is not the critical feature of the noises discussed by Bryan". However, it is pointed out by Leventhall [3] that "Bryan's subjects had long term exposure in real settings, whilst Goldstein and Kjellberg's listened to 10 second samples in the laboratory, so removing any temporal growth of annoyance from the response" [3]. A study of the annoyance caused by combinations of a few tones at different levels and frequency separations may provide a better insight into the influence of spectrum in determining annoyance.

In addition to the subjective assessment of annoyance, researchers have proposed objective assessment methods that can predict the degree of annoyance for practical purposes. Berglund *et al.* [11] used three community noises (with considerable energy at higher frequencies) and mixed the noises at different levels to create combinations of noises. The measurement of annoyance for separate noises and for combined noises showed that the annoyance of a combined noise was different from the arithmetical sum or the arithmetical mean of the individual components. They used different objective methods in order to correlate the evaluation with subjective measurements, and showed that the loudest component model, which states that the only effective parameter was the loudest of the component noises, gave the best results for practical purposes. Poulsen and Mortensen [4] measured annoyance using field recordings of actual noises with intense low frequency components. The noises were presented to general subjects and to a special group of subjects, who were known to be disturbed by low frequency noise. The annoyance ratings were measured for different assumed circumstances, such as day, evening and night times. The results were then evaluated by objective methods derived from standards and limiting criteria of different countries, some of which are shown in Fig. 1. It has been shown that the method derived from the Danish standard, which is based on low frequency A-weighting, has the best correlation with the subjective evaluations.

The performance of these objective methods however, has not been evaluated in the previous studies for tones just above the threshold and for combinations of two tones, which may cause a "throbbing". As it has been suggested that low frequency tones just above the threshold have been found to cause annoyance, and "throbbing" increases the annoyance [12,13], the prediction from the objective methods has to be investigated for pure tones and combinations of two tones.

In the light of the above discussion, the objectives of the present study were set to investigate annoyance of low frequency tonal sounds and combinations of two tones at low sound pressure levels experimentally, and also to examine some of the objective methods that can give best relation with the subjective assessment of the annoyance obtained from the experiments.

2. EXPERIMENTAL METHOD

2.1 Experimental Room

The experiment was conducted in a soundproof cabin of size 1.8 × 1.2 × 2.3 m. The schematic drawing of the plan view of the experimental cabin is shown in Fig. 2a. The floor of the room was carpeted and the walls and ceilings were covered with 55 mm thick sound absorbent cloth. Four speakers (Yamaha, YST 800) were used to generate the test sounds. The subjects were placed 1 m in front of the speakers in a sitting position as shown in Fig. 2b. They were given a clipboard, a pen and few pages to read or write inside the cabin. Although the cabin had an exhaust fan, it was not used during the experiment. An electric bulb was used to illuminate the room. The experimental cabin had a 450 mm × 450 mm

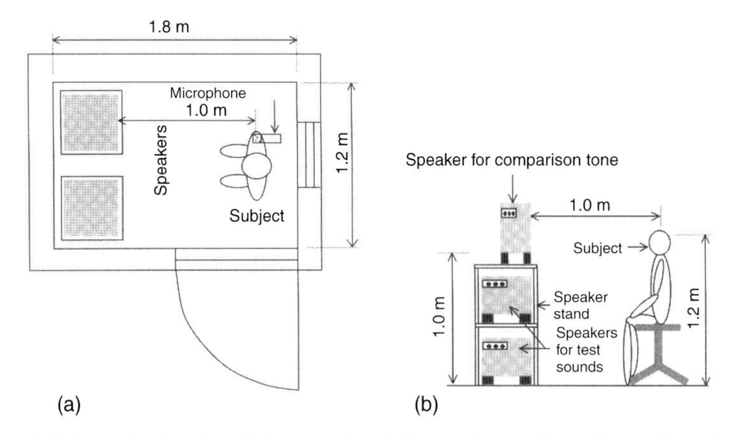

Figure 2. Schematic drawing of the experimental setup inside the cabin, a. Plan view; b. Elevation showing speakers location and subjects' position.

opening covered by two layers of glass which could be used to observe the subjects during the experiment.

Fig. 3 shows the background noise measured inside the cabin at the position of the microphone shown in Fig. 2a. The background noise was below the average threshold defined in ISO 389-7 [5] in the frequency range of the test sounds. The noise exceeded the average threshold only above 160 Hz, and it was not considered to affect the experiment significantly.

The experimental setup was arranged in such a way that in the case of any discomfort, the subjects could turn off the whole system by pressing one switch. The switch was kept within reach of the subjects, and the exit door was also located within reach of the subjects. The subjects were instructed either to turn off the system and leave the cabin, or to leave the cabin immediately in case of any discomfort. However, the situation did not arise at any time during the experiment.

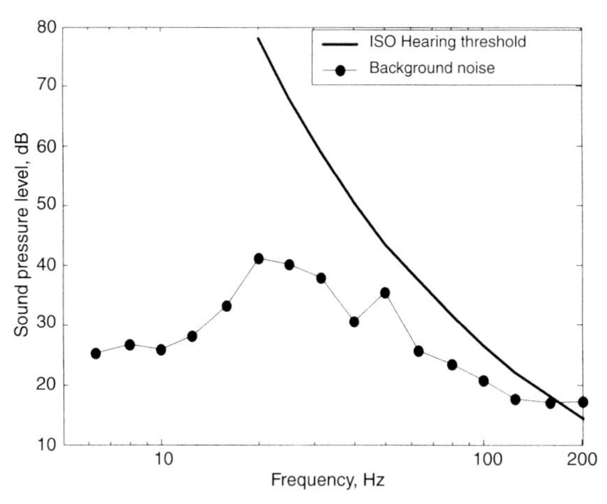

Figure 3. Background noise inside the cabin at the location of microphone position in Fig. 2a. Hearing threshold curve defined in ISO 389-7 is also shown.

2.2 Subjects

Sixteen students, 14 males and 2 females, aged from 22 to 31 years were asked to be the subjects. Nine of the subjects were Japanese (JS), and 7 subjects were foreign students (FS) studying in Japan.

2.3 Test Sounds

The experiments were conducted with two types of test sounds: pure tones and combined tones. Three frequencies, 31.5, 50 and 80 Hz, were selected for the pure tones, and four levels of each frequency were considered. The lowest sound pressure level, SPL 1, was selected at 6 dB above the threshold specified in ISO 389-7 [5] for the pure tone at 31.5

Hz, and the level at other frequencies were selected in such a way that the loudness levels specified in ISO 226 [14] were nearly equal at all the frequencies. The next level, SPL 2, was formed by adding 5 dB to SPL 1 at 31.5 Hz, and by increasing the sound pressure level at other frequencies to such a level that the loudness level at these frequencies were nearly equal to that at 31.5 Hz. The same procedure was adopted to form SPL 3 and SPL 4. The designed sound pressure levels for all the pure tones used in the experiment are shown in Fig. 4.

For the combined tones, the tonal sounds at levels of SPL 1 to SPL 4 at frequencies of 31.5, 50 and 80 Hz were mixed with a tonal sound at 40 Hz at a constant level. The level of the 40 Hz was kept 5 dB above the threshold level of individual subjects at 40 Hz measured prior to the measurement of annoyance in order to ensure that the tone was clearly audible for all the subjects. The tonal sounds at levels of SPL 1 to SPL 4 at frequencies of 31.5, 50 and 80 Hz were considered as main tones and the tonal sound at 40 Hz was considered as additional tone. In summary, 24 test sounds, which consists of 12 pure tones and 12 combined tones, were used in the experiment.

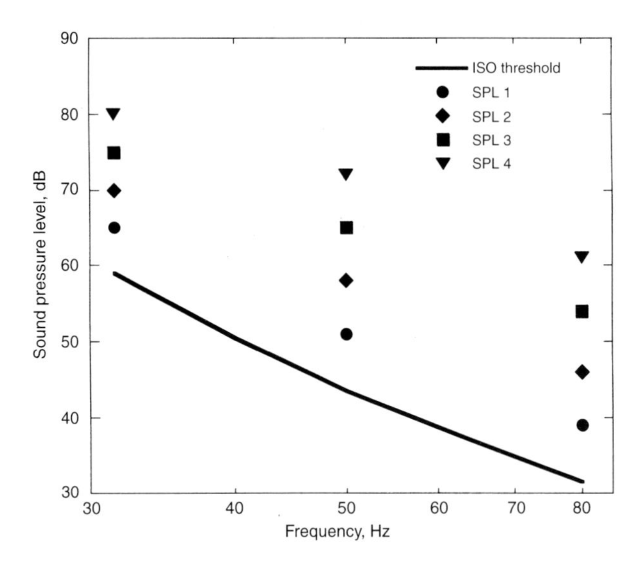

Figure 4. Designed four levels of pure tones used in the experiment. Hearing threshold curve defined in ISO 389-7 is also shown.

2.4 Method of Annoyance Measurement

Two scales were used in order to record the annoyance rating. The first scale was a graphical scale, this was a linear scale with a length of 150 mm marked with "not annoying at all" at one end and "very annoying" at the other end, assuming the environment of the experiment room as not annoying at all. This scale is referred to as Graphical Scale in this paper. Moller [6] used a graphical scale to measure annoyance of low frequency tones and concluded that the graphical scale was practicable for comparative measurements.

The second scale was a pure tone at 63 Hz: the subjects were asked to adjust its intensity so that it was perceived as equally annoying as the sound under test. This scale is referred to here as Comparison Tone.

The term FUKAISA was used in the measurement for the Japanese subjects. The nearest translation of FUKAISA in English may be unpleasantness. Kuwano *et al.* [15] reported that "the impressions of loudness, annoyance and unpleasantness of synthesized sounds are similar to each other and mainly determined by loudness level... On the other hand, in the case of recorded road traffic noise, there was some difference between loudness and annoyance judgments" (cf. Inukai *et al.* [8]). Therefore, it was considered in this study that the impression of FUKAISA is similar to that of annoyance.

2.5 Experimental Procedure

Instructions to the subjects were given orally and the same information was provided to them in written form. The instructions were given in Japanese to the Japanese subjects, and in English to rest of the subjects.

The test sound, generated from a personal computer and controlled by the experimenter, was presented to the subject for one minute. After this one minute, the sound was turned off and the subjects were asked to answer the question 'How annoyed do you feel?' and mark the annoyance rating on the Graphical Scale. It is noted that Moller [6] used three exposure times, 30 seconds, 3 minutes and 15 minutes, in order to investigate dependence of annoyance rating on exposure time and concluded that the results were independent of the exposure time. After the subjects determined the annoyance rating with the Graphical Scale, they were instructed to turn on the Comparison Tone, and to adjust its intensity so that the tone was perceived equally annoying as the test sound. The Comparison Tone was generated by another speaker (Yamaha, YST 350) kept inside the experimental room and at a distance of 1m in front of the subjects as shown in Fig. 2b. The source of the Comparison Tone was a function oscillator (NF Electronic Instruments, E-1011A) kept near to the subjects. The subjects could operate the function oscillator by themselves whenever required. Once the recording of the signal was over, the subjects were asked to turn off the Comparison Tone. The same procedure was repeated for all the test sounds.

The experiments for the pure tone and combined tones were conducted on two separate days for each subject. On the first day, the experiments were conducted in two parts. In the first part, measurements of thresholds of pure tones at 1/3 octave band center frequencies from

31.5 Hz to 80 Hz were made. The thresholds were measured by the indirect method of adjustment where the subjects did not have direct control over the sound pressure levels. The experimenter adjusted the sound pressure level according to the subject's indication. Two repetitions were made for each signal.

In the second part of the first day, which started after a gap of 10-15 minutes, the experiments for the annoyance of pure tones were conducted. In this part, twelve test sounds of pure tones were presented to the subject with a five minute break after every four test sounds. The same procedure was repeated for all sixteen subjects. In order to avoid carry-over effects, the order of test sounds for twelve subjects was determined from a 12 × 12 Latin Square which ensured that no test sound was preceded by any other test sound for more than one subject. For the remaining four subjects, a random order of presentation was used. The experiments on pure tones for all sixteen subjects were completed in three days.

On the second day, which started two weeks after the completion of the experiments on pure tones, measurements for twelve combined tones were carried out. Additionally, the measurement for one test sound of a pure tone at 40 Hz and a repeated measurement of a pure tone test sound of 50 Hz at SPL 4 were also carried out. The level of the pure tone at 40 Hz was the same level as that of the additional tone in the combined tones. The measurement of SPL 4 at 50 Hz was carried out in order to verify whether any systematic difference was produced by conducting experiments on two separate days. The repeated measurement of 50 Hz was conducted at the beginning of the second day for each subject. For the remaining sounds, a similar procedure to that used for the pure tones was used to ensure that no test sound was preceded by same test sound for two subjects. The experiments on combined tones were also completed in three days.

2.6 Recording of the Signal

The measurement of the sounds was made with a microphone (Rion NA-18) and the data was recorded in a PC. The sound pressure level used in this paper is the average sound pressure level measured for 20 seconds. The measured sound pressure levels of pure tones and combined tones differed from the designed sound pressure level by ±1 dB.

3. RESULTS AND DISCUSSIONS

3.1 Perception Threshold

The median of the perception thresholds of all the 16 subjects along with the inter- quartile ranges for the pure tones measured inside the cabin are shown in Fig. 5. The thresholds were measured at 1/3 octave band center frequencies from 31.5 Hz to 80 Hz. The median thresholds measured in this experiment are compared with the hearing thresholds defined in ISO 389-7 (1996). The comparison showed that the measured thresholds were higher than the ISO threshold at all frequencies used in this experiment. One possible reason for the higher threshold may be because of the fact that method of adjustment was used to record the threshold. The lowest sound pressure level, SPL 1 (shown by cross mark in Fig. 5), used in the experiment was below the median threshold at 50 Hz (by 0.3 dB), therefore, the measurement of annoyance for this condition was problematic for some subjects. However, for all other input stimuli, the relatively high threshold of the subjects should not have caused any difficulties in the measurements of annoyance. The median thresholds are comparable, with the threshold measured by Inukai *et al.* [8] for 27 subjects aged from 19 to 62 years.

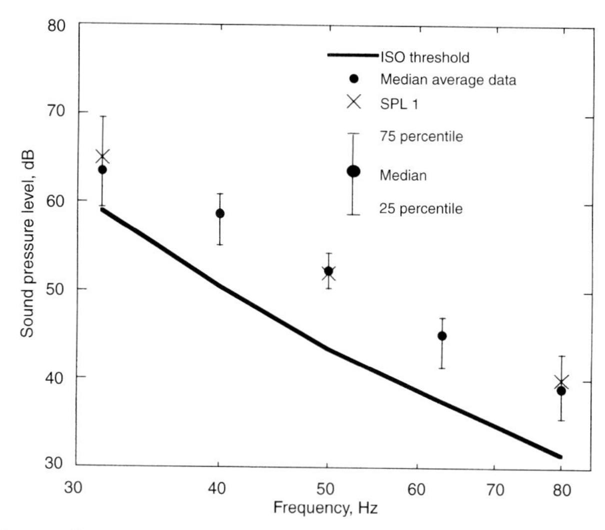

Figure 5. The median perception thresholds and the inter-quartile ranges. The lowest sound pressure level, SPL 1, used in the measurement of annoyance and hearing threshold curve defined in ISO 389-7 are also shown.

3.2 Annoyance

3.2.1 Effect of measurement on two separate days

The measurements of annoyance of the test sound of 50 Hz pure tone at SPL 4 conducted in first and second session were compared by using Wilcoxon signed ranks test. The test showed no systematic variation produced by taking measurements on two separate days for both the Graphical Scale and Comparison Tone ($p>0.05$). In this paper, the results obtained with the Graphical Scale are presented. The results for the Comparison Tone will be published in Subedi *et al.* [16] and are not discussed here.

3.2.2 Measurement for two groups of subjects

Since all the subjects did not speak English, the languages of instruction to subjects were different: Japanese instruction for Japanese subjects (JS) and English instruction for foreign subjects (FS). As Guski *et al.* [17] argued that the subjects from different language groups might "stress different facets of noise annoyance", it was considered necessary to investigate any systematic variation in the measured results between the two groups.

The annoyance ratings measured were separated into two groups according to the language of instructions, and Mann-Whitney U test was performed to investigate variation among the results from the two groups. The test showed that there was no significant difference in the annoyance rating among the subjects in two groups ($p > 0.05$) except for SPL 2 at 31.5 Hz. Although the test showed variation for SPL 2 at 31.5 Hz, the data were treated as those from one group in all the cases.

3.2.3 Annoyance of pure tones

The median annoyance ratings for pure tones for all the subject are shown in Fig. 6. The results are plotted for sound pressure levels above the measured median perception threshold shown in Fig. 5. As seen clearly in the figure, the rate of increase in annoyance was higher for lower frequencies. The results obtained in this study indicated that even a small change in sound pressure level above threshold could cause a relatively large change in the annoyance at lower frequencies.

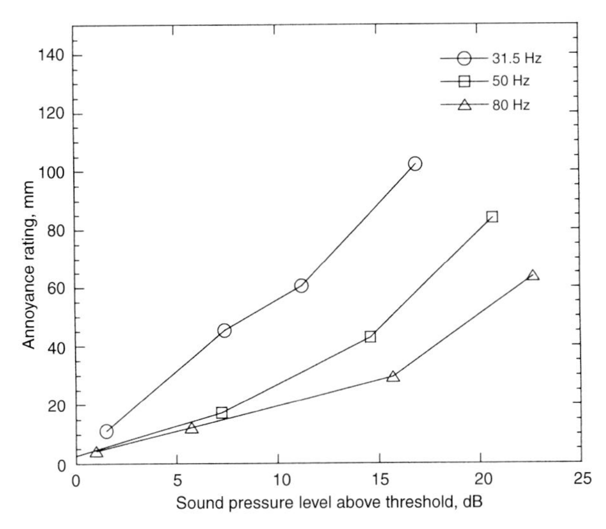

Figure 6. Increase in annoyance rating with increase in the sound pressure level above threshold for pure tones.

Similar results were obtained for frequencies of 31.5 Hz and below by Moller [6] and Adresen and Moller [7]. It was found in this study that a similar trend was observed at frequencies up to 80 Hz. The equal unpleasantness contour from 10 Hz to 500 Hz obtained by Inukai *et al.* [8] showed a similar trend also; the lowest ranked "somewhat unpleasant" contour was about 20 dB above the average threshold defined in ISO 389-7 [5]. In the present study, a similar trend was observed at sound pressure levels just above the perception thresholds.

3.2.4. Annoyance of combined tones

The results of pure tones and combined tones for all the subjects plotted together are shown in Figs. 7a-c for 31.5, 50 and 80 Hz, respectively. In the figures, the abscissa is the sound pressure levels of the tone for pure tones and the 1/3 octave band sound pressure level of the main tone for the combined tones. It should be noted that the median sound pressure level for 40 Hz pure tone measured separately was 64.6 dB and that the median annoyance rating of this tone for all subjects was 29 mm.

The figures compare increases in annoyance with increases in sound pressure levels for the pure tones to increases in annoyance with increasing the level of main tones for the combined tones when the additional tone at 40 Hz at constant level was added. As shown in the Fig. 7a, at lower sound pressure levels of the main tone at 31.5 Hz, the annoyance rating of the combined tone tended to be higher than that of the pure tone because of the addition of 40 Hz tone. It can be seen in the figure that with the increases in the level of main tone, the effect of 40 Hz tone tended to be reduced. At the highest level of the main tone used, the annoyance rating for the combined tone approached the annoyance rating of the pure tone. Similar effects can be seen for 50 Hz and 80 Hz as shown in Figs. 7b and 7c. However, the reduction in the effect of the additional tone with increasing the level of main tone was not similar at all frequencies.

Wilcoxon signed ranks tests were carried out in order to compare the measured results for the pure tones and the combined tones statistically. The results are listed in Table I. The statistical results showed that for SPL 1 to SPL3 at 31.5 Hz, the differences in the two results are statistically significant ($p < 0.05$), which indicate that the effect of additional tone was significant at these levels. However, for SPL 4 at 31.5 Hz, the test showed that the difference produced by addition of 40 Hz was not statistically significant ($p > 0.05$). Similar comparisons for 50 Hz showed that there was no significant effect of additional tone for SPL 3 and SPL 4, while the effect of additional tone was statistically significant for all sound pressure levels in the case of 80 Hz.

These results may be explained by the partial masking effect, which Zwicker and Fastl [18] described as the phenomenon by which the loudness of a test tone was reduced but not completely masked in the

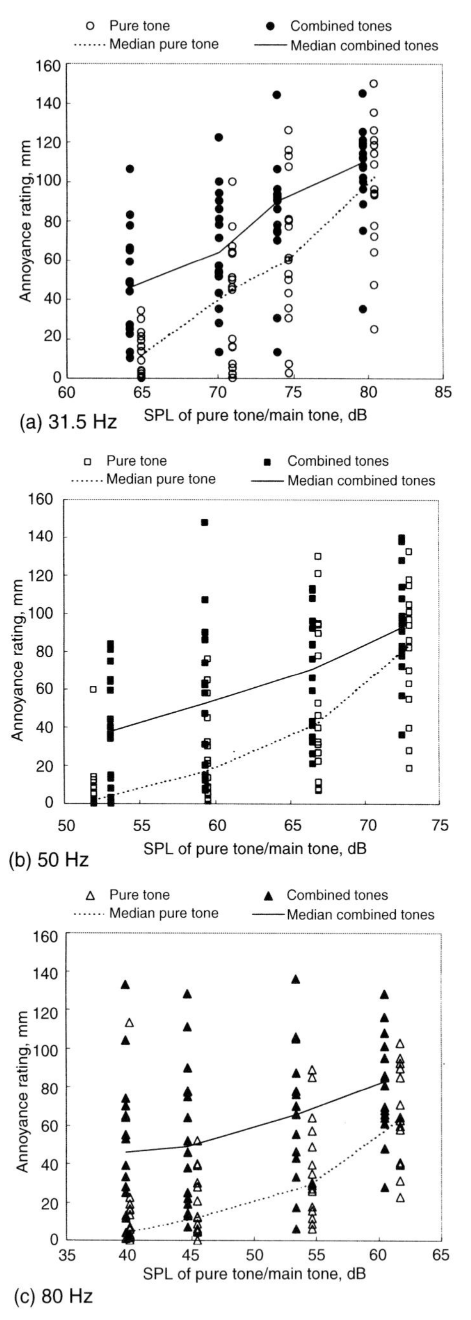

Figure 7. Annoyance rating for pure tones and combined tones. Main tones in the combined tones are: a. 31.5 Hz; b. 50 Hz; c. 80 Hz (SPL of pure tone at 45 dB is increased by 0.8 dB in the figure for the sake of clarity). (SPL: sound pressure level)

presence of a masker sound. For example, the tone at 31.5 Hz at SPL 1 might have partially masked the additional tone at 40 Hz. As the level of 31.5 Hz tone increased, the partial masking effect increased. This might be the reason that the rate of increase in annoyance of the combined tones was less than that of the pure tone. At the level of SPL 4, the tone at 31.5 Hz might have masked the effect of 40 Hz tone almost completely, and no significant effect of additional tone was observed. In the case of 50 Hz, which is also at one 1/3 octave band away from the additional tone at 40 Hz, the results showed that no significant effect of 40 Hz was present at SPL 3 and SPL 4. Although these results might imply that the tone at 50 Hz is more effective than the tone at 31.5 Hz in masking the additional tone at 40 Hz, this may be only because of the difference in sound pressure levels above the hearing threshold. As seen in Fig. 6, the sound pressure level above threshold for SPL 4 is higher by 5 dB at 50 Hz compared to that at 31.5 Hz. As the frequency separation of the two tones increased, relatively large sound pressure level was required to completely mask the effect of the tone at 40 Hz. This might be the reason that even at the largest sound pressure level at 80 Hz, the effect of additional tone can be seen in the annoyance rating.

Table 1. Wilcoxon non-parametric test on pure tones and combined tones

Main tone		Wilcoxon test	
Frequency	**Level**	**p**	**Effect of 40 Hz tone**
31.5 Hz	SPL 1	0.001	Yes
31.5 Hz	SPL 2	0.008	Yes
31.5 Hz	SPL 3	0.044	Yes
31.5 Hz	SPL 4	0.320	No
50 Hz	SPL 1	0.001	Yes
50 Hz	SPL 2	0.024	Yes
50 Hz	SPL 3	0.215	No
50 Hz	SPL 4	0.171	No
80 Hz	SPL 1	0.001	Yes
80 Hz	SPL 2	0.001	Yes
80 Hz	SPL 3	0.011	Yes
80 Hz	SPL 4	0.030	Yes

4. OBJECTIVE METHODS OF EVALUATION OF ANNOYANCE

4.1 Evaluation Models

It is often desirable to find a single index for a noise spectrum that corresponds to the subjective response to noise. The index obtained from three models, namely Moore's loudness model, low frequency A-weighting and the total energy summation model, have been evaluated here.

Loudness level [19] has been one of the widely used indices for the assessment of noise. Loudness is closely related to the physical characteristics of the noise, while annoyance is moderated by many other personal and social factors. Hence, the contribution of loudness to annoyance is only a part of the annoyance observed in real cases. However, Goldstein and Kjellberg [10] showed that the subjective responses of loudness and annoyance do not differ significantly for real sounds over a wide frequency range from 25 to 12000 Hz. Andresen and Moller [7] compared equal annoyance contours and equal loudness contours and concluded that the two sets of curves are remarkably similar in their general shape. However, they also mentioned that the annoyance level was higher than the loudness level at lower frequencies compared to their relation at 1000 Hz. Inukai *et al.* [8] m easured equal unpleasantness sound pressure levels of pure tones from 10 Hz to 500 Hz and compared their results with equal loudness contours. They showed that the gradients of equal unpleasantness contours differed with the gradients of equal loudness contours for frequencies below 100 Hz. In this study, loudness level obtained from Moore's loudness model [20] has been used as an index.

Low frequency A-weighting was proposed by Vercammen [21]. The method was derived from the A-weighting filter, which is one of the popular indices to evaluate community noises. The low frequency A-weighting is the A-weighting filter applied at 1/3 octave band center frequencies for sounds below 160 Hz. Previous studies at low frequencies showed that the A-weighting filter did not give a good correlation with the measurement of annoyance when large amounts of energy were present at low frequencies [9-10, 22-23]. However, Poulsen and Mortensen [4] showed that the low frequency A-weighting had a good correlation with the laboratory measurements of annoyance for noises with strong low frequency components.

The total energy summation model is the summation of energies across the whole frequency range, and the summation is made only for sound pressure levels above the hearing threshold. The energy summation has the limitation that the contribution to the auditory sensation from energies at different frequencies is not similar. However, this fact is not taken into account in the total energy summation model. Despite this limitation, this model has been evaluated here, as the frequency range used in this study is only from 31.5 Hz to 80 Hz. Furthermore, being one of the simplest models to use, this model may be attractive for practical purposes.

4.2 Application of Models and Discussion

The annoyance ratings obtained from the measurements in this study are plotted against the loudness level in phon obtained from Moore's loudness model in Fig. 8. As seen in the figure, the plot of loudness obtained from Moore's loudness model vs. annoyance rating shows scattered results. Although a clear trend was observed for 50 and 80 Hz with the correlation coefficient equal, to 0.9, the correlation coefficient decreased to 0.78 when the results for 31.5 Hz were included. For the same loudness level, the pure tones and combined tones at 31.5 Hz tended to be judged with higher annoyance rating than those at other frequencies.

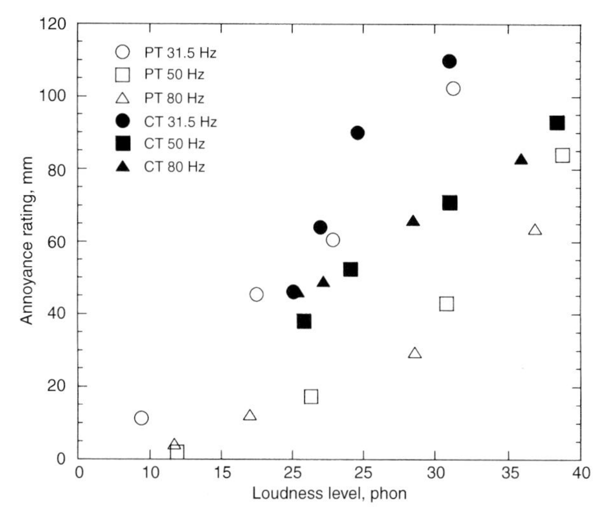

Figure 8. Loudness level (obtained from Moore's loudness model [20]) vs. annoyance rating for pure tones and combined tones (PT: Pure tone; CT: Combined tones).

The low frequency A-weighted sound pressure levels obtained from low frequency A-weighting are compared with the measured annoyance ratings in Fig. 9. The unit of weighted sound pressure is given as dBA(L) in order to show that the dB values are A-weighted, and the methodology is used only for 1/3 octave band sound pressure levels below 160 Hz. The correlation coefficient in this case was equal to 0.93.

The annoyance ratings plotted against total sound pressure level obtained from the total energy summation model are shown in Fig. 10. The total sound pressure level is the sound pressure level of the tone in the case of pure tone and the energy summation of intensity of two tones in the case of combined tones. As seen in the figure, although the trend was reasonable at higher sound pressure levels, the results were scattered at low sound pressure levels. The correlation coefficient was equal to 0.83.

The results showed that the low frequency A-weighting gave the best result among the three methods used here. Poulsen and Morensen [4] also reached a similar conclusion. They compared the low frequency A-weighting with different methods derived from low frequency noise standards of different countries, some of which are shown in Fig. 1. They used low frequency noise from real situations containing energies in wide frequency ranges. The present study showed that the index obtained from this method had the best correlation for the pure tones and the combined tones in the low frequency range at a level just above the average threshold.

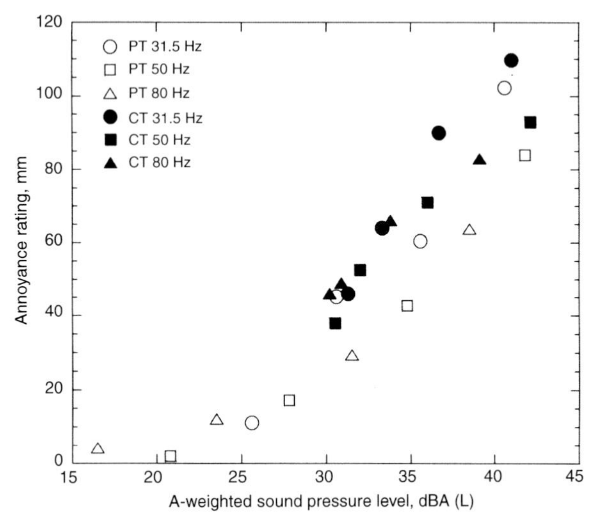

Figure 9. Low frequency A-weighted sound pressure level vs. annoyance rating (PT: Pure tone; CT: Combined tones).

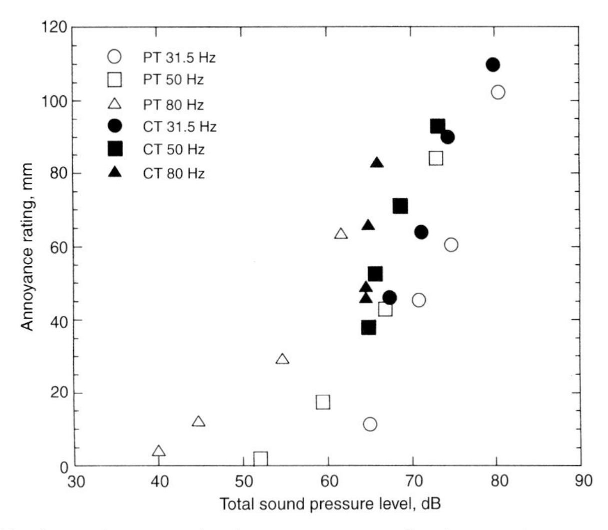

Figure 10. Total sound pressure level vs. annoyance rating for pure tones and combined tones (PT: Pure tone; CT: Combined tones).

Although the change in the annoyance of two tones could be explained by partial masking effects, the loudness level calculated from Moore's loudness model, which takes into account of partial masking phenomenon, had poor correlation with the annoyance rating. This might indicate that the annoyance differs from loudness level at lower frequencies. This hypothesis is in agreement with the results from Andresen and Moller [7] that the annoyance level was higher than the loudness level at lower frequencies.

5. CONCLUSIONS

Annoyance of pure tones and combined tones were measured in a laboratory. The pure tones were at four sound pressure levels, from about 6 dB to 24 dB above the average hearing threshold, of three low frequency tones at frequencies of 31.5, 50 and 80 Hz, while the combined tones were combinations of two tones: the four levels of 31.5, 50 and 80 Hz tones with a constant level of 40 Hz tone. The results from the experiment were compared with the evaluation by three objective methods: the loudness level calculated from Moore's loudness model, the low frequency A-weighted sound pressure level and the total energy summation. The main findings of this study are summarized as follows:

1. The rate of increase in annoyance of pure tones with increasing the sound pressure level was higher for lower frequencies in the frequency range of 31.5 Hz to 80 Hz. A similar trend was observed for equal unpleasant levels in a previous study. This study showed that the trend is observed at sound pressure levels just above the perception thresholds.
2. The results indicated that the total annoyance for complex sound may be dependent not only on level of each component, but on their relative level and separation in frequency.
3. Among the three objective methods, the low frequency A-weighted sound pressure level derived from the low frequency A-weighting gave the best correlation with the annoyance of pure tones and combined tones. This result was consistent with previous findings that the low frequency A-weighting gave the best, corre1ation with low frequency noises with energy in wide frequency ranges.

REFERENCES

1. Broner, N., The effect of low frequency noise on people — A review, *Journal of Sound and Vibration*, 1978, *58(4)*, 483-500.

2. Berglund, B. and Hassmen, P., Sources and effects of low-frequency noise, *Journal of the Acoustical Society of America*, 1996, *99(5)*, 2985-3002.

3. Leventhall, G., A review of published research on low frequency noise and its effects, *Report for Defra,* 2003.

4. Poulsen, T. and Mortensen, R., Laboratory evaluation of annoyance of low frequency noise, Working report No. 1, Danish Environmental Protection Agency, Danish Ministry of the Environment, 2000.

5. International Organization for Standardization, Acoustics-Reference zero for the calibration of audiometric equipment — part 7 : Reference threshold of hearing under free field and diffuse-field listening condition, ISO 389-7, 1996.

6. Moller, H., Annoyance of audible infrasound, *Journal of Low Frequency Noise and Vibration,* 1987, *6(1),* 1-17.

7. Andresen, J. and Moller, H., Equal annoyance contours for infrasonic frequencies, *Journal of Low Frequency Noise and Vibration,* 1984, *3(3),* 1-8.

8. Inukai, Y., Nakamura N. and Taya, H., Unpleasantness and acceptable limits of low frequency sound, *Journal of Low Frequency Noise, Vibration and Active Control,* 2000, *19(3),* 135-140.

9. Bryan, M.E., Low frequency noise annoyance, in: Tempest, W., ed., *Infrasound and Low Frequency Vibration,* 1976, Academic press, London, 65-96.

10. Goldstein, M. and Kjellberg, A., Annoyance and low frequency noise with different slopes of the frequency spectrum, *Journal of Low Frequency Noise and Vibration,* 4(2), 1985, 43-51.

11. Berglund, B., Beglund, U., Goldstein, M. and Lindvall, T., Loudness (or annoyance) summation of combined community noises, *The Journal of Acoustical Society of America,* 1981, *70(6),* 1628-1634.

12. Broner, N. and Leventhall, H. G., Low frequency noise annoyance assessment by low frequency noise rating (LFNR) curves, *Journal of Low Frequency Noise and Vibration,* 1983, *2(1),* 20-28.

13. Persson, K. and Rylander, R., Disturbance from low-frequency noise in the environment: *A survey among the local environmental health authorities in Sweden,* 1988, *121(2),* 339-345.

14. International Organization for Standardization, Acoustics — Normal equal-loudness level contours, ISO 226, 1987.

15. Kuwano, S., Fasti, H. and Namba S., Loudness, annoyance and unpleasantness of amplitude modulated sound, *Inter-noise,* 1999, *99,* 1195-1200.

16. Subedi, J. K., Ishihara, M., Yamaguchi, H. and Matsumoto, Y., Annoyance of low frequency pure tones and effect of one additional tone, *Proceeding of JSCE 54th Annual Conference,* 2004 (to be published).

17. Guski, R., Felscher-Suhr, U. and Schuemer, R., The concept of noise annoyance: how international experts see it, *Journal of Sound and Vibration,* 1999, *223(4),* 513-527.

18. Zwicker, E. and Fastl, H., *Pyscho-acoustics - Facts and Models,* Second updated edn., Springer-Verlag, Germany, 1999.

19. International Organization for Standardization, Acoustics — Method for calculating loudness level, ISO 532, 1975.

20. Moore, B.C.J., Glasberg, B.R. and Baer, T., Model for the prediction of thresholds, loudness, and partial loudness, *Journal of Audio Engineering Society,* 1997, *45(4),* 224-239.

21. Vercammen, M. L. S., Low-frequency noise limits, *Journal of Low Frequency Noise and Vibration,* 1992, *11,* 7-13.

22. Kjellberg, A., Goldstein, M. and Gamberale, F., An assessment of dB(A) for predicting loudness and annoyance of noise containing low frequency components, *Journal of Low Frequency Noise and Vibration,* 1984, *3(3),* 10-16.

23. Moller, H., Comments to: Infrasound in residential area — case study, *Journal of Low Frequency Noise and Vibration,* 1995, *14(2),* 105-107.

LFN and the A-weighting

Piet Sloven
Noise Section, Enviromental Protection Agency Rijnmond, Netherlands

ABSTRACT
Low frequency noise, which can arise from a variety of environmental sources, frequently gives rise to complaints of noise annoyance, which are not necessarily predicted on the basis of conventional dB(A) assessment. This paper, which is particularly concerned with the noise arising from music with a heavy 'beat', considers the extent to which dB(A) underestimates low frequency noise and proposes a rating procedure based on the use of dB(A) and dB(C) measurements.

INTRODUCTION
Many people dealing with noise problems talk about decibels but mean A-weighted decibels, the db(A). However, in order to describe the disturbance arising from low-frequency noise the dB(A) is unsatisfactory (Sloven, 2003). The weighting formula fails and others, such as dB(C) and even dB(G) are considered for the very lowest frequencies. In many cases the problem of Low Frequency Noise (LFN) is dealt with by using a dB(lin) system with 1/3 octave band filters to measure the spectrum levels.

However, in practice it is not a happy situation to have an LFN 'world' which only few specialists understand and which is quite different from the 'world' of higher frequency noise measurements. There are many situations where noise (and possibly vibration) occur in both LFN and 'normal' ranges and there is a clear need to bring both regions together. In order to make progress with the evaluation of LFN it it important to make this clear to the much larger group of people who use dB(A).

Many workers who use dB(A) are familiar with the relationships between dB(A) and, for example, annoyance for different kinds of noise source such as trains, road traffic and aeroplanes. The need is for simple measurement tools which be used quickly and reliably. This is the thinking behind the present paper.

It is perhaps useful, in relation to LFN to remember the World Health Organisation definition that health is not simply the absence of illness,but a state of complete physical, mental and social beneficence. Vulnerable people must not be excluded.

LFN AND A-WEIGHTED STANDARDS IN DIFFERENT CONDITIONS
It is possible to compare the 'LFN-climate' by comparison with other 'noise climates' in the light of available data and some standards.

a) In the Netherlands the use of the Environmental Quality Measure (EQM = MKM = 'Milieukwaliteitsmaat') is widespread. It is an expression used to compare different kinds of noises, The assessments of noise, in percentage terms, play an important role.
b) Although LFN is dealt with as a problem inside houses, it is of course possi- ble to express noise inside and outside. In the Netherlands the mean difference is 22dB(A) - with large variations (Breugelmans et al 2004).
c) Thanks to the problems arising from airports much more is known about the problem of sleep disturbance by noise (Houthuijs 2004).
d) The relations between loudness and sound pressure level for pure tones are well known. The dB(A) can be compared with the level at 1000 Hz and the loudness of any other frequency.
e) The stress levels induced by LFN are considered to be as severe as those due to night-time aircraft noise.

Table I. Description of Noise Climates

a Class	b description of acoustic quality	c L24h+ out side dB(A)	d % ann-oyed	e % seve-ley ann'd	f Lnight out-side db(A)	g Lnight in-side dB(A)	h Lnight LFN inside db(lin)	i Severe sleep Dist
I	Excellent	45	5%	2%			47	
II	Good	50	10%	4%	40	18	50	4%
IIIa	Tolerable fair,passable	55	16%	6%	45	23	53	7%
IIIb	sober (modest?)	60	24%	10%	50	28	56	11%
IIIc	unfavourable weak	65	33%	15%	55	33	59	18%
IV	bad	70	50%	30%	60	38	62	27%
V	Miserable						65	

NOTES

a) and b) In the Dutch guidelines for noise between houses five classes are mentioned (NEN 1070). Class III is 'Moderate' . The dB(U) (universal, united) guidelines were developed in the Netherlands in the 1990's with 8 classes.
c) L24h+ = maximum of LAeq 07-19h, LAeq 19-23h + 5, LAeq 23-07h + 10 dB(A)
d) and e) are estimations, from comparison figures for LFN around airports.

f) Lnight = L24h+ - 10 dB(A) in cases where the night dominates annoyance.

h) LFN. At night often the 1/3 octave of 50 Hz dominates. The figures given here are loudness in phons.

i) Severe = remembered waking three times or more. In Class I there is no problem. Non-desired noise is seldom heard, Annoyance and severe annoyance are <5% and <2% respectively. In Class II there are 5-10% annoyed and 2-4% severely annoyed. In Class IV occupants need to adapt to the noise climate to a considerable extent, with annoyance and severe annoyance arising in 25-50 and 10-30% of cases respectively. In Class V the climate inside houses is often disturbed and it is hard for the occupants to adapt.

These classifications can be used in many LFN situations. They can be combined with the figures from the LFN 'licence-curve' (Sloven, 2000) and in that way a single figure can be used to identify the acoustical description of the noise environment.. This provides a valuable means of communication, bringing LFN 'down to earth'. In the situation where the 50Hz band dominates the 1/3 octave spectrum,the 'licence-curve' level is 47dB, this fits in well with the figures for noise and sleep disturbance.

COMMUNITY RESPONSES

Table II below (From ISO 1996) shows the estimated community response to noise when criteria are exceeded.

Table II. Description of Community Responses to LFN

Acoustical Climate	dB excess of LFN (50Hz)	Community Response Category	Community Response Description
Excellent or good	0	None	No observed reaction
Tolerable	3	Little	Sporadic Complaints
Sober	6	Medium	Widespread Complaints
Weak	9	Strong	Threats of Community Action
Bad & Miserable	12	Very Strong	Vigorous Community Action

Note that in the ISO table differences extend up to 20dB(A) while at LFN, here loudness is of more importance and the table only runs to 12 dB. A little more LFN gives a greater reaction than the same increase in dB(A)

THE IMPORTANCE OF 50HZ

It is interesting to consider possible explanations of the fact that the 1/3 octave of 50 Hz is so often the trigger of LFN disturbance and complaint.

At the lower LFN frequencies, the sound pressure level needs to be (very) high before it is possible to hear and/or feel the sound. The frequency of 50 Hz seems to be the first (i.e. lowest) which occurs in houses at an observable level. In my experience this is often the case, see also Mirowska (1997) for some further data illustrating this point (Fig 10 at p 91).

This indicates that 50Hz is the lowest familiar frequency and is widespread. It is noticed because it is often heard, lower frequencies will be neglected in terms of perception because they provide no extra information. It is found by investigators in this area that in many cases of low frequency complaints the first frequency to exceed the threshold, or the frequency with the highest excess above the threshold is 50Hz.

THE SIGNIFICANCE OF THE C-WEIGHTING

At the present time, in the LFN world, we consider the spectral content of noises and look at the individual band levels, this appears to be our best available method. However for higher frequencies we make use of single figure ratings, in particular the dB(A) .

There is certainly a need for a simple, preferably single figure rating system for LFN, since such a figure is the most effective communication tool, much more effective (and comprehensible) than a set of 1/3 octave band levels. What we need is a) a simple expression, b) one that is easy to calculate and c) something as easily recognisable as dB(A). In the case of LFN the possibilities seem to be, firstly, dB(C), and in rare cases dB(G).

In my working experience the notion of LFN as a specific aspect of noise nuisance in its own right is growing quickly. My simple rule is to express measurements in db(A) and dB(C) as well, Very often you can just use the former - hence business as usual - and sometime you find an indication that you have an LFN problem to deal with.

AIRCRAFT NOISE

A level LCMax outside houses of 75dB(C) corresponds with annoyance suffered by 25% of the residents, and high levels of course give more, the C-weighted levels providing a fair relationship (Miller 1998). An increase of 10 dB(C) corresponds to some 80% annoyed. This finding is in accordance with the data in table I and II above that relatively small differences in dB(lin) give substantial increases in annoyance.

FESTIVALS AS NOISE SOURCES

Due to the problems arising from musical festivals in the Netherlands and in Belgium the C-weighting has become of huge importance. These festivities often take place at weekends and also run on into the night. They are held in the open air and the heavy bass beat or rhythm gives rise to LFN audible over considerable distances. More and more of the organisers are becoming aware that the lower frequencies are very important in terms of disturbance. The implementation of C-weighted monitoring and the use of C-weighted level defined licences are a part of the solution. In some cases all noise in the octaves of 32Hz and lower is forbidden, the participants do not hear the difference but they just miss some of the sensation in their stomachs ....

At longer distances the control of noise levels at the lowest frequencies has a major effect (Sloven,2002) and shows that reductions in low frequency content at source have an important role in reducing annoyance. Local authorities are becoming more aware of these factors and are conscious of the importance of the weighting problem in measurements, LFN is being recognised as a significant issue.

THE GERMAN PROPOSAL

In the German Standard DIN 45680 the C-weighted noise level plays an important part in the evaluation of LFN. A 20 dB difference is used as a criterion (between L(C)eq and L(A)eq, or between L(C)max and L(A)max). See also (Sloven,2000)

INDUSTRIAL NOISE

In this situation the use of dB(C) is not novel at all. there were attempts as far back as 1971 to incorporate this approach in order to minimise LFN problems (see Hessler 2004). In many cases noise engineers have much data available on the spectra from a variety of industrial sources and can use this to estimate octave band, and in turn dB(C) (or dB(G)) levels from the dB(A) data.

THE EXTENT TO WHICH DB(A) UNDERESTIMATES LFN

It has long been established that the psychophysical parameter of loudness tells us far more about the annoyance generated by a noise than does the parameter of sound pressure level, it is reasonable to expect that the SPL is not at all likely to be a good annoyance indicator in every circumstance. The curves of loudness vs SPL (Phons vs dB) show up the differences especially in the LFN range of the spectrum. This effect is not too evident in the highest two octaves, 250 and 125 Hz of the spectrum, it is greater in the the middle part, 63 and 32 Hz, and is very evident in the lowest bands of 16 and 8 Hz. The names audio, centre and infra are often used to describe these regions and in that way to distinguish LFN (broad band) from infrasound (Sloven,2000 and Berglund et al, 1996). The degree of underestimation is about 6 phons for the centre-LFN bands (equivalent to about 10 dB). A similar addition of 5 to 10 dB has been put forward by Vercammen in 1992 and by others much earlier in 1972 see (Berglund st al 1996) . To deal with this discrepancy we must make changes, I would propose changes in the A-weighting system.

ADDITIONAL 'LOADINGS' FOR LFN

It is a well-established practice to 'load', ie increase the measured level by a number of decibels, any type of noise which causes extra annoyance. The increased level is then used in comparing the noise to the permitted or standard value. This practice has been used with dB(A) for such factors as impulsiveness, tonality or musical nature.

The extent of this 'loading' may not always be the same, it could depend on the time of day for example, or the degree of impulsiveness (where it might range over 2 - 14 dB(A)). A very straightforward LFN loading is used in the Dutch community of Nijmegen during the monitoring of noise from musical events; 5dB(A) is added in the case of audible LFN, this is quite a good system. Hence the formula is simply;

$$L(\text{LFN rating}) = LAeq + \text{Loading}$$

where the value of the loading term could depend on the LFN content of the noise.

The above equation is sometimes written more simply as;

$$L = A + E$$

where "E" is the "Extra Loading"

More complex loadings have been proposed, for example the following has been put forward for shooting noise;

$$L = A + c + b(C\text{-}A)(A\text{-}a)$$

where a, b, and are numerical constants (Vos, 2001)

The expression (C-A)>15 has been widely used as an identifier of LFN (see Holmberg et al 1997)

With the above in mind, an equation has been derived for LFN, and is put forward as follows;

$$L = A + (0.84A\text{-}29.44) + 0.04(C\text{-}A)(41\text{-}A)$$

where A and C are the A- and C- weighted noise levels respectively and L is the rating level (loaded) in dB(A). This formula is valid in cases of LFN type Q, night and evening, inside houses, and is based on a comparison with the Dutch standard of 25 dB(A) as a a good environmental standard. The heart of the formula is the value 'C-A'= 21dB giving 'normal' LFN, a value derived from experience.

The results of the formula are set out in Table III below which show that when (C - A) is small the loadings are very small, as (C - A) is increased, then the loading become larger.

Table III. Loading as a function of dB(A) and dB(C) levels.

C-A A	16	21	26
16	0	5	10
21	1	5	9
26	2	5	8

The advantages of the method are:-

a) Simplicity. It is always possible to express measurements and calculations in db(A) and dB(C) and most sound level meters have the required weighting networks built-in.
b) It is not necessary to know if you are dealing with a situation where one 1/3 octave band level dominates, or a broad-band noise where it does not.
c) It is flexible, further development is possible.
d) It connects A-weighted measurements with LFN.
e) When LFN is being measured in rooms, it is often found that there are considerable differences between 'room-centre' and 'corner average' data, this is particularly evident in the case of 1/3 octave measurements. It is found that 'C - A' measurements are much less dependent on microphone position, a distinct advantage in many cases.

RECOMMENDATIONS

1. It would be helpful to have more certain information about the figures for annoyance and sleep disturbance in specific LFN-dominated cases. This will be difficult because most research is done in laboratories where people seldom have time to habituate, or the subjects are young. People under 25 years respond differently from older subjects. Differences of 10dB(A) to get the same degree of annoyance are common (Berg 1999). However available results are at least good indicators of the disturbances, when combined with the steepness of the relation 'dB vs disturbance' it will provide information in other cases.
2. Experience the use and usefulness of C- and A- weighting figures as much as possible This is a helpful way to improve ones understanding of LFN.
3. Let us find a forum to exchange data on practical LFN-cases. In any one country there are few studies per year, Learning from experience is valuable.

REFERENCES

Berg M van den,(1999) Lecture *'Differences in annoyance due to sex and age'*

Berglund B P, Hassmen R F, & Soames J,(l996) *J Acoust Soc Am,* Vol. 99, pp. 2986-2990

Breugelmans O et al (2004) *Tussenrapportage Monitoring (Gezondheidskundige Evaluatie Schipol)* (In Dutch, Noise and Health)

Hessler G F (2004) *J Low Freq Noise, Vib and Active Control,* Vol. 24

Holmberg K U, Landstrom A S & Kjellberg (1997) *J Low Freq Noise Vib and Active Control,* Vol. 15, No. 2, pp. 81-87.

Houthuijs D (RIVM) 2004 *(Beoordelingskader Gezondheid en Milieu: Nachtelijk geluid vanVliegverkeer rond Schipol en slaapverstoring),* (In Dutch, about health and environment at airport)

ISO 1996-1987 (1987) *'Description and measurement of environmental noise'.* ISO Geneva.

Miller N P, (1998) *Internoise 'LFN' from aircraft start and takeoff'*

Mirowksa M, (1998) *J Low Freq Noise Vib and Active Control,* Vol. 17, No. 3, pp. 119-126.

Sloven P (2000) *Low Freq Noise Vib and Active Control,* Vol. 20, No. 2, pp, 77-84.

Sloven P (2002) *Proc Low Frequency 2002,* York, Multi-Science Pub Co Ltd, Brentwood, Essex, U.K., pp. 269-284

Sloven P (2003) *'Effects of LFN on humans',* ICBEN 2003, Rotterdam.

Vercammen M L S, (1992) *J Low Freq Noise Vib and Active. Control,* Vol. 11, No. 1, pp. 7-13

Vos J (2001) *J Acoust Soc Am.,* Vol. 109, No. 1, pp. 244-253.

Low frequency noise annoyance and the negotiation challenge for environmental officers and sufferers

Stephen Benton
Human Factors Research Group, Department of Psychology, University of Westminster, U.K.

INTRODUCTION

The common theme to which many Low Frequency Noise (LFN) researchers subscribe to is that here is a phenomenon that is consistently under-rated in terms of its status as an environmental pollutant (Benton & Leventhall, 1994, Persson-Waye 1995, Bengtsson et al, 2000). The difficulties surrounding the development of an effective and systematic approach to the quantification of LFN incidence and associated impact have centred upon source detection, identification, location and annoyance loading. The quantification of each and all of these aspects is often complicated by the combination of significant 'individual differences' in sensitivity to LFN and the relatively low sound pressure levels (SPLs) which can be associated with disturbance, annoyance and stress (Persson-Waye et al, 2000). The overall LFN context within which environmental health and related agencies undertake an assessment and development of a solution is further complicated by the impact that these measurement factors have upon the sufferers' behaviour.

ASSESSMENT PROTOCOLS

Whilst research has sought to clarify issues within each of these areas as separate influences, in practice, Environmental Health Professionals (EHPs) engaged in resolving noise complaints, work with composite problems produced by the interaction of all three areas. The restricted development of an effective and systematic LFN 'complaint-handling' methodology has meant that when initiating case assessments, EHPs are usually reliant upon existing dB(A) guided protocols. The often reported limited effectiveness of this protocol may be in part, illustrated by a comparison of the dB(A) filter characteristics with another weighting network, (db(C)), as shown in figure 1.

The dB(A), gradually reduces the significance of frequencies below 1000Hz, until at 10Hz the attenuation is 70dB. The C-weighting is flat to within 1dB down to about 50Hz and then drops by 3dB at 31.5Hz and

14dB at 10Hz. Many researchers have drawn attention to the inaccuracies associated with the measurement of environmental noise using dB (A) as it incorporates an over reliance upon the mean hearing sensitivity curve and loudness as predictors of noise annoyance, Leventhall et al (2003). Even in the instances where LFN has been identified as a potential source, problems can occur as in many cases the SPL involved may be low, relative to the mean hearing threshold. (Leventhall et al, 2005).

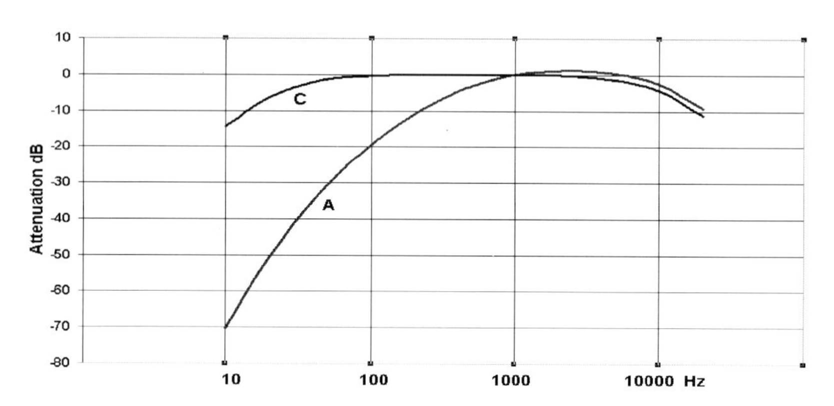

Figure 1. Sound level meter weighting curves – A and C

ANNOYANCE THRESHOLDS AND VARIABILITY

Measurement of low frequency hearing sensitivity has provided weighting networks, such as dB(A), with a sensitivity curve, (such as that shown in Figure 2) against which filtering characteristics can be shaped and estimates of both loudness and annoyance developed.

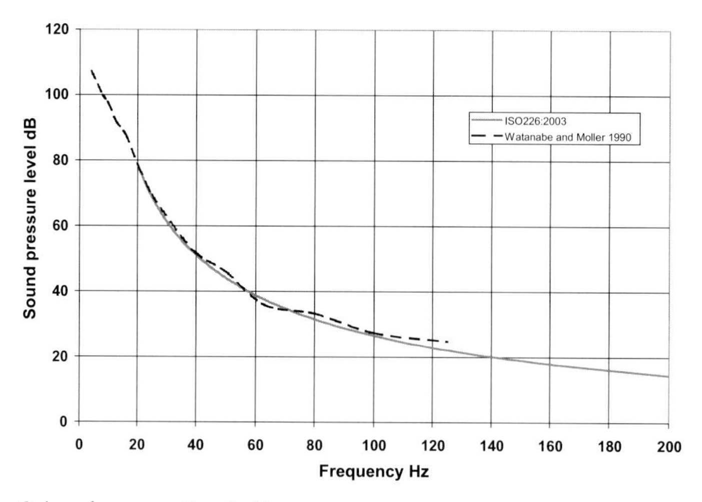

Figure 2. Low frequency threshold

The average perception threshold shown in Fig 2 contains two overlapping studies. The threshold above 20Hz is from ISO226 (ISO:226, 2003), whilst that from 4Hz to 125Hz is from Danish work (Watanabe and Møller, 1990b). The threshold varies from 107dB at 4Hz to 14dB at 200Hz and is 97dB at 10Hz.

However, even in this area of relatively robust measurement, the description of sensitively needs some conditionality. For example, the threshold values shown in Fig 2 are median values, for which 50% of the test subjects (who were typically young adults) are less sensitive and 50% more sensitive. The standard deviations of threshold measurements are about 6dB, which leads to 16% of the population at least 6dB more sensitive than the median and about 2% at least 12dB more sensitive than the median. Thus, it is possible that the EHP may encounter individuals with the equivalent of two standard deviations away from the mean threshold. Another, and related factor, within this area is the assumption that as individuals age increases their hearing sensitivity will generally decrease, relative to the mean threshold curve, the consequences of which finding a role within the initial relationship and assessment phase between the EHP and the LFN sufferer. The thresholds shown in Fig 2 are for young adults and although hearing generally does deteriorate with age, the main effect is at higher frequencies. A Netherlands study (N S G, 1999; Sloven, 2001; van den Berg and Passchier-Vermeer, 1999) defines the threshold for the 10% most sensitive 50 – 60 year olds as a criterion for noise assessment. These thresholds are about 3dB higher than those of ISO226 as in Table 1, but ISO 7029 (ISO7029, 2000), which deals with the statistics of the threshold in the frequency range from 8000Hz down to 125Hz, shows that, at 125Hz, 10% of 60 year old males have at least 4dB greater hearing sensitivity than the median young adult , shown in Table 1. There is clearly sufficient variation in hearing thresholds to require caution in using the median threshold to assess a noise problem (Leventhall et al, 2005).

While the overall shape of human hearing thresholds show a decline in detection sensitivity for the low frequencies there is a more rapid growth in annoyance as the level increases at low frequencies, shown in Fig 3 (Møller, 1987) At the lowest frequencies, which are on the right of the figure, the level must be greater for the sound to be perceived, but the annoyance range at 4Hz is covered in about 10dB, compared with 40dB at 31.5Hz (Leventhall et al, 2005).

The capability of EHPs' to provide effective solutions may be compromised by a combination of influences, which form a particularly complex problem environment. The overall LFN environment (to include sufferers responses to the problem) is one that is frequently shaped indirectly by the assessment protocols as much as by the direct

Table 1. NSG reference curve (From Leventhall et al, 2005)
Low frequency hearing threshold for levels for 50% and 10% of the population.
(NSG reference curve in bold)

	Otologically Unselected Population 50 –60 years		**Otologically Selected Young adults (ISO 226)**	
Freq Hz	**50% dB**	**10% dB**	**50% dB**	**10% dB**
10	103	92	96	89
12.5	99	88	92	85
16	95	84	88	81
20	85	74	78	71
25	75	64	66	59
31.5	66	55	59	52
40	58	46	51	43
50	51	39	44	36
63	45	33	38	30
80	39	27	32	24
100	34	22	27	19
125	29	18	22	15
160	25	14	18	11
200	22	10	15	7

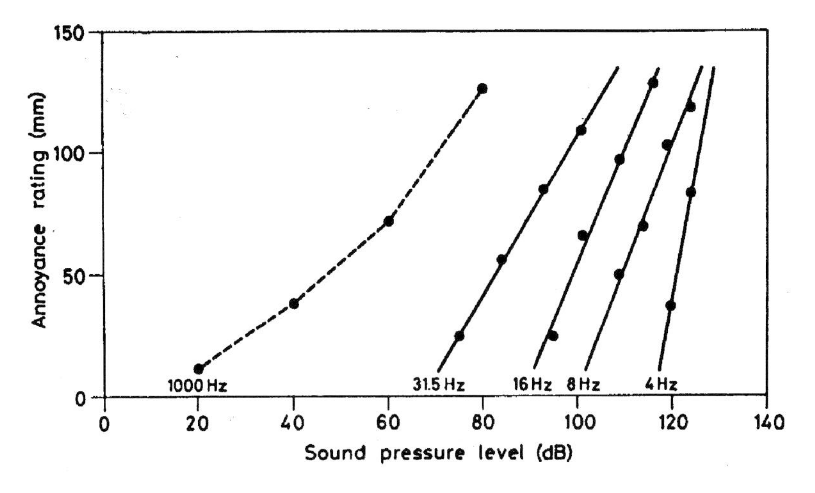

Figure 3. Annoyance rating, showing rapid growth at low frequencies

impact of the LFN stimulus. An example of this 'shaping', common to many of the most intransigent LFN problems, is that sufferers experience EHP interventions as contributing to a decline in quality of life and increases in annoyance (Guest, 2000). Failed resolution of the situation combines with the physical and psychological features of the acoustic signature to create second order stress effects. These effects may develop over time acting to intensify the noise impact value beyond that likely to be associated with the acoustic signature in a neutral context, or experimental conditions, and act to undermine cognitive and emotional stability, which underwrite individuals' capacity to cope, and to therein re establish quality of the subjective environment.

THE PSYCHO-ACOUSTIC ENVIRONMENT

For those individuals trapped within the experience of living with LFN and the consequent impact upon their well being and quality of health, the failure by EHP professionals to identify and locate the source forms yet another component of irritation rather than a sense of support. Issues surrounding the application of inappropriate assessment procedures and the particular problems associated with low frequency source location form a significant context within which an individual's experience is formed. The issues are wide-ranging and frequently interactive. Perhaps the noise can be identified, yet the A-weighted Sound Pressure Level (SPL) is too low to be classified as a statutory nuisance; maybe the individual complainant is the only person in the area that is concerned by the noise. Often it is reported that they are the only members in the household suffering from the noise. Sometimes no others can hear it, and no measurement can detect it. Sufferers may be found to have tinnitus, which is frequently taken as an end to the acoustic problem. It is likely that the category of long term (possible chronic) LFN sufferer hosts a mixture of etiologies wherein failure to satisfy presenting behaviours will consistently play a fundamental role in the degree to which a sustainable solution is found for both the sufferer and the EHP. It is also the case that, from this category of sufferer, will emerge a disproportionate number of long term noise complaints and a familiar cluster of symptoms (Møller and Lydolf, 2002) that include; sleep interference/insomnia, headaches, poor concentration and mood swings.

It seems that important and practical questions need to be asked concerning the status of LFN complaints and the composition and proportion of effective EHP interventions, as LFN complainants tend to remain within the EHP system longer than most other noise related complaints. Importantly, the duration of exposure to the perceived noise, acoustic assessment protocols and perceived EHP behaviours result in an interpersonal environment, which is extremely difficult to manage using, established protocols. The longer the duration the more likely that failed diagnostic and assessment procedures will engender or intensify failed coping. In brief, for a small yet significant number of LFN complaints, involvement with the EHP system is likely to further undermine behavioural (and often failing) coping behaviours, exacerbate stress symptoms, and fuel interpersonal mis perception and miscommunication between parties concerned. No standard or set of standard criteria are available for the EHP, which when deployed have the capacity to address the particular interpersonal demands, engendered by the LFN complaint. Without an effective 'out' strategy, for both sufferer and the EHP pressure for resolution can fall upon the interpersonal as much as the acoustic and psycho acoustic as both parties find themselves trapped in a system without an 'out' strategy.

ASSESSMENT ENVIRONMENT AND THE NON-STANDARD COMPLAINT

Most assessments of LFN, as an environmental pollutant, have necessarily centred on comparisons made against other noise impact criteria. Such assessments of impact are guided by reference to a number of established impact criteria and subjective indices, which include, speech intelligibility (ISO, 1975) annoyance, sleep deprivation and performance degradation (Cooper, and Quick, 1999). Each of these categories has seen the development of empirically based protocols which, under well defined exposure and stimulus conditions, have led to the production of criteria designed to protect health and the quality of life as captured within increasingly internationalized standards. This trend is in response to the rapid growth in technological developments (ranging from heavy duty transport to home used items, from industrial to individual scale applications e.g. air conditioning). All this has occurred against a background trend of continually increasing developments in transport infrastructure, with attendant concerns over the widening environmental impact of noise (Kalveram, 2000).

Clear-cut procedures of assessment and weighting are now available for a number of key 'noise impact' categories including annoyance, under specific circumstances (Leventhall, et al 2003). The widespread application of standards and measurement techniques are an indication of the extent to which subjective and physical attributes of a stimulus 'impact' have been reliably correlated. However, as proponents of separate or discrete weighting networks for LFN are likely to note, this reliability has not been extended to include the effective resolution of LFN complaints, with principal difficulties associated with the measuring impact of exposure. This is particularly key to assessments of LFN impact as the consequences of exposure may well be expressed the disruption of cognitive and behaviourally based coping strategies. This disruption may well then act, as consequent driver of behavioural dysfunction (e.g. failed coping/reducing quality of life).

EMPIRICAL MEASURES: AS A FORMAT FOR RESOLUTION

It is the case that, in general, empirical findings have provided a raft of guidelines, objective measurement and procedural bodies of evidence, which provide the EHO with ways in which to define, quantify and categorise the physical noise signature, and therein to assign an impact rating (e.g. an annoyance weighting) and complainants' experience and associated 'annoyance' behaviour. Importantly, an expected outcome being the effective quantification of the environment and problem enabling all parties to the complaint to identify the source, or likely source, to agree upon existent and perceived levels of the noise and identify effective and explicit steps towards reduction of the impact and a solution. As this process unfolds the complainants perception of the problem and resultant behaviours will frequently be validated. This level of understanding and co-operation forms an essential part of the puzzle for the complainant, as it serves to validate, in explicit terms, their personal experience. The increased access to and sharing in, professional and expert explanations of the physical parameters contribute to regaining a sense of control over their personal and interpersonal environment.

LOW FREQUENCY NOISE: THE EROSION OF THE INTERPERSONAL.

However, before the complainant is able to access this route leading towards a reassertion of personal control and therein coping, they will need to be able to co-opt concepts that allow them to explain both the 'behaviour' of the noise and their experience of 'their' noise. This explanation will need to offer connection points for the relevant professional's technical body of knowledge and experience. In principle, expertise provided by the EHP, drawn from their technical knowledge and personal experience of the environment, acts to form an assessment path that leads to a common language of representation and through which explanations will draw upon shared meaning and understanding. This expertise, although based upon explicit operations, will carry a high degree of tacit validation for the complainants experience and symptoms. From this the complainants are able to seek and achieve a degree of consensus and support for their situation, symptoms and anxieties. The form of correspondence between complainant's personal experience and this (EHP) expertise may be summarized as shown in Figure 4 below

EXPLICIT TO TACIT: NON LFN ROUTE

The EHP intervention (measurement) yields evidence, which encourages the formation of a descriptive language based upon statements of fact that does not contradict complainants' personal knowledge, in a manner that invalidates their experience, and which improves the actual and perceived quality of information. The improved quality of information serves to confirm the personal body of knowledge (evidence), which reinforces the construct validity underlying the language used (statements) in the language used to describe their symptoms, supporting correspondence with their personal experience. The correspondence between explicit and the tacit creates effective problem solving, coping and resolution.

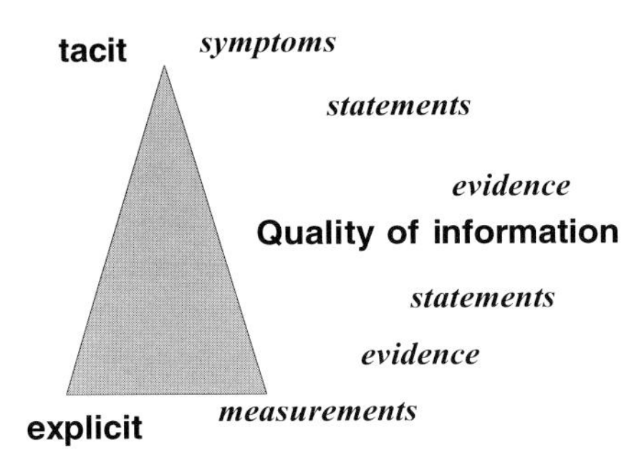

Figure 4. Correspondence: balanced upon secure measurement

SPECIFIC SYMPTOMS

Complainants report that one of the most debilitating aspects of noise is its 'intrusiveness'. They lose control over the quality of sound in their personal environment. Where noise is involved it seems that one person's choice of sound, or activity is another person's noise, and loss of personal space. For low-level LFN the situation is further complicated by the high degree of intrusiveness apparently associated with near threshold sound pressure levels. A number of studies have investigated the relationship between individual differences in the and how they impact upon the relationship between personality and task processing demands. Individual differences in sensitivity to noise in general, and LFN in particular, have been identified, for example the scores on work related tasks have shown differential responses under specific LFN experimental conditions. The work reported by (Persson-Waye et al., (2002) 32 subjects were exposed to moderate levels of low frequency noise during mentally demanding tasks. The work demands weighted the coping processes in order to measure LFN effects in terms of stress and annoyance, and the influence on the secretion of cortisol. Subjects were exposed for a period of two hours to ventilation noise, with dominant low frequencies (low frequency noise condition) or a flat frequency spectrum (reference noise), both at 40dBA level. Subjects were categorized as being either ‘high- or low-sensitive’ to noise in general, or low frequency noise in particular, based upon scores from self-report questionnaires. While results showed that cortisol concentrations during the task were not significantly modulated by the noises, or related to noise sensitivity alone. The normal circadian decline in cortisol concentration was, however, significantly reduced in subjects rated as ‘high-sensitive to noise’ in general, when they were exposed to the low frequency noise. This noise was rated as more annoying and more disruptive to working capacity than the reference noise.

The role of personality and individual differences in individual’s repsonse to noise has led to some contradictory and confusing results. The review of this issue conducted by Belojevic (Belojevic et al., 2003) covered a twelve year period of research into the role of neuroticism, extraversion and general noise sensitivity during task performance under various noise conditions. The relevant model of individual differences indicated that persons scored as ‘neurotic’ might show enhanced “arousability” i.e. their arousal level increases more in stress. Additional unfavorable factors for neurotics were ‘worry’ and ‘anxiety’, which tended to inhibit their coping successfully with noise, or some other stressors during mental performance.In numerous experiments, introverts have shown higher sensitivity to noise than extraverts under conditions of mental-loading performance, while extraverts often coped with a boring task by requesting short periods of noise during performance! The analysis of correlation results regularly revealed a highly significant negative relationship between extraversion and noise annoyance during task performance. Many studies of performance under noise conditions were found to show that individuals with higher ‘self rated’ scores for noise sensitivity achieved poorer work results compared to those scored as less sensitive to noise. Belojevic et al (2003) suggest that those scoring higher on the ‘stable’ personality dimension, with extravert tendencies and with a relatively lower subjective noise sensitivity may be expected to cope better with noise during mental performance tasks, compared to people that scored higher on the introvert personality dimension. The range of factors, which are likely to influence an individual’s predisposition to annoyance and/ or stress from exposure to noise, are unlikely to be amenable to single measures of impact or single and specific features which comprise individual differences.

LFN: THE UNWANTED CHARACTER

The emerging picture from research into LFN effects is one that emphasises the role of an interaction between noise character and individual differences in producing high impact noise with low sound pressure levels. Recent work investigated the properties of LFN character

(Pawlaczyk-Luzszynska et al., 2004). The experiments showed that LFN, defined as broadband noise with dominant content of low frequencies (10-250 Hz), differed in its perceived nature and impact from other noises at comparable levels. The study assessed the influence of LFN on human mental performance using 193 male paid volunteers as subjects. They performed standardized tests: the Signal Detection Test (test I), the Stroop Colour-Word Test (test II), and two sub-tests of the General Aptitude Test Battery, i.e. the Math Reasoning Test (test III) and the Comparing of Names Test (test IV). The experimental design employed three different acoustic conditions. These conditions were; background laboratory noise of about 30 dB (A), LFN and a broadband noise at comparable dB (A) levels of 50 dB. Subjects were assigned randomly to the experimental conditions.

After the test session, the subjects completed a questionnaire aimed at rating the subjective annoyance of exposure conditions during the tasks, and assessing individual sensitivity to noise in general and LFN in particular. The main effects of exposure and/or noise sensitivity on the tests results or their interactions were found in three of the four tests performed (tests I, II and IV). The tendency toward weaker results in low frequency noise compared to other conditions being observed in persons classified in test II as more sensitive to LFN (higher value of reading interference). The significant effect of both exposure and sensitivity to noise on annoyance rating during test performance was also noted. The annoyance of LFN and the reference noise was rated higher than that of background noise. LFN at 50 dB(A) could be perceived as annoying and adversely affecting mental performance (concentration and visual perception), particularly in persons sensitive to LFN and particularly in those persons self-rated as sensitive to LFN. As the subjective 'loudness' of the noise had been matched, the authors argued that the 'character' of the noise carried a form of added value in terms of its capacity to annoy, in a manner that would not have been predicted by the single measure of sound pressure values.

Corresponding evidence of the complex nature of annoyance was found in a pilot study reported by Broner (Broner, 2004). This paper describes how annoyance effects for LFN noise 'character' behaved in a manner contrary to that predicted by loudness values. Subjects listened to stimuli with prominent low frequency spectral peaks for an hour. Loudness and annoyance ratings were elicited using a method of Magnitude Estimation. The findings showed that, at lower frequencies, individuals rate of habituation to perceived loudness was more rapid than that for Annoyance. Broner argued that the basic assumption upon which many noise assessment metrics are founded is flawed and that a non-linear relationship can exist between annoyance and loudness. In this instance, as frequency decreased to below 50Hz, the relationship was indeed inverse. These studies appear to offer a degree of support to Benton and Leventhall's findings based upon an investigation into the impact of noise character upon performance and associated subjective states (Benton and Leventhall, 1986). The experiment compared the impact of loudness pure tones centred at 40 Hz and 100 Hz (both modulated at 1 Hz) and a narrow band noise centred at 70 Hz, all at a level of 25 dB above the individual hearing threshold, and recorded traffic noise (90 dB Lin), matched for loudness, and a silent control condition. They found that the tones centred at 40 Hz and 100 Hz caused more errors in a dual task situation, i.e. when the subjects performed two tasks in parallel, compared with the scores during traffic noise and silence. The effects were especially pronounced during the last ten minutes of the total 30-minute exposure. Further support for LFN intrusive and interference effects upon performance is given by Benton and Robinson (1993) where it was found that under conditions of narrow band low frequency noise at 70 dBC or 95 dBC, subjects made more semantic and spelling errors on a proof reading task. The subjects also rated the low frequency noise as more annoying than two other noise conditions (speech and white noise, 20-20k Hz) matched for loudness against the narrow band low frequency noise. It would appear that both these studies lend support to the view that LFN places an extra degree of demand upon individuals processing and coping strategies.

The common element from much of the research conducted into these psycho-physiological parameters confirms the central role of the 'unwantedness' or 'intrusiveness' of the noise. These attributes have an established role in the growth of stress and confirm the essential role played by psychosocial factors in interacting with other factors (including personality) in shaping how well individuals are able to deal with competing demands. Kalverman (2002) points out that much psycho acoustical noise research has limitations, because it is based upon the correlation between annoyance ratings and physical measurements of sound energy, with a subsequent correlation of annoyance and sound level. He further suggests an "ecological" approach to noise research, which emphasises the psychological functions of sounds. Annoyance originates from acoustical signals which are not compatible with, or which disturb, these psychological functions. Kalveram has extended his approach to include "psycho-biological" effects. Within this model annoyance conveys a "possible loss of biological and behavioural fitness" where such a loss of fitness would result from an imbalance between the psychological and bio/physiological mechanisms of coping, evidenced by the break down in cognitive selectivity and habituation (Callan and Hennessey, 1989). The formation of effective behavioural coping strategies is an integral part of the adaptability of these processes, as they act to match resource with demand (Lazarus and Folkman,

1984). Effective balance between capacity and demand may prompt experiences that correspond to effective habituation while an imbalance may correspond to sensitisation.

PERSONAL EXPERIENCE: A VALID PART OF THE RESOLUTION

The common theme across studies assessing the subjective impact of LFN has been the tendency for subjects'/sufferers' annoyance to increase and the quality of lives to degrade, over time. The polarization of positions taken between the EHP and sufferer seem to intensify and the consequence is a series of behaviours that are reflective of failing negotiation. Positions are taken, options become restricted and the focus is aimed at justifying the differences of view rather than in generating common ground and new ways of solving the problem The fact of the 'problem' recedes as the mutual invalidation of the sufferers personal experience and the EHP's professional knowledge serves to inhibit each parties capacity to work with the 'opposing terms of reference'. The sufferer now needs to find ways to justify both that they are having a noise problem and the way in which they are coping with it. Co-developing effective personal coping strategies for individuals may offer a targeted solution to many sufferers, in order to achieve this, a detailed examination of their 'experience' is a prerequisite. In order to accomplish this re-focus the tacit information available from the sufferers experience needs to play a central role in the assessment and language between the parties. Their experience is often the only 'evidence' that validates the sufferer's view of the problem. Figure 5 highlights some examples of the tacit elements, which characterize this problem environment.

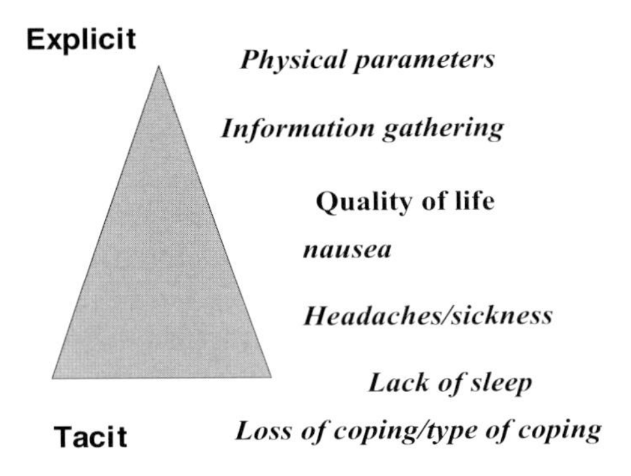

Figure 5. Tacit to Explicit: The LFN Route

The sufferers experience yields evidence that encourages the formation of a descriptive language that blends the personal with the factual in a manner that does not invalidate the individuals' experience. From within this framework the rationale for a variety of behaviours will emerge. Each aspect designed to gain information and solve the noise problem thereby reestablishing effective coping. Statements describing the noise characteristics and its impact are attempts to make sense of experience that in the absence of demonstrably effective assessment practice is the primary source of coherent information. The tacit start-point may be symptoms, which align with failing or failed coping leading to an erosion of maintenance behaviours such as relaxation and sleep. These symptoms interact aggressively over time to undermine sufferers' personal resources to cope. Some common symptoms reported are headaches, nausea and fatigue, each of which can undermine the experience of quality of life. These elements typically result in a narrowing of attentional fields and tend to inhibit the generation and elaboration of alternative explanations. Information gathering, becomes one of the coping mechanisms and is directed towards eliciting confirming evidence for personal experience of the physical environment and may prompt a route towards building correspondence, which effectively excludes the EHP's body of knowledge. Individuals' coping behaviour now actively shapes the type of physical evidence that is rated as important and moreover how it fits into the puzzle of LFN.

BEHAVIOURAL

The need for EHP's to address the psychological impact of noise is an integral element in initiating steps that will build an effective intervention and resolution.

Addressing individuals' experience of the noise is pivotal, as LFN appears to operate at the margins of processes that underpin coping.

The subjective problems associated with the physical impact of the stimulus occupy one level of psychoacoustics; this can be assessed in relative terms of interference, loudness and pitch (intrusiveness) and to some degree annoyance. However, there also exists a secondary subjective impact and this originates from the methods of assessment themselves, a form of supra stimulus impact, not of the noise but rather of the behavioural context generated around the exposure and one that comes to ***define*** the experience for the complainant.

It is in this area that LFN is particularly likely to be problematic for the EHP. The area of incongruence between the complainant's experience and the EHP's findings may be stark; the resolution of this difference however is fundamental to any subsequent and enduring complainant resolution. Many times initial assessments may at best downplay the noise as a problem or at worse measurements taken fail to substantiate even the existence of a noise. All of which is compounded by the fact that usually by the time the EHP is called in individuals would already be experiencing unwanted subjective effects and started to develop personal coping strategies.

OBJECTIVE AND SUBJECTIVE PARAMETERS:

For LFN cases, where a widespread and standardized measurement protocol has yet to be established, there is a high probability that a mismatch between objective measurements and contextual psychological aspects will characterize the situation. In order to proper evaluate the type and degree of behavioural coping, which may also presents as part of the problem, it may be of use to provide for an improved, behaviourally based approach to the problem, one that includes the physical, psychological and the contextual. By the time the EHP arrives the noise impact has already registered within the contextual, influencing the quality of statements used and therein the communication style and content. The feature here being that the complainant has developed a way of making sense of their experience within the terms of reference accessible to them and consistent with the overall behavioural context (e.g. their understanding of how and why the noise behaves the way it does). The rules and justification for a complainant's personalized context may be difficult to communicate and are often supported by perception of implicit relationships rather than the explicit and objective ones. This form of implicit knowledge (experience that is difficult to communicate and demonstrate to others because it has been thoroughly internalized) is characteristic of expert knowledge in that it has tacit validity. While tacit knowledge is frequently valid and leads to coherent personal judgments they are notoriously prone to confirmation bias (Kaufman, 1990). The individual tends to seek for confirming instances, in order to sustain an internally consistent evaluation of their experience, while disregarding or downgrading instances (findings) that run counter to their evaluations. In this way the tacit assessment criteria are sustained.

This is not to suggest that dissociation from reality has occurred. Far from it, it is just as likely that the experience has been interpreted in a manner that would be common to others if they had shared the same route to acquiring it. Many complainants report that EHP's undervalue the distress that they are experiencing and this often becomes a confounding aspect in subsequent exchanges. This situation may be an example of mis communication that results from the schism between the explicit and tacit bodies of knowledge and consequent perspectives. One way forward could be to elicit, explore and profile the tacit properties (those cognitive, emotional and behavioural adjustments that sufferers have had to make in order to cope with the noise) that underpin their form of correspondence. The context of experience for the complainant fails to be translated into accessible concepts likely to be achieved through the application of objective assessments. The common ground sought by the EHO, from which to 'make sense' of the complaint, fails to hold as the evidence does not support the context within which the complaint is observed.

TACIT KNOWLEDGE AND THE INTERVENTION PROCESS: A SKILLS SHIFT

How to improve the quality of information to be exchanged between the complainant and the EHP? If this process could be enhanced then the development of neutral (if not objective) ground and a commonly accessible frame of reference (a combination of the physical, psychological and contextual) could ensue. In order to up grade the level of information available to both parties it will be necessary to inhibit the interpersonal disablers, which commonly act to distort communication between EHP's and LFN sufferers. Perhaps lessons could be learnt from the field of negotiation and conflict resolution where parties are commonly derailed by persistent and antagonistic perceptions of what counts and what doesn't. An illustrative summary is shown in table1 below

Table 1. Strategies for Improved Resolution (Fisher and Ury, 1997)

Separate People from the problem	(avoid being dragged into contests of will, focus on improving the quality of information)
Focus on Interests NOT positions	(interests, tacit information)
Generate variety of alternative solutions	(problem solving perspective for mutual gain)
Seek objective measures/standards	(or at least agreed criteria)

The above strategies represent a segment of an approach designed to guide negotiations, where the answer cannot be achieved by simple, and no matter how clear, the reference to 'facts'. Particular behaviours/skills support these strategies, examples of which are shown in table 2.

Table 2. Skills Sets for LFN Resolution?

Suspend position taking	(Information gathering, types of questions, active listening)
Suspect selectivity	(switch perspectives, what is blocking alternative perspectives)
Empathy	(try on their point of view, explore personal circumstances and challenge personal preconceptions)
Mutual holders of a problem	(focus on tacit and explicit information) Blaming will induce defense, confirm positions and assumptions

These strategies are able to support a move towards a coherent approach and an improvement in the resolution of LFN complaints, resolution that currently is dominated by reliance upon single measure assessments. Where the physical attributes available are often confounded by some assessment practices, the non-physical and contextual attributes may offer a useful set of secondary indicators available for inclusion in the overall appraisal of the complaint. Under conditions where one set of findings is indeterminate in regards to solving the problem, and responding with a series of dis confirmation of the complainant's experience the logical consequence of available analysis, subjective symptoms combined with the behavioural context may prompt alternative interpretations and generate intervention options.

CONCLUSION:

The behavioural context associated with LFN exposure in the environment offers a rich source of tacit information and may provide routes towards enhanced practice for the EHP. The structured collection of such information would provide a valuable source for developing 'common ground' between the complainant and the EHP, an important element in finding a sustainable and practical base for improved coping, reduced isolation and stress while encouraging improved resolution for the complainant and EHP.

REFERENCES

Benton,S. and Leventhall, H.G. (1994). The Role of 'Background Stressors' in the Formation of Annoyance and Stress Responses. *Journal of Low Frequency Noise and Vibration*; 13.

Benton, S., and Leventhall, H. G. (1986): Experiments into the impact of low level, low frequency noise upon human behaviour. *Jnl Low Freq Noise Vibn,* Vol. 5, pp. 143-162.

Belojevic, G., Jakovljevic, B., and Slepcevic, V. (2003): Noise and Mental Performance: Personality Attributes and Noise Sensitivity. *Noise and Health,* Vol. 6, pp. 67-89.

Bengtsson,J., Persson Waye, K., Kjellberg,A., and Benton, S. (2000) *Proceedings of the 9th International Meeting. Low Frequency Noise and Vibration.* ISBN no. 87-90834-06-2.

Broner, N. (2004): Low Frequency Noise assessment-A New Insight? *Proceedings 11th International Meeting: Low Frequency Noise and Vibration and its Control.* Maastricht, 2004, pp. 27-34.

Callan, V. J., and Hennessey, J. F. (1989): Strategies for coping with infertility. *British Journal of Medical Psychology*, Vol. 62, pp. 343-354.

Cooper, C. and Quick,J. (1999) *Stress and Strain*. Oxford: Health Press.

ISO (1974) *Assessment of Noise with Respect to its Effects on the Intelligibility of Speech*. ISO Technical Report TR 3352-1974.

Fisher, R. and Ury, W. (1997) *Getting to YES. Negotiating an agreement without giving in*. Arrow Books Ltd.

Guest, H. (2002): Inadequate standards currently applied by local authorities to determine statutory nuisance from LF and infrasound. *10th International Meeting Low Frequency Noise and Vibration and its Control*. York UK (Editor: H G Leventhall), pp. 61-68.

Kalveram, K. T. (2000): How acoustical noise can cause physiological and psychological reactions. *5th Int Symp Transport Noise and Vibration*. June 2000, St. Petersburg, Russia.

Kaufman, B.E. (1990) A New Theory of Satificing. *Journal of Behavioural Economics*. Spring, pp. 35-51.

Lazarus, R. S., and Folkman, S. (1984): *Stress, appraisal, & coping*. New York: Springer Publishing Co.

Leventhall, H. G., Benton, S., and Pelmear, P. (2003): *A review of published research on low frequency noise and its effects*. Prepared for Defra http://www.defra.gov.uk/environment/noise/research/lowfrequency/index.htm

Leventhall. H.G., Bentons,S. and Robertson, D. (2005) *A review prepared for Defra: Coping Strategies for Low Frequency Noise* Report number: NANR 125

Møller, H. (1987): Annoyance of audible infrasound. *Jnl Low Freq Noise Vibn*, Vol. 6, pp. 1-17.

Møller, H., and Lydolf, M. (2002): A questionnaire survey of complaints of infrasound and low frequency noise. *Jnl Low Freq Noise Vibn*, Vol. 21, pp. 53 – 65.

Pawlaczyk-Luzszynska, M., Dudarewicz, A., Waszkowska, M., Szymczak, W., and Sliwinska-Kowalska, M. (2004): The impact of low frequency noise on the cognitive functions in humans. *Proc. 11th International Meeting on Low Frequency Noise and Vibration and is Control.* Maastricht, pp. 251 – 260.

Persson-Waye, K. (1995). Doctoral thesis., Goteborgs University, ISBN 91-628-1516-4. *On the Effects of Environmental Low Frequency Noise.*

Persson-Waye, K., Rylander, R., Bengtsson, J., Clow,A., Hucklebridge, F., and Evans, P. (2000). *Proceedings of the 9th International Meeting. Low Frequency Noise and Vibration.* ISBN no. 87-90834-06-2.

Persson-Waye, K., Bengtsson, J., Rylander, R., Hucklebridge, F., Evans, P., and Clow, A. (2002): Low frequency noise enhances cortisol among noise sensitive subjects during work performance. *Life Sciences,* Vol. 70, pp. 745-758.

Sloven, P. (2001): A structured approach to LFS complaints in the Rotterdam region of the Netherlands. *Jnl Low Freq Noise Vibn,* Vol. 20, pp. 75-84.

Van den Berg, G. P., and Passchier-Vermeer, W. (1999): Assessment of low frequency noise complaints. *Proc Internoise'99*, Fort Lauderdale.

Watanabe, T., and Møller, H. (1990b): Low frequency hearing thresholds in pressure field and free field. *Jnl Low Freq Noise Vibn,* Vol. 9, pp. 106-115.

Chapter 3. Physiological effects of low frequency noise

In addition to annoyance and sleep disturbance, low frequency noise can cause physiological effects and has been suspected to interfere with brain function and mental activity and it can also induce body surface induced vibration. These effects are discussed in the papers in this chapter.

High level, low frequency noise can induce vibration of the body surface and there have been suggestions that this can lead to illness. In this paper the authors report on vibration measurements on a nine test subjects at locations on their chest, abdomen and head, when subjected to tonal noise between 20 and 50 Hz and sound pressure levels from 100 to 110 dB. A number of conclusions are outlined regarding the effect of frequency, the locations of highest vibration level and the linear increase of vibration level with sound level.

The authors measured low frequency vibration the chest, abdomen and head of nine test subjects and asked the subjects to rate the sensation they felt at each location. It was found that the sensation at all locations including the head correlated strongly with the body vibration (and sound pressure level of the noise stimulus) and not with the head vibration.

Six male subjects were exposed to airborne white noise, two pure tones (31.5 and 50 Hz) and five complex noises made up of pure tones at levels ranging from 90 to 100 dB for each tone used. It was found that the vibration levels measured on the forehead, chest and abdomen increased linearly with increasing sound pressure level in the range considered.

Blast noise equates to impulsive low frequency noise for which frequencies below 10 Hz are the most important in terms of affecting communities at distances larger than 1 km. However at distances greater than 300 m from the blast site, the effects of the blast sound are small and generally limited to rattling windows and building fittings. Closer than 300 m, complaints of sleep disturbance will occur and at distances closer than 50 m people report oppressive or oscillatory feelings. The paper is included in Chapter 2.

5. An investigation on the physiological and psychological effects of infrasound on persons.

Changes in blood pressure, heart rate and subjective feelings were measured after one hour exposures to low frequency sound (2 Hz and 4 Hz) for six male and four female test subjects. Subjects reported feeling headachy, fretful and tired and increases in the systolic blood pressure and heart rate, with the extent of the effects being very subject dependent.

6. Effects on spatial skills after exposure to low frequency noise.

A study of spatial skills was undertaken using twenty seven male and twenty seven female test subjects in which subjects were asked to perform a mental rotation task after being exposed to 21 Hz low frequency noise for 20 minutes. Statistical analysis showed that a noise level of 81 dB(A) resulted in significant post exposure effects but exposure to 77 dB(A) or 86 dB(A) generated no significant effects.

7. Does low frequency noise at moderate levels affect human mental performance.

191 male test subjects were used to determine the influence of low frequency noise of 50 dB(A) on mental performance including attention, visual perception and logical reasoning. No effects were observed.

Some characteristics of human body surface vibration induced by low frequency noise

Yukio Takahashi, Kazuo Kanada and Yoshiharu Yonekawa
National Institute of Industrial Health, 6-21-1, Nagao, Tama-ku, Kawasaki 214-8585, Japan, email: takahay@niih.go.jp

ABSTRACT
The body surface vibration induced by low frequency noise (noise-induced vibration) was measured at the forehead, the anterior chest and the anterior abdomen. At all the measuring locations, the increase steps in the vibration acceleration levels of the noise-induced vibrations was in good agreement with the increase steps, in the sound pressure levels of the noise stimuli. The vibration acceleration level measured at the forehead was found to increase suddenly at around 31.5-40 Hz, while the acceleration levels measured at the chest and abdomen increased with frequency at approximately constant rates in the 20- to 50-Hz range. Our results showed no clear evidence of the effect of posture or bilateral asymmetry in the noise-induced vibration. We found that the noise-induced vibrations measured at the chest and abdomen were correlated negatively with the body fat percentage.

INTRODUCTION
Low frequency noise, that is, noise below 100 Hz, is widely generated in living and working environments [1]. One of the typical sources of low frequency noise in the living environment is built-in air-conditioning system in a house, which generates low-level but continuous low frequency noise. In the working environment, on the other hand, various machines such as blowers, exhaust fans, air compressors and the like generate high-level low frequency noise, the sound pressure level of which occasionally exceeds 100 dB (SPL).

Mechanical vibration is induced in the human body when a person is exposed to high-level low frequency noise [2, 3]. The level of this vibration (noise-induced vibration) is not especially high, but many previous studies suggest that noise-induced vibration plays an important role in the human psychological and perceptual responses to low frequency noise. For example, Yamada et al. reported that deaf persons perceived low frequency noise through sensing vibration in the chest [4]. Takahashi et al. showed that subjective ratings of vibratory sensation perceived in the chest and abdomen were closely correlated with the vibration induced at each corresponding location [5].

Recent studies by Castelo Branco et al. suggest that noise-induced vibration is associated with adverse health effects caused by low frequency noise [6-8]. They examined workers who had been exposed to high-level low frequency noise for more than 10 years and found some pathological changes in the workers' bodies, such as pericardial thickening, pulmonary fibrosis and so on. They called the pathological changes 'vibroacoustic disease' and presumed that long-term exposure to the noise-induced vibration was one of the causes. Many pathological changes were found in the chest, which is where researchers would expect noise-induced vibration to be induced at a comparatively high level [2, 3]. Clarifying the characteristics of noise-induced vibration should be helpful to clinicians attempting to assess low frequency noise from a medical viewpoint.

In our previous study we measured noise-induced vibration at the chest and abdomen of sitting subjects [2]. We found that the vibration acceleration levels of the noise-induced vibration measured at the chest were higher than those measured at the abdomen. We also found that the increase steps in the noise-induced vibration were in good agreement with the increase steps in the sound pressure levels of the noise stimuli. However, we wanted to make more precise measurements, because the previous results were obtained under experimental conditions with too many variables, such as a wide range of the subject's age (24-57 yr).

The aim of the present study was to clarify the detailed characteristics of noise-induced vibration. We limited the subjects' age to fall within a narrow range and carried out the measurements in a test chamber where temperature and humidity were controlled. The forehead was newly included in the measuring locations, in addition to the chest and abdomen. We measured the noise-induced vibration not only with subjects sitting but also with them standing, which allowed us to examine characteristics such as position-dependence, effects of subject's posture, bilateral asymmetry, effects of the subject's physical constitution etc.

SUBJECTS

Nine subjects participated in the measurements; All of the subjects were healthy male students, and their ages ranged from 21 to 24 yrs (mean = 22.6, SD = 1.0). Their height and weight were 173.0 ± 3.5 cm and 65.8 ± 4.5 kg (mean ± SD), respectively.

This study was approved by the ethics committee of the National Institute of Industrial Health, and informed consent was obtained from each subject before the measurement.

MEASUREMENT METHODS

The methods for measuring noise-induced vibration, which are briefly described in the following paragraphs, were almost the same as the methods used in our previous study [2]. The measurements were carried out in a test chamber with a capacity of about 25 m^3 [9] in winter (a dry season in Japan). The temperature in the test chamber was initially set at 25°C and, if the subject complained of discomfort, it was adjusted within the 23-27°C range. The humidity in the test chamber was kept at 40% by a humidifier.

Fifteen kinds of low frequency noise stimuli (5 frequencies x 3 sound pressure levels) were used. All of them were pure tones; we used frequencies of 20, 25, 31.5, 40 and 50 Hz and sound pressure levels of 100, 105 and 110 dB (SPL). No noise stimulus in the frequency range higher than 50 Hz was used, because the spatial uniformity of the sound pressure levels in the test chamber would deteriorate above 50 Hz [9]. To detect only those noise-induced vibrations which were higher in level than those inherent in the human body, we avoided using noise stimuli lower than 20 Hz and the sound pressure levels of the noise stimuli were set to be sufficiently high. A function generator (HP3314A, Hewlett Packard, USA) generated sinusoidal signals as the source of the noise stimuli. The signals were amplified by power amplifiers (PC4002M, Yamaha, Japan) and then reproduced by 12 loudspeakers (TL-1801, Pioneer, Japan) installed in the wall in front of the subject (Fig. 1). The levels of the higher harmonics of the noise stimuli had proved adequately low when the stimuli were reproduced at levels of 110 dB or less [9] . The sound pressure level was adjusted so that the desired level could be measured at the centre of the test chamber, at 100 cm height (corresponding to the position of the sitting subject's chest) or at 120 cm height (corresponding to the position of the standing subject's abdomen), without a subject present.

Noise-induced vibrations were measured at 5 measuring locations on the body surface: at the forehead (2 cm above the level of the eyebrow, and on the midline), the right anterior chest (2 cm above the right nipple), the left anterior chest (2 cm above the left nipple), the right anterior abdomen (5 cm below the pit of the stomach, and 5 cm to the right of the midline) and the left anterior abdomen (5 cm below the pit of the stomach, and 5 cm to the left of the midline). When an accelerometer is attached on the body surface, the accelerometer and the local tissue around it may form an additive local mechanical system and result in a difference between the accelerometer measurement and the acceleration of the body surface. To minimize this change around each measuring location, it is necessary to use an accelerometer which is as small and lightweight as possible [10]. Hence, we used a small (3.56 mm x 6.86 mm x 3.56 mm) and lightweight (0.5 g) accelerometer (EGA-126-l0D, Entran, USA) as a vibration detector. The accelerometer, which was attached at each measuring location with double-sided adhesive tape and no other supporting material, detected noise-induced vibration perpendicular to the body surface. After being amplified by a strain amplifier (6M92, NEC San-ei Instruments, Japan), the detected vibration was recorded on DAT with a multi-channel data recorder (PC216Ax, Sony Precision Technology, Japan). Simultaneous measurements at 5 measuring locations were performed with 5 identical measuring sets, each of which consisted of an accelerometer and a strain amplifier.

Frequency-analysis by an FFT analyzer (HP3566A, Hewlett Packard, USA) yielded the power spectrum of the noise-induced vibration detected at each measuring location. Then, the frequency component corresponding to the noise stimulus was transformed to a vibration acceleration level (VAL) defined as

$$\text{Vibration acceleration level (VAL)} = 20\text{x}\log 10(a_{meas} / a_{ref}),$$

where a_{meas} was a measured acceleration (m/s^2(r.m.s.)) and a_{ref} was the reference acceleration equal to 10^{-6} m/s^2 [11]. In this transformation, we did not separate the inherent vibration from the total vibration, because the phase relationship between the inherent vibration and the noise-induced vibration was unknown. It should be noted that in this paper, the VALs of the measured noise-induced vibration were probably contaminated by those of the inherent vibration and that the degree of contamination was more significant at lower frequencies [2].

Each measurement consisted of two sessions. In the first session the subject sat on a stool at the centre of the test chamber (Fig. 1), and in the second session he stood at the same place without any supporting

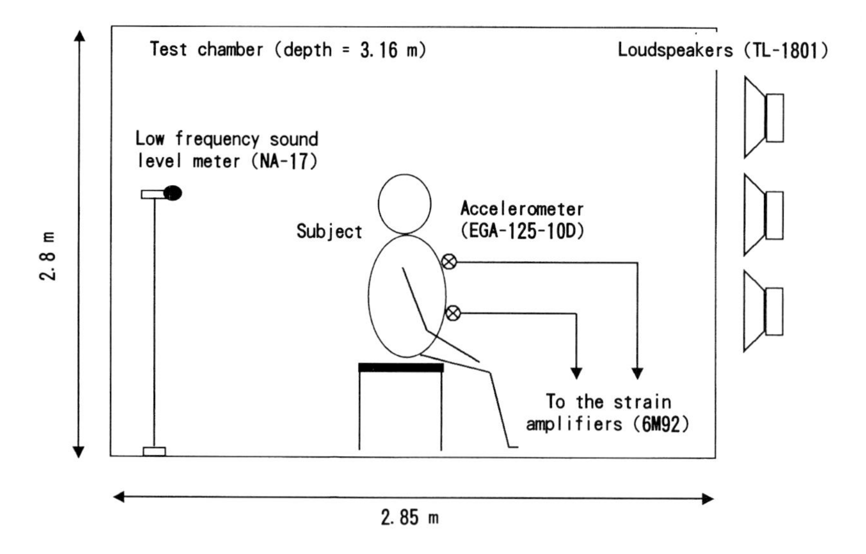

Figure 1. The experimental apparatus in the test chamber. The subject sat on a stool at the centre of the test chamber stood at the same place after removing the stool.

instruments. At the beginning of each session, the inherent body surface vibration with no noise stimulus was recorded (1 min). Next, a rest period with no noise stimulus (1 min) and an exposure period with a noise stimulus (1 min) when the noise-induced vibrations were recorded were continued alternately. The subject, who wore no clothes on the upper half of the body to allow the accelerometers to be attached, was instructed to keep his upper body erect during an exposure period. Fifteen kinds of noise stimuli were presented in random order for every subject and every session.

We used a statistics software package (SPSS for Windows 10.0J, SPSS Japan, Japan) to evaluate the effect of the subject's posture, bilateral asymmetry and physical constitution. In the statistical analysis of the effect of posture, we used ANOVA to examine the statistical significance of the difference between the noise-induced vibrations for the sitting subjects and those for the standing subjects. We also used ANOVA to examine the bilateral asymmetry between the noise-induced vibrations measured at bilaterally symmetrical measuring locations. To determine the effect of the subject's physical constitution, we used the statistics package to calculate Pearson's correlation coefficients between the VALs of the noise-induced vibration and the subject's body fat percentage, which was estimated by a body-fat meter (TBF-501, Tanita, Japan).

RESULTS

Figure 2 shows the VALs of the noise-induced vibration (means ± SD) measured at five measuring locations, together with the VALs of the inherent vibration. Solid lines in the figure represent the VALs for the sitting subjects and dotted lines represent those for the standing subjects. For simplicity, only the upward SDs are shown for the sitting subjects and only the downward SDs are shown for the standing subjects. At all the measuring locations, the VALs of the noise-induced vibration increased with the frequency and sound pressure level of the noise stimulus, while the VALs of the inherent vibration decreased as the frequency of the noise stimulus increased. This is because the inherent vibration originating in vital activities such as the heartbeat is distributed mainly in the lower frequency range. The highest VAL of the noise-induced vibration (about 107 dB with 50 Hz, 110 dB noise stimulus) was measured at the chest. The noise-induced vibrations measured at the abdomen were at slightly lower levels than those measured at the chest, and the noise-induced vibrations measured at the forehead were at clearly lower levels. At all the measuring locations, the increase step in the VALs of the noise-induced vibrations was found to be about 5dB, which was in good agreement with the increase step (5dB(SPL)) in the sound pressure levels of the noise stimuli.

The VALs of the noise-induced vibration measured at the chest were found to increase with frequency at an approximately constant rate. A similar tendency was found in the noise-induced vibration measured at

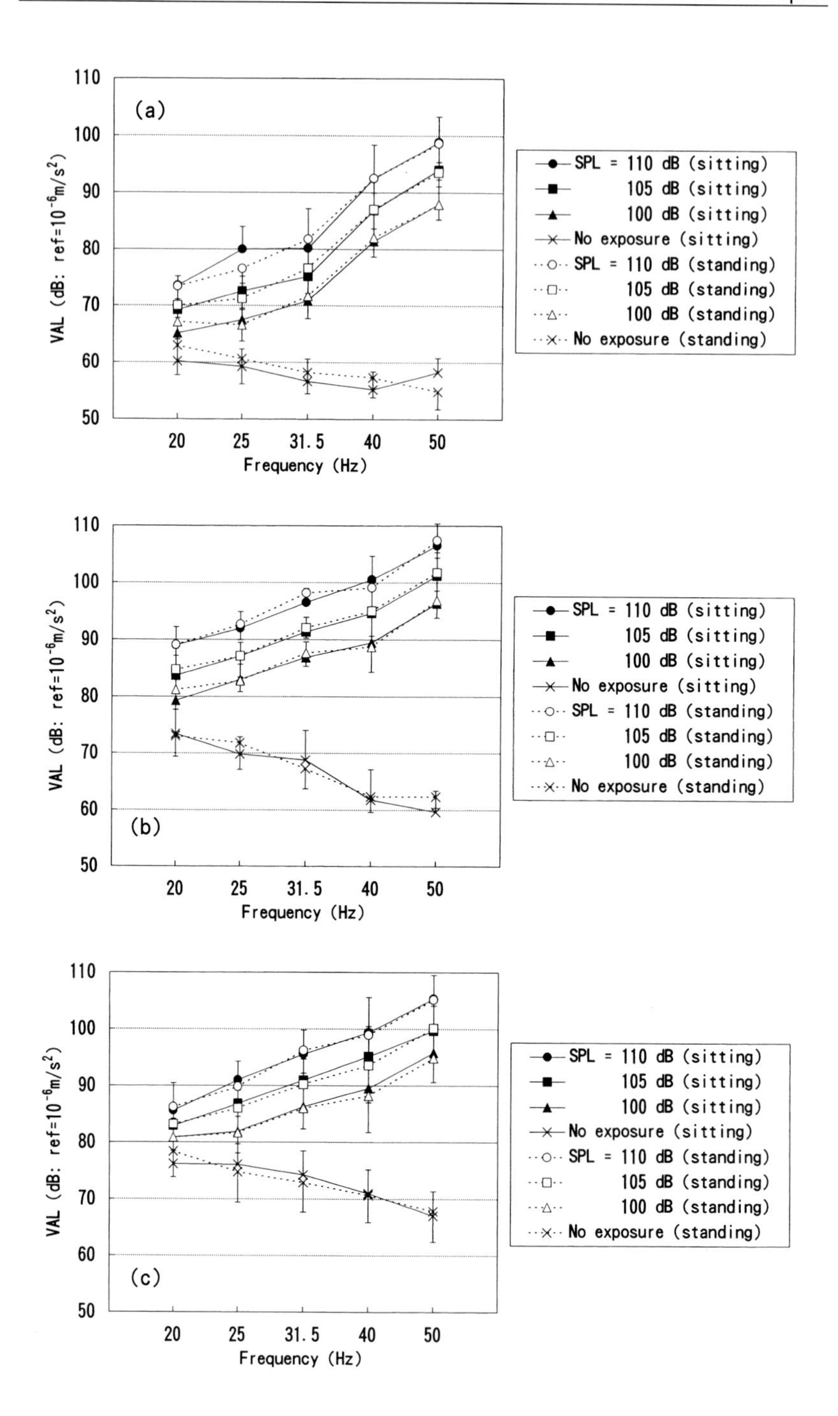

Figure 2. *continued opposite*

the abdomen. At the forehead, in contrast with the other measuring locations, the VALs of the noise-induced vibration were very low at lower frequencies (20-25 Hz), but they rose suddenly around 31.5-40 Hz (Fig. 2(a)). As a result, the highest rate of increase with frequency in the 20- to 50-Hz range was measured at the forehead.

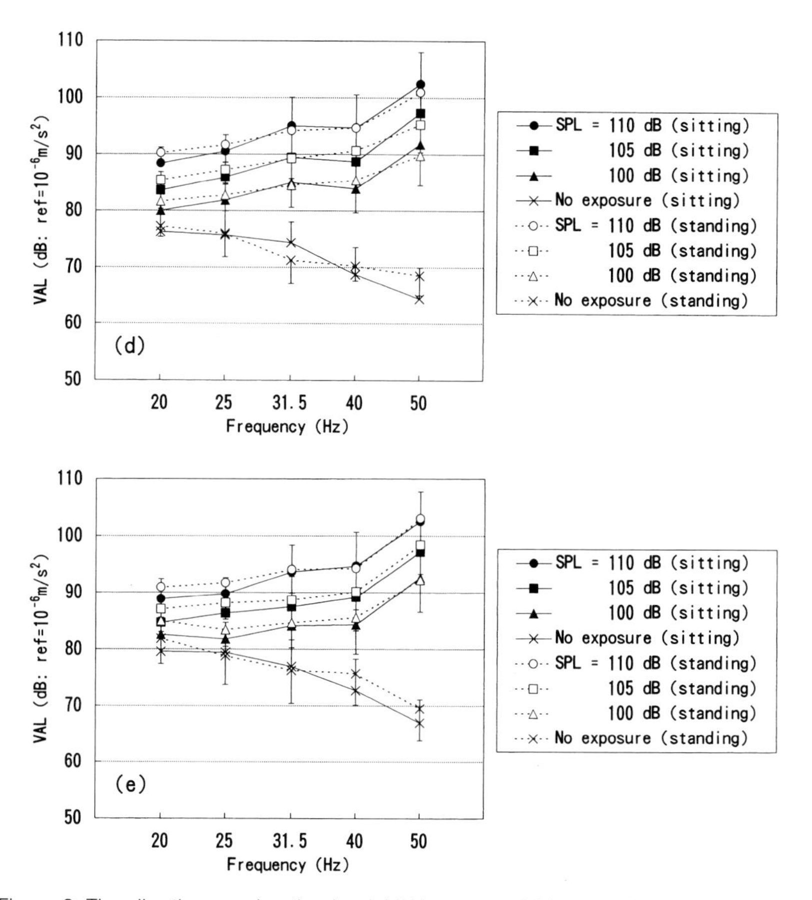

Figure 2. The vibration acceleration level (VAL) measured (a) at the forehead, (b) at the right chest, (c) at the left chest, (d) at the right abdomen and (e) at the left abdomen. Solid lines represent the VALs for the sitting subjects, and dotted lines represent those for the standing subjects.

Table I summarizes the statistical significance of the difference between the VALs for the sitting subjects and those for the standing subjects. The effect of posture was not statistically significant at any of the measuring locations.

Table 1. Statistical significance of the difference between the vibration acceleration levels for the sitting subject and those for the standing subject. No significant difference was found.

Measuring locations		p-values
Head		0.912
Chest	Right	0.308
	Left	0.508
Abdomen	Right	0.569
	Left	0.058

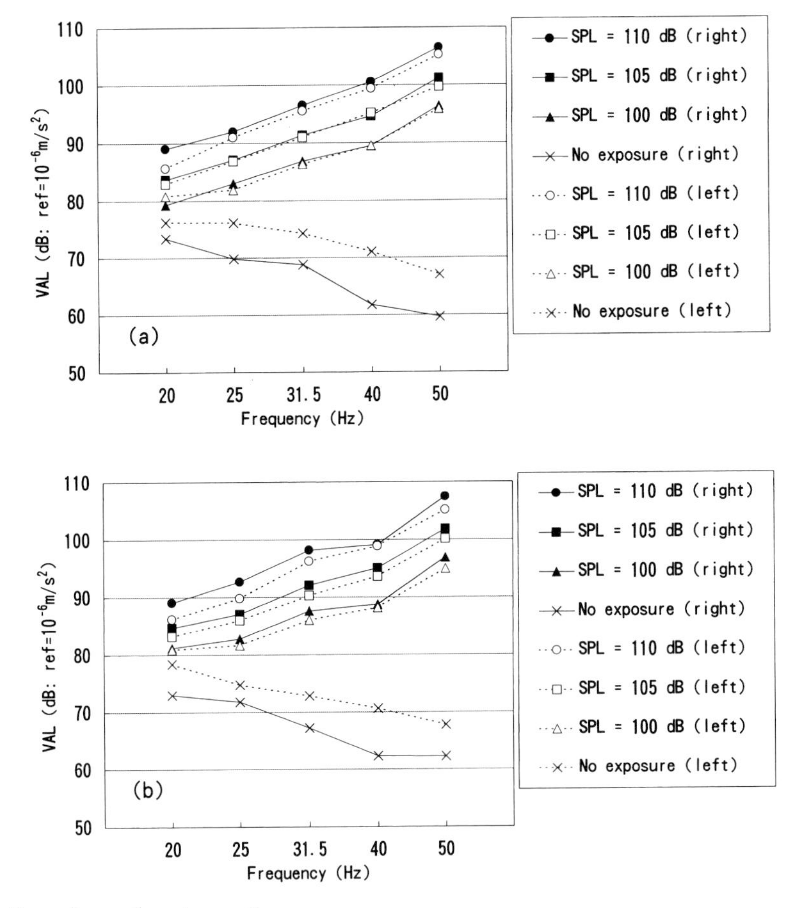

Figure 3. *continued opposite*

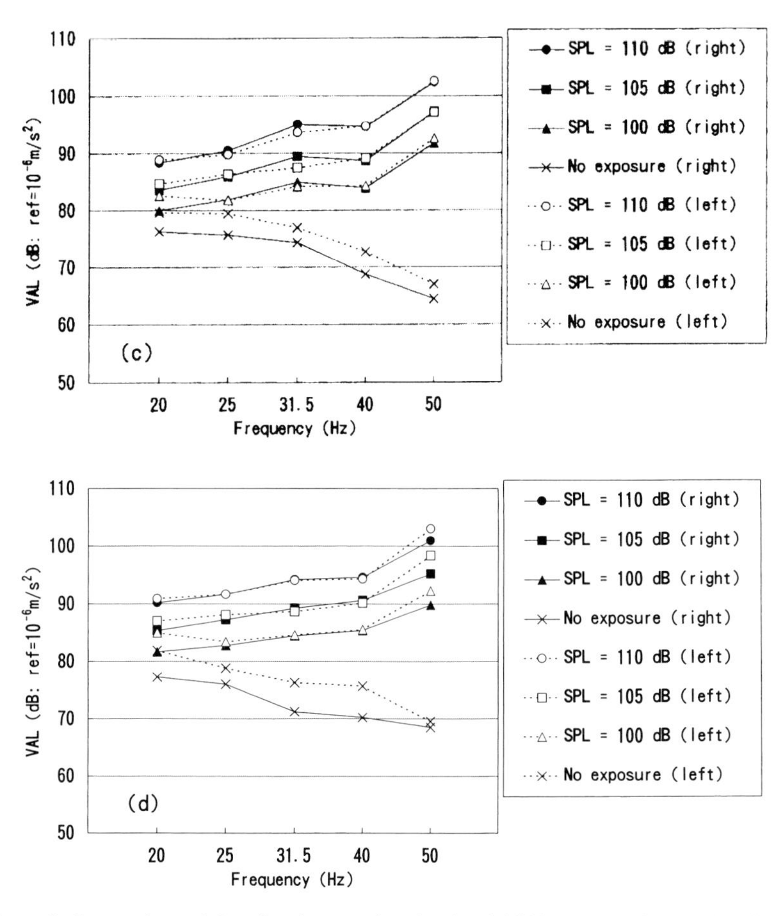

Figure 3. Comparison of the vibration acceleration level (VAL) measured (a) at the chest of the sitting subject, (b) at the chest of the standing subject, (c) at the abdomen of the sitting subject and (d) at the abdomen of the standing subject. Solid lines represent the VALs measured on the right side of the body, and dotted lines represent those measured on the left side of the body.

Figure 3 compares the VALs of the noise-induced vibrations measured at bilaterally symmetrical measuring locations. Solid lines in the figure represent the VAL measured on the right side of the body, and dotted lines represent those measured on the left side of the body. As seen in Table II, significant bilateral asymmetry was found at the chest of standing subjects ($p = 0.003$); however, no statistically significant difference was found at the other measuring locations.

Table II. Statistical significance of the difference between the vibration acceleration levels measured at the bilaterally symmetrical locations. (**) represents a significant difference at the level of $p < 0.01$ (two- sided) .

Measuring locations	**p-values**	
	For sitting subjects	**For standing subjects**
Chest	0.174	0.003**
Abdomen	0.883	0.120 .

Figure 4 shows examples of the correlation between the VAL of the noise-induced vibration and the body fat percentage. Tables III and IV summarize the Pearson's correlation coefficients and the statistical significance of the correlation. The VALs of the noise-induced vibrations measured at the chest and abdomen, as a whole, tended to correlate negatively with the body fat percentage. At the forehead, on the other hand, no significant correlation was verified for any of the noise stimuli.

Table III. Pearson's correlation coefficients between the vibration acceleration levels and the body fat percentage. (*) represents a significant difference at the level of $p < 0.05$ (two-sided) and (**) at the level of $p < 0.01$ (two-sided). The blank cell in the table represents no significance.

Noise stimuli		**Pearson's correlation coefficients**			
		Chest		**Abdomen**	
Frequency (Hz)	**SPL(dB)**	**Right**	**Left**	**Right**	**Left**
20	100	-	-	-0.731 (p=0.025)*	-
	105	-	-	-0.702 (p=0.035)*	-
	110	-0.711 (p=0.032)	-	-0.714 (p=0.031)*	-
25	100	-	-	-	-
	105	-	-	-	-
	110	-	-	-	-
31.5	100	-	-	-	-
	105	-	-	-	-
	110	-	-	-	-
40	100	-	-	-0.735 (p=0.024)*	-
	105	-	-	-0.784 (p=0.012)*	-
	110	-	-	-0.716 (p=0.030)*	-
50	100	-0.797 (p=0.010)*	-0.786 (p=0.012)*	-	-0.726 (p=0.027)*
	105	-0.871 (p=0.002)*	-0.741 (p=0.022)*	-	-0.808 (p=0.008)**
	110	-0.864 (p=0.003)*	-0.762 (p=0.017)*	-	-0.671 (p=0.048)*

Table IV. Pearson's correlation coefficients between the vibration acceleration levels and the body fat percentage. (*) represents a significant difference at the level of $p < 0.05$ (two-sided) and (**) at the level of $p < 0.01$ (two-sided). The blank cell in the table represents no significance.

Noise stimuli		Pearson's correlation coefficients			
		Chest		Abdomen	
Frequency (Hz)	SPL(dB)	Right	Left	Right	Left
20	100	-0.837 (p=0.005)**	-	-	-
	105	-0.776 (p=0.014)*	-	-0.833 (p=0.005)**	-
	110	-0.755 (p=0.019)*	-	-0.811 (p=0.008)**	-
25	100	-	-	-	-
	105	-	-	-0.678 (p=0.045)*	-
	110	-	-	-	-
31.5	100	-	-	-	-
	105	-	-	-	-
	110	-	-	-	-
40	100	-	-	-	-
	105	-	-	-0.758 (p=0.018)*	-
	110	-	-	-	-
50	100	-	-0.762 (p=0.017)*	-	-0.739 (p=0.023)*
	105	-	-0.731 (p=0.025)*	-	-0.761 (p=0.017)*
	110	-	-0.757 (p=0.018)*	-0.706 (p=0.034)*	-0.795 (p=0.011)*

DISCUSSION

In this study we measured noise-induced vibration at the forehead, the chest and the abdomen. At all the measuring locations, the increase step in the VALs of the noise-induced vibration corresponded well to the increase step in the sound pressure levels of the low frequency noise stimuli (Fig. 2). This linear response to the noise stimuli is consistent with our previous results [2]. In general, the wavelength of low frequency noise is sufficiently longer than the size of a human being. For example, provided that the sound velocity in the atmosphere is 340 m/s, the wavelength of the noise with a frequency of 50 Hz or less is 6.8 m or more. Hence, for a human being, exposure to low frequency noise with frequencies below 50 Hz is similar to periodic fluctuation in the atmospheric pressure, and it is reasonable to hypothesize that noise-induced vibration is isotropic and generated mainly by the difference in the atmospheric pressures between the inside and the outside of the body. Because this pressure difference is considered to change approximately in proportion to the sound pressure level of the low frequency noise stimulus, the linear response of the noise-induced vibration can be elucidated by the above hypothesis. Thus, in the limited range of frequency and sound pressure level, the human body is considered to be a mechanical

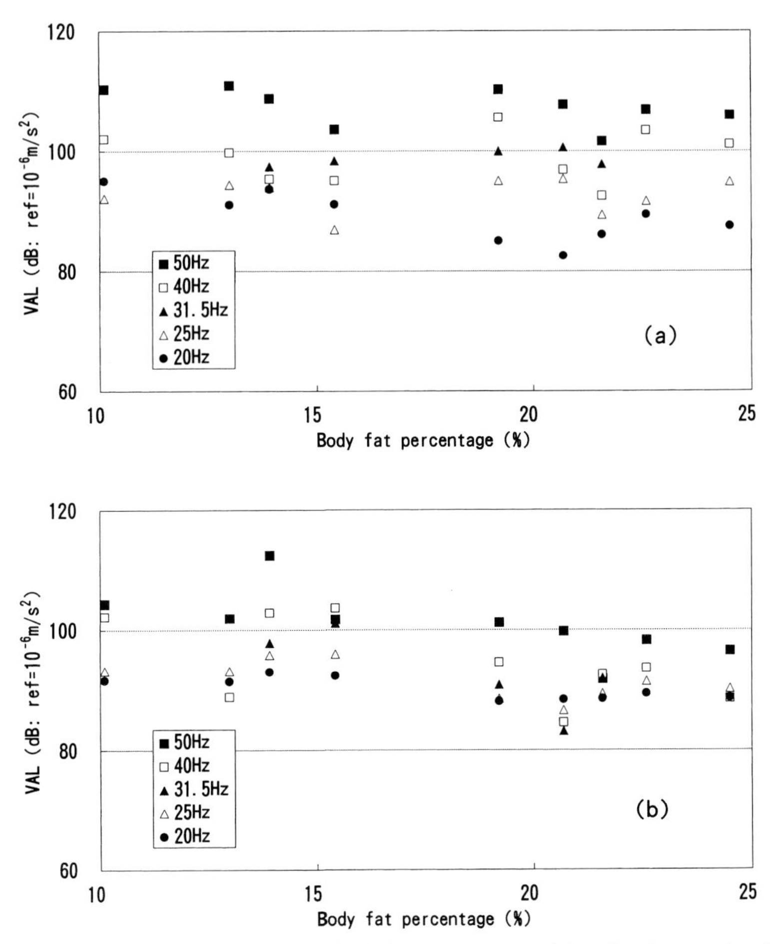

Figure 4. The correlation between the body fat percentage and the vibration acceleration levels (VAL) measured (a) at the right chest of the standing subject and (b) at the right abdomen of the standing subject. The sound pressure levels of the noise stimuli are 110 dB(SPL) in all cases.

system that responds linearly to airborne vibrating stimuli such as low frequency noise.

The bilateral asymmetry in the noise-induced vibration was found to be statistically significant only at the chest of the standing subject (Table II), with higher-level noise-induced vibration measured at the right chest. Wodicka et al. introduced sound stimuli from the mouth to the lung and measured the chest wall vibration [12, 13]. They found that the right chest wall was excited at higher levels than the left chest wall and presumed that this was due to the bilateral asymmetry in the anatomical structures in the human body. They used broadband sound stimuli with frequencies higher than 100 Hz and vibrated the intrathoracic structures. In our study, on the other hand, we used pure tonal stimuli in the 20- to 50-Hz range and vibrated the whole body from the atmosphere. Nevertheless, the chest wall vibrations measured in both studies were generated by an essentially identical source, which is the pressure difference between the inside and the outside of the body. Therefore, it is possible that the low frequency noise stimuli induced the bilaterally asymmetrical vibration in the chest in both cases. The anatomical structures in the abdomen are also organized to be bilaterally asymmetrical, which implies that the abdominal vibration would also be bilaterally asymmetrical, but we could not verify a statistically significant bilateral asymmetry in the noise-induced vibration at the abdomen. Wodicka et al. used noise stimuli in a frequency range higher and wider than ours. Further studies using noise stimuli in a wider range of frequency are needed to verify the bilateral asymmetry in the noise-induced vibration.

The mechanical characteristics of the anatomical structures in a sitting person are different from those in a standing person. For example, the intraabdominal pressure in the body part around the hips is presumed to be at a higher level for a sitting person than for a standing person. The change in the intraabdominal pressure causes a change in the pressure difference between the inside and the outside of the body. which may result in a change in the magnitude of the noise-induced vibration. In this study, however, the effect of posture on the noise-induced vibration was not statistically significant (Table I). We speculate that this is due to the fact that the anatomical structures around the five measuring locations, none of which were proximal to the lower trunk, hardly changed their mechanical characteristics when the subject changed his posture. To verify the effect of posture on the noise-induced vibration, one needs to measure the body surface vibration at the lower abdomen or the lower back.

We found that the VALs of the noise-induced vibrations measured at the chest and abdomen tended to correlate negatively with the body fat percentage (Tables III and IV). This is consistent with our previous results that the magnitude of the noise-induced vibration measured at the abdomen was correlated negatively with BMI (Body Mass Index) [2]. An increase in the body fat results in thickening the body tissue under the skin and changes in the mechanical characteristics of the body tissue, including a reduction of stiffness in the body tissue. These changes in the body tissue interfere with the propagation of vibration in the human body and result in the negative correlation we found. The VAL of the noise-induced vibration measured at the forehead showed no significant correlation with the body fat percentage. We believe the vibration induced at the forehead is less influenced by the change in the mechanical characteristics of the body tissue than the vibrations induced at the chest and abdomen because the body tissue under the skin at the forehead is scarcer than that under the skin at the chest and abdomen. This may be speculated to be a reason why we could find neither negative nor positive significant correlation between the vibration induced at the forehead and the body fat percentage.

One new finding in this study was that the VAL of the noise-induced vibration measured at the forehead increased steeply around 31.5-40 Hz, in contrast with approximately constant rates of increase with frequency found at the chest and abdomen. Kobayashi et al. vibrated the human head using a vibrating table and found that the vibration transmissibility in the human head had a resonance-like peak around 50-80 Hz [14]. In our study, on the other hand, the human head was vibrated by high-level low frequency noise stimuli with frequencies below 50 Hz. In spite of the different vibrating methods, the results of both studies correspond well qualitatively in the frequency range just below 50 Hz. Because the body tissue between the skin at the forehead and the skull is scarce, the characteristics of the skull may dominate over the characteristics of the vibration induced at the forehead. Some resonance frequencies of the skull were found in the frequency range around 1 kHz or higher [15, 16] but, to the authors' knowledge, they were not measured in the frequency range below 100 Hz. Although we have not drawn a specific conclusion because of the small amount of available experimental data, we believe resonance-like vibration may be induced in the human head at frequencies around 50 Hz by airborne vibrating stimuli such as high-level low frequency noise.

In this study we failed to sufficiently control some of the experimental conditions. For example, only male subjects participated, and the measurements were conducted under limited experimental conditions, such as using a narrow range of frequency and a narrow range of sound pressure level. Hence, many characteristics of the human body surface vibration remain unknown. In addition, we still do not know how the body surface vibration is related to the vibration induced in the inner body, which needs to be investigated for assessing low frequency noise from a medical viewpoint. Future studies of noise-induced vibration should be conducted under more various experimental conditions, with an additional goal of establishing the relationship between the body surface vibration and the vibration induced in the inner body.

ACKNOWLEDGMENTS

This study was supported in part by a fund from the Environment Agency of Japan.

REFERENCES

1. Berglund B., Hassmén P., Job R.F.S.,(1996) Sources and effects of low frequency noise. *J. Acoust. Soc. Am.*, Vol. 99, No. 5, pp. 2985-3002.

2. Takahashi Y., Yonekawa Y., Kanada K. and Maeda S., (1999) A pilot study on the human body vibration induced by low frequency noise. Ind. *Health*, Vol. 37, No. 1, pp. 28-35.

3. Smith S.D.,(2002) Characterizing the effects of airborne vibration on human body vibration response. *Aviat. Space. Environ. Med.*, Vol. 73, No. 1, pp. 36-45.

4. Yamada S., Ikuji M., Fujikata S., Watanabe T, and Kosaka T., (1983) Body sensation of low frequency noise of ordinary persons and profoundly deaf persons. *J. Low Freq., Noise & Vib.*, Vol. 2, No. 3, pp. 32-36.

5. Takahashi Y, and Yonekawa Y., (2001) Relationship between vibratory sensation and the human body surface vibration induced by low frequency noise. *Proc. Inter-Noise 2001*, pp. 1079-1082.

6. Castelo Branco N.A.A. and Rodriguez E., (1999) The vibroacoustic disease – An emerging pathology. *Aviat. Space. Environ. Med.*, Vol. 70, No. 3, Pt 2, Al-A6.

7. Alves-Pereira M., 1999 Noise-induced extra-aural pathology: A review and commentary. *Aviat. Space. Environ. Med.*, Vol. 70, No. 3, Pt 2, A7-A21.

8. Castelo Branco N.A.A., (1999) The clinical stages of vibroacoustic disease. *Aviat. Space. Environ. Med.*, Vol. 70, No. 3, Pt 2, A32-A39.

9. Takahashi Y., Yonekawa Y., Kanada K. and Maeda S., (1997) An infrasound experiment system for industrial hygiene. *Ind. Health,* Vol. 35, No. 4, pp. 480-488.

10. Kitazaki S. and Griffin M.J., (1995) A data correction method for surface measurement of vibration on the human body. *J. Biomech.*, Vol. 28, No. 7, pp. 885-890.

11. International Organization for Standardization, (1990) Human response to vibration – Measuring instrumentation. ISO 8041.

12. Wodicka G.R., DeFrain P.D. and Kraman S.S., (1994) Bilateral asymmetry of respiratory acoustic transmission. *Med. Biol. Eng. Comput.*, Vol. 32, No. 5, pp. 489-494.

13. Pasterkamp H., Patel S, and Wodicka GR., (1997) Asymmetry of respiratory sounds and thoracic transmission. *Med. Biol. Eng. Comput.*, Vol. 36, No. 2, pp. 103-106.

14. Kobayashi F., Nakagawa T., Kanada S., Sakakibara H., Miyao M., Yamanaka K, and Yamada S., (1981) *Measurement of human head vibration. Ind. Health,* Vol. 19, No. 3, pp. 191-201.

15. Håkansson B., Brandt A., and Carlsson P., (1994) Resonance frequencies of the human skull in vivo. *J. Acoust. Soc. Am.*, Vol. 95, No. 3, pp. 1474-1481.

16. Khalil T.B., Viano D.C, and Smith D.L., (1979) Experimental analysis of the vibrational characteristics of the human skull. *J. Sound Vib.*, Vol. 63, No. 3, pp. 351-376.

The relationship between vibratory sensation and body surface vibration induced by low-frequency noise

Yukio Takahashi, Kazuo Kanada and Yoshiharu Yonekawa
National Institute of Industrial Health, 6-21-1, Nagao, Tama-ku, Kawasaki 214-8585, Japan e-mail: takahay@niih.go.jp

ABSTRACT
Human body surface vibration induced by low-frequency noise was measured at the forehead, the chest and the abdomen. At the same time, subjects rated their vibratory sensation at each of these locations. The relationship between the measured vibration on the body surface and the rated vibratory sensation was examined, revealing that the vibratory sensations perceived in the chest and abdomen correlated closely with the vibration acceleration levels of the body surface vibration. This suggested that a person exposed to low-frequency noise perceives vibration at the chest or abdomen by sensing the mechanical vibration that the noise induces in the body. At the head, on the other hand, it was found that the vibratory sensation correlated comparably with the vibration acceleration level of the body surface vibration and the sound pressure level of the noise stimulus. This finding suggested that the mechanism of perception of vibration in the head is different from that of the perception of vibratory sensation in the chest and the abdomen.

1. INTRODUCTION
Low-frequency noise, defined as noise in the frequency range below 100 Hz, is widely prevalentl in living and working environments [1, 2]. Low-frequency noise is not of high loudness, because the hearing threshold levels of human beings are quite high at frequencies below 100 Hz [3, 4]. However, low-frequency noise is well known to cause annoyance and discomfort [5-7]. In addition, people perceive vibration when they are exposed to low-frequency noise [1] . The apparent presence of vibratory sensation is a characteristic of low-frequency noise, and many studies have investigated this phenomenon. For example, Inukai et al. performed factor analysis on subjects' impressions of low-frequency noise and found that vibration was one of the main factors in human psychological responses to low-frequency noise [8]. Nakamura and Tokita found that the threshold contour for inducing vibratory sensation was lowest around 60 Hz, where the threshold sound pressure level was about 70

dB(SPL) [9].

In almost all previous studies. researchers focused on the relationship between vibratory sensation and the acoustic properties of low-frequency noise, such as frequency and sound pressure level. There is no doubt that such acoustic properties are the most important factors in the stimulation of the human auditory organs. However, it is uncertain whether or not the vibratory sensation is caused directly by stimulation of the auditory organs, though the possibility is not excluded that the increased pressure on the ear resulting from low-frequency noise contributes to the vibratory sensation.

Low-frequency noise induces mechanical vibration in the human body, but the level of this vibration is not especially high [10, 11]. It is considered reasonable that the vibratory sensation is indeed the body sensing actual vibration of its tissues (noise-induced vibration). Nevertheless, to the authors' knowledge, few studies have investigated vibratory sensation in association with noise-induced vibration.

The aim of this study was to clarify the relationship between noise-induced vibration and vibratory sensation induced by low-frequency noise. We measured the noise-induced vibration at five locations on the body surface of male subjects. While being exposed to low-frequency noise, each subject rated the intensity of each vibratory sensation on a scale of 1 to 3. The ratings correlated with the measured noise-induced vibration. It was expected that changes in the subject's posture would influence the mechanical characteristics of the body and might therefore change that relationship. To verify such a postural effect, the measurements were carried out with each subject in two postures: a sitting position and a standing position.

2. SUBJECTS

Nine healthy male subjects whose ages ranged from 21 to 24 yr (mean = 22.6, SD = 1.0) participated in the experiment. Each subject's hearing ability was confirmed to be normal in the 125- to 8000-Hz range by means of an audiometer (AA 73A, Rion, Japan). This study was approved by the ethical inquiry committee of the National Institute of Industrial Health, and informed consent was obtained from each subject before the measurements were taken.

3. METHODS FOR MEASURING BODY SURFACE VIBRATION

The subjects were exposed to low-frequency noise stimuli in a soundproof test chamber with a capacity of about 25 m^3 [12]. Background noise in the test chamber proved to be lower than 30 dB(A), which was sufficiently low for our study. Fifteen kinds of low-frequency noise stimuli (five frequencies and three sound pressure levels) were used. The noise stimuli were pure tones with frequencies of 20, 25, 31.5, 40 and 50 Hz and sound pressure levels of 100, 105 and 110 dB(SPL). No noise stimulus in the frequency range above 50 Hz was used because such frequencies would reduce the spatial uniformity of the sound pressure levels in the test chamber [12]. On the other hand, to detect noise-induced vibrations that were at higher levels than vibrations inherent in the human body, noise stimuli lower than 20 Hz were not used and the sound pressure levels of the noise stimuli were set sufficiently high [10]. Sinusoidal signals generated by a function generator (HP3314A, Hewlett Packard, USA) were used as the sources of the noise stimuli. After power amplification, the sinusoidal signals were fed to 12 loudspeakers (TL-1801, Pioneer, Japan) installed in the wall in front of the subject (Fig. 1), which reproduced the low-frequency noise stimuli. With the subject absent from the chamber, the gains of the power amplifiers (PC4002M, Yamaha, Japan) were adjusted so that the desired sound pressure level could be measured at the center of the chamber at a height of either 100 cm (corresponding to the chest of the sitting subject) or 120 cm (corresponding to the chest of the standing subject). The levels of higher harmonics of the noise stimuli proved sufficiently low when the stimulus was reproduced at levels up to 110 dB(SPL) [12]. During the measurement, a low-frequency sound level meter (NA-17, Rion, Japan) monitored the sound pressure levels of the noise stimuli.

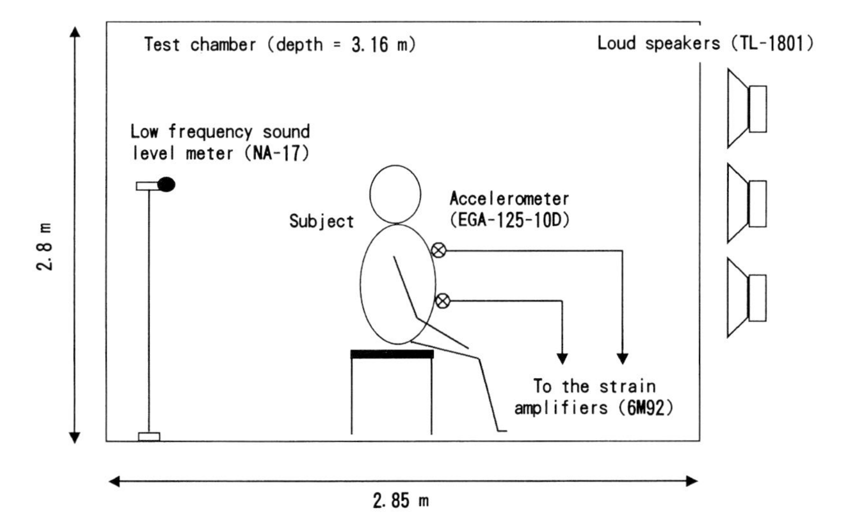

Figure 1. The experimental apparatus with the subject in a sitting position.

Noise-induced vibration was measured at five locations on the body surface; the forehead (2 cm above the level of the eyebrow and on the midline), the right anterior chest (2 cm above the right nipple), the left anterior chest (2 cm above the left nipple), the right anterior abdomen (5 cm below the pit of the stomach and 5 cm to the right of the midline) and the left abdomen (5 cm below the pit of the stomach and 5 cm to the left of the midline). Generally, the attachment of an accelerometer to the body can add an additional mechanical system to the body tissue, causing a difference between the acceleration of the accelerometer and that of the body tissue at the site of measurement. To reduce this undesirable effect [13], we used a lightweight miniature accelerometer (EGA- 125- 10D, Entran, USA) as a vibration detector. With an accelerometer attached to each measuring location by double-sided adhesive tape and no other supporting material, we detected noise-induced vibration perpendicular to the body surface. After amplification and low-pass filtering (cutoff frequency = 100 Hz) by a strain amplifier (6M92, NEC San-ei Instruments, Japan), the output signal of the accelerometer was recorded on DAT (digital audio tape) by a multi-channel data recorder (PC216Ax, Sony Precision Technology, Japan). Using five identical measuring sets, each consisting of an accelerometer and a strain amplifier, the noise-induced vibrations were measured simultaneously at all five locations.

Analysis by means of an FFT analyzer (HP3566A, Hewlett Packard, USA) yielded the power spectrum of the noise-induced vibration measured at each location for each stimulus. The frequency component corresponding to the noise stimulus was transformed to a vibration acceleration level (VAL) defined as:-

Vibration acceleration level (VAL) = $20\text{x}\log_{10}(a_{meas} / a_{ref})$ [dB], where a_{meas} was the measured acceleration (m/s^2(r.m.s.)) and a_{ref} was the reference acceleration equal to 10^{-6} m/s^2 [14]. To eliminate any effects of transient vibrations corresponding to the beginning or end of each l-min noise exposure, the first and last 10 seconds of each data recording were disregarded and only the remaining 40 seconds were analyzed. We did not separate the inherent vibration from the total measured vibration, for two reasons: (1) the phase relationship between the noise-induced vibration and the inherent vibration was unknown and (2) it was considered that the inherent vibration also contributed to the vibratory sensation.

4. METHODS FOR MEASURING VIBRATORY SENSATION

The subject rated vibratory sensation in three areas: the head, the chest and the abdomen. It should be noted that we instructed each subject to rate the sensation not merely at each measuring location (e.g., the forehead), but in the entirety of each part of the body being measured (e.g., the whole head). The subject rated each vibratory sensation as either 1, 2 or 3, corresponding to three grades: 'not sensed', 'mildly sensed' or 'strongly sensed'. To record his ratings, the subject used an LED display (labeled 1-3) built into a small response box inside the chamber. Because no reference noise was presented in the measurement, each vibratory sensation was rated independently of the others.

5. MEASUREMENT PROCEDURES

The measurements were carried out in winter (a dry season in Japan). The temperature at the subject's position in the test chamber was initially set at 25°C although it was adjusted, if he complained, within the range of 23-27°C to keep him comfortable and prevent him from sweating. The humidity in the chamber was maintained at 40% with a humidifier throughout the measurements. Before the measurements began, the subject spent at least 10 min in the test chamber to become adjusted to these conditions. The subject wore no clothes on the upper half of the body, to allow the accelerometers to be attached . The measurements were conducted in two sessions. In the first, the subject sat on a stool in the center of the chamber (Fig. 1). In the second, the stool was removed and the subject stood in the same place. At the beginning of each session, the inherent vibration was recorded (1 min) with no noise stimulus. Then, a rest period (1 min) with no noise stimulus was followed by an exposure period (1 min), during which the subject was exposed to a noise stimulus and the noise-induced vibration was recorded. The rest and exposure periods continued alternately. In each rest period, the subject rated the vibratory sensation he had just perceived (during the preceding exposure period) at each of the three areas of the body. Fifteen kinds of noise stimuli were presented in random order for each subject and in each session. The subject was instructed to keep his upper body erect during exposure to a noise stimulus, regardless of his posture. The second session started immediately after the first session ended.

6. RESULTS

The VALs of the noise-induced vibration measured at all of the measuring locations were reported in another article [15]. At all locations, it was found that the VALs increased with the frequency and sound pressure level of the noise stimulus. These VALs increased at roughly 5 dB increments, which corresponded well to the increments in the sound pressure levels of the noise stimuli. No clear bilateral asymmetry was found between the VALs measured at the right side of the body and those measured at the left side. In addition, no clear effect was observed between the VALs measured while the subject was sitting and those measured while the subject was standing. In Figs. 2 (for the subjects sitting) and 3 (for the subjects standing), the mean rating scores of the degrees of vibratory sensations are shown as a function of the

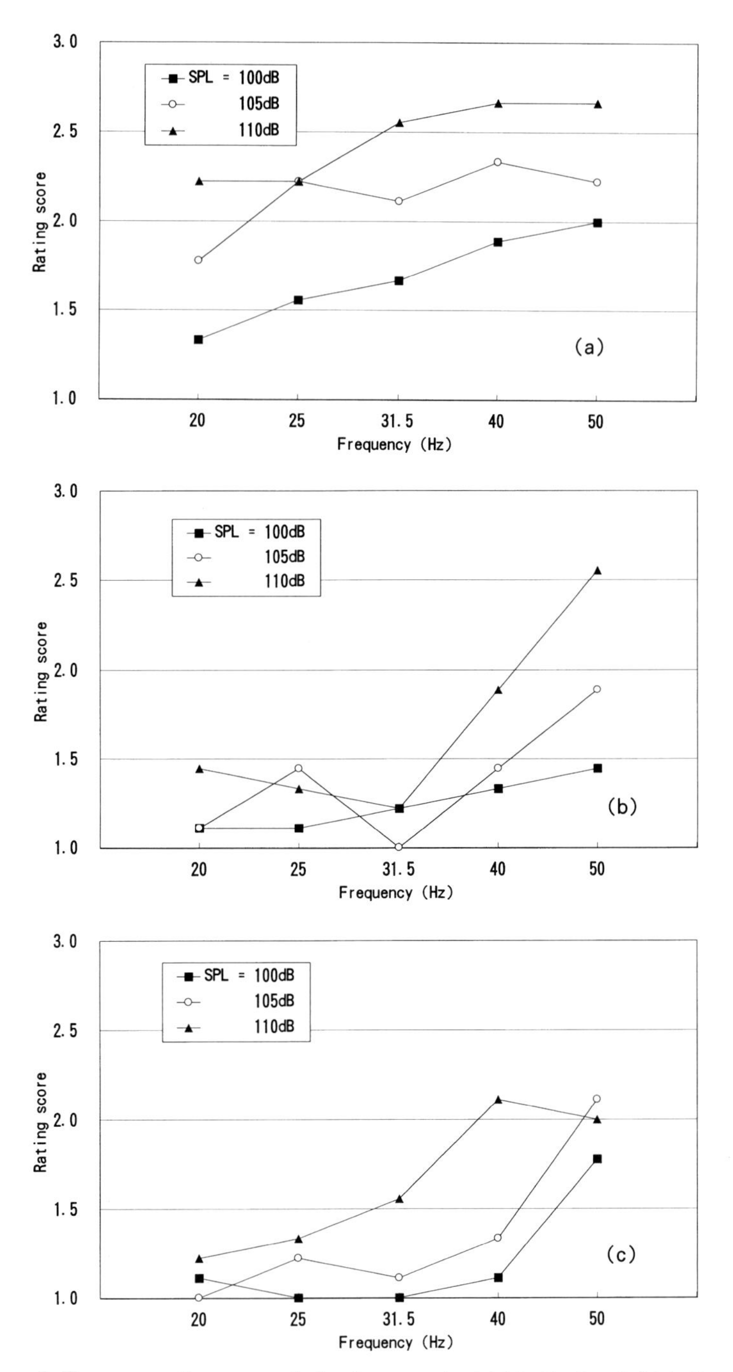

Figure 2. The mean rating score of vibration perceived (a) in the head, (b) in the chest and (c) in the abdomen of the standing subject.

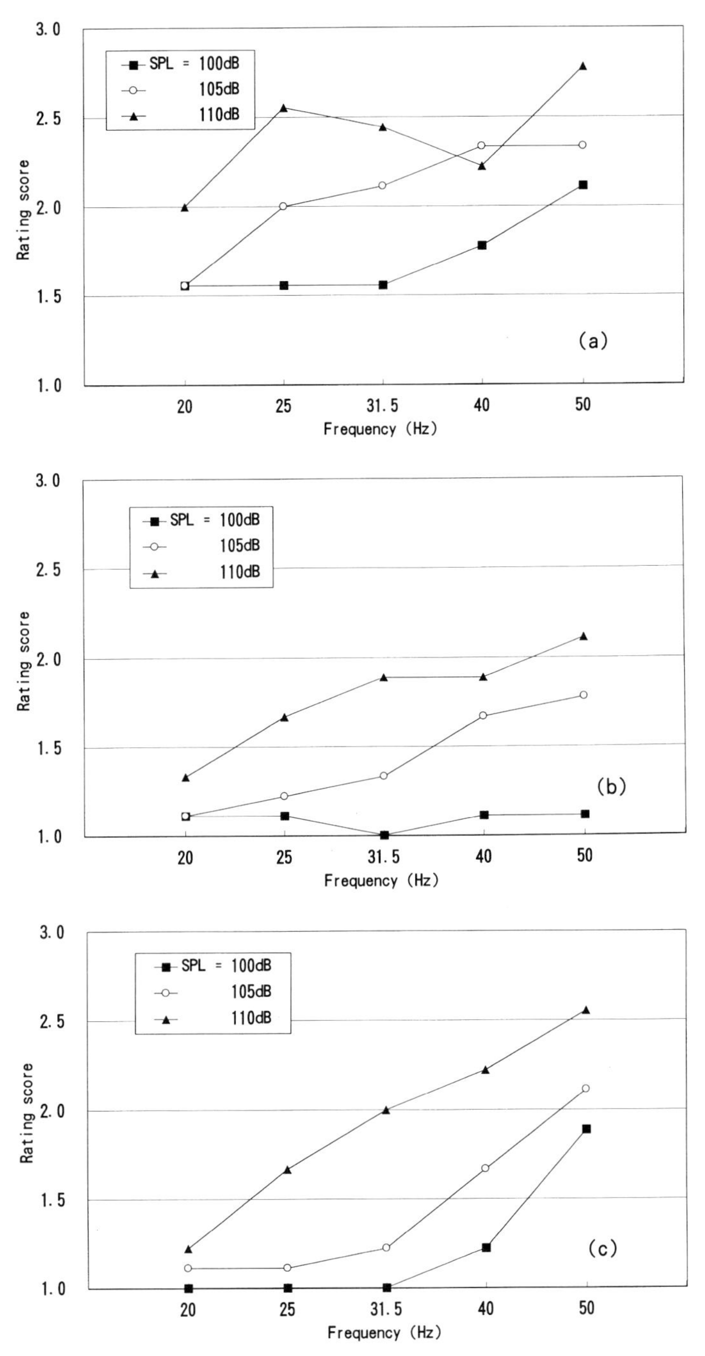

Figure 3. The mean rating score of vibration percieved (a) in the head, (b) in the chest and (c) in the abdomen of the standing subject.

frequency of the noise stimulus. All of the vibratory sensations tended to be rated highly at higher frequencies and at higher sound pressure levels. It should be noted that the vibratory sensation in the head was rated highly even at 20 and 25 Hz, whereas the perceptions in the other two parts were very low at those frequencies.

Figures 4 (for the subjects sitting) and 5 (for the subjects standing) show the correlations between the mean rating scores and the mean VALs of the noise-induced vibration measured at each corresponding location. In the correlations obtained at the chest and the abdomen, two VALs—those measured at both the right and left sides of the body — are plotted to one rating score. A solid line and a formula incorporated in the figure represent a regression line calculated with all the VALs plotted in each figure, regardless of the frequency of the noise stimulus. The gentlest gradient of the regression line was found at the head, whereas the steepest was found at the abdomen. This showed that the vibratory sensation in the abdomen was very sensitive to changes in the magnitude of the noise-induced vibration.

With a statistics software package (SPSS 10.0J for Windows, SPSS Japan, Japan), we calculated Pearson's correlation coefficients for these frequency-independent correlations (Table .I). The correlation coefficients obtained at the chest and the abdomen were both higher than 0.8, while that obtained at the head was lower. At all of the measuring locations and for both postures, these frequency-independent correlations were found to be statistically significant ($p<0.01$). Examination of the correlations shows that each subgroup categorized by frequency had a different feature of correlation from the other. Among the three locations, the frequency-dependence was the most apparent at the head, which was the main reason why the data plotted in the correlation at the head appeared to be scattered. At the chest and abdomen, on the other hand, the frequency-dependences in the correlations were found to be weaker than at the head.

Table 1. Pearson's correlation coefficients between the mean rating score of the vibratory sensation and the mean VAL of the noise-induced vibration measured at the corresponding location. The coefficients were calculated independently of the frequency of the noise stimulus. (*) represents statistical significance at the level of $p<0.05$, and (**) represents $p<0.01$.

Measuring locations	**Pearson's correlation coefficients (on sitting / on standing)**	
	For VAL	**For SPL**
Forehead	0.763** / 0.760**	0.830** / 0.735**
Chest	0.808** / 0.841**	0.465 / 0.796**
Abdomen	0.894** / 0.930**	0.459 / 0.582*

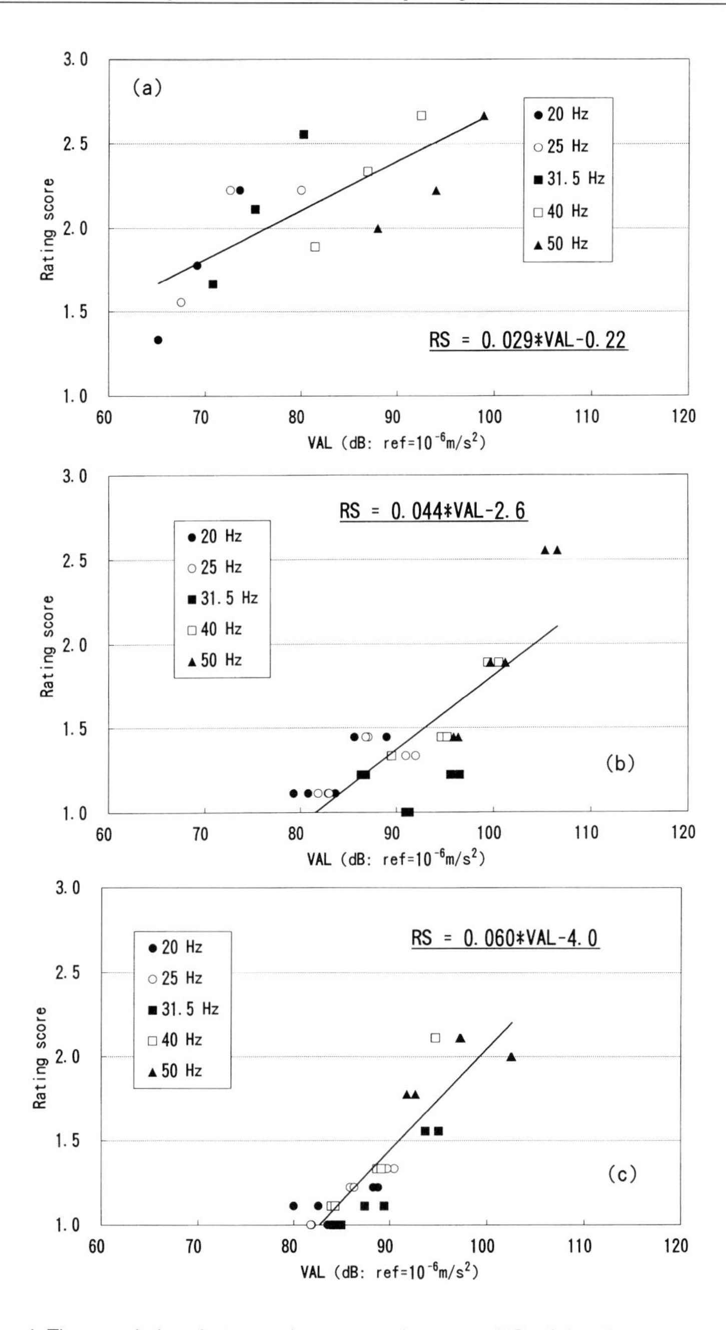

Figure 4. The correlations between the mean rating score (RS) of the vibratory sensation and the mean VAL measured (a) at the head, (b) at the chest and (c) at the abdomen of the sitting subject. At the chest and abdomen, two VALs, those measured at the right and left sides of the body, are plotted to one rating score.

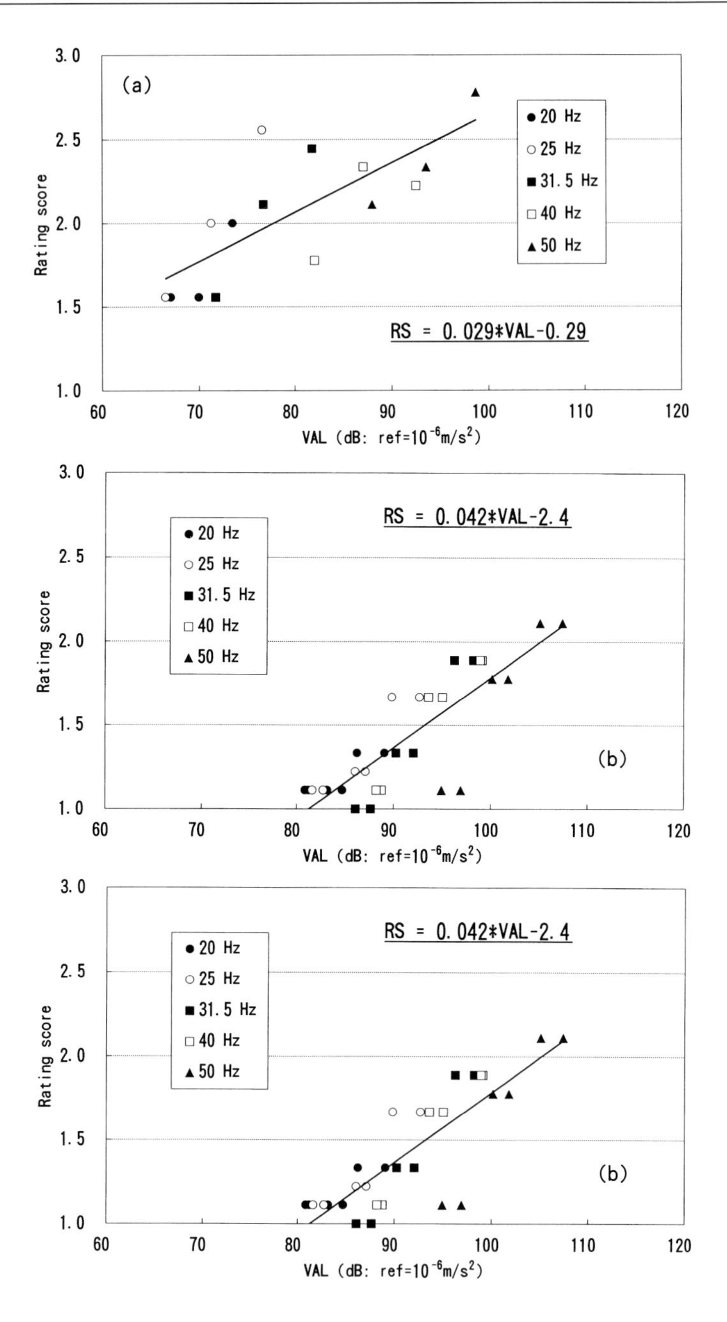

Figure 5. The correlations between the mean rating score (RS) of the vibratory sensation and the mean VAL measured (a) at the head, (b) at the chest and (c) at the abdomen of the standing subject. At the chest and abdomen, two VALs, those measured at the right and left sides of the body, are plotted to one rating score.

No clear effect of the subject's posture was found in the Pearson's correlation coefficients at any of the measuring locations (Table I). At the abdomen (Figs. 4(c) and 5(c)), however, the gradient of the regression line was found to be clearly higher for the subject standing (0.089) than for the subject sitting (0.060). This difference suggested that standing, as opposed to sitting, increased sensitivity to changes in the magnitude of mechanical vibration induced on the abdomen. At the head and the chest, no clear difference was found between the gradients of the regression lines for the subjects while sitting versus standing. This suggested that the vibratory sensation in these parts was independent of the subject's posture.

For comparison, Figs. 6 (for the subjects sitting) and 7 (for the subjects standing) show the correlations between the mean rating scores of the vibratory sensations and the sound pressure levels of the noise stimuli, which were obtained by the same manner as described above. The Pearson's coefficients calculated for these correlations are also listed in Table I At the chest and the abdomen, the coefficients were smaller than those obtained for the correlation with the VAL of the noise-induced vibration. This suggested that the vibratory sensations in these parts were more closely related to actual vibration induced on the body surface than to acoustic stimulation by low-frequency noise. At the head, on the contrary, the coefficients for the correlation with the sound pressure level were nearly equal to those for the correlation with the VAL. This suggested that actual body vibration and acoustic stimulation contributed comparably to the perception of vibratory sensation in the head.

7. DISCUSSION

If we hypothesize that the threshold VAL for inducing vibratory sensation corresponds to a score of 1 ('not sensed'), we see consistently in Figs. 4 and 5 that the threshold VAL for the vibratory sensation in the chest or the abdomen is about 82 dB. The figures show also that the threshold VAL for the vibratory sensation in the head is much lower than that. However, these prudent considerations may mislead us. It should be noted that the subjects were exposed only to high-level low-frequency stimuli in the 100- to 110-dB(SPL) range. The exposure to stimuli in such a limited range might bias a subject's judgment in rating vibratory sensation. Hence, the threshold VAL for inducing a vibratory sensation will not be discussed in this article.

The results of this study suggested that at the chest and the abdomen, vibratory sensation induced by low-frequency noise was closely related to mechanical vibration induced on the body surface. Yamada et al. reported that deaf persons perceived low-frequency noise by sensing vibration on the chest [16]. Our findings are consistent with their results and show the possibility that persons with normal hearing perceive low-

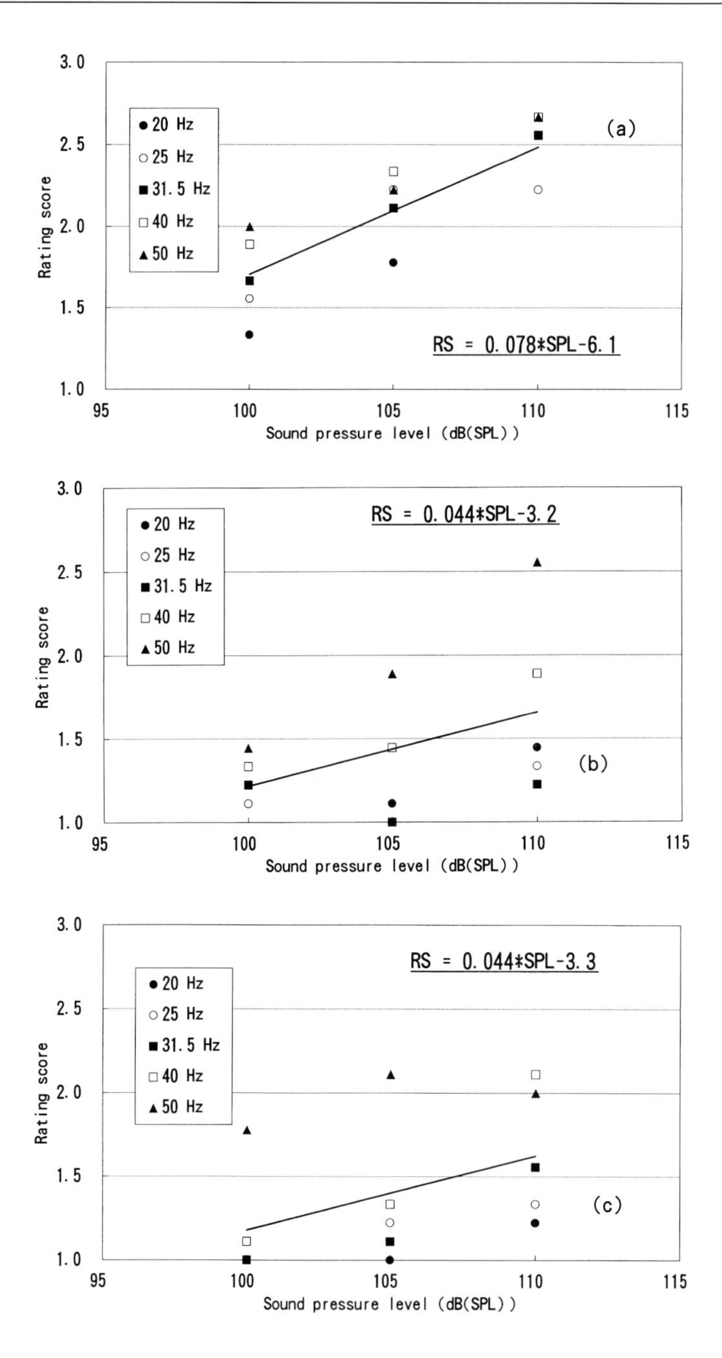

Figure 6. The correlations between the mean rating score (RS) of the vibratory sensation and the sound pressure level of the low-frequency noise stimulus. The correlations were obtained (a) at the head, (b) at the chest and (c) at the abdomen of the sitting subject.

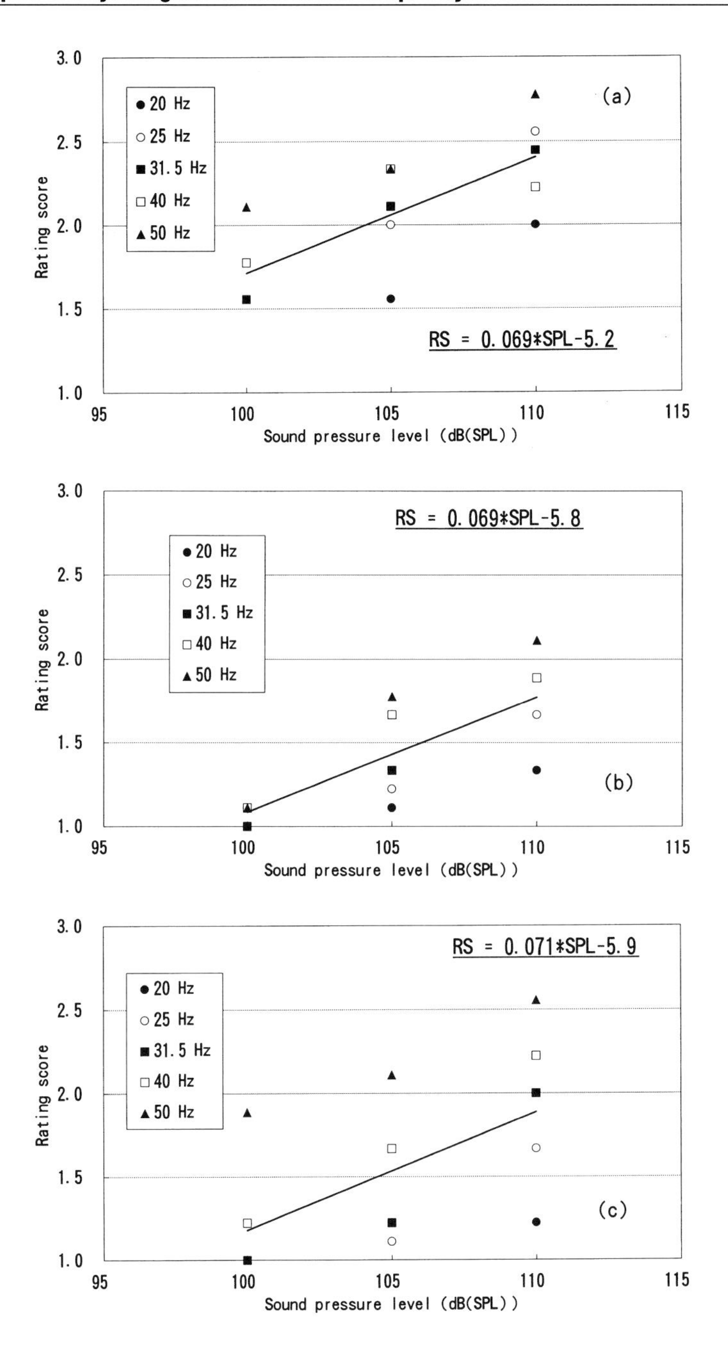

Figure 7. The correlations between the mean rating score (RS) of the vibratory sensation and the sound pressure level of the low-frequency noise stimulus. The correlations were obtained (a) at the head, (b) at the chest and (c) at the abdomen of the standing subject

frequency noise through a mechanism similar to that which deaf persons use. Four types of mechanoreceptive channels are known to mediate somatic sensation in the glabrous skin [17], as are three types in hairy skin [18]. Although the dominant type of mechanoreceptive channel has yet to be identified, it is considered that mechanoreception in the human body plays an important role in perceiving vibration induced by low-frequency noise.

In contrast with the vibratory sensations in the chest and abdomen, those in the head correlated well not only with the VAL of the noise-induced vibration but also with the sound pressure level of the low-frequency noise stimulus. This suggests that not only the actual body vibration but also acoustic stimulation of the auditory organs contribute to vibratory sensations in the head. After the measurements, many subjects reported that they felt vibration not on the surface of the head but inside the head. Also, a study by Møller showed that one of the main responses to low-frequency noise was an oppressive sensation on the ear [7]. One speculation is that high-level low-frequency noise induces periodic pressure changes in the tissue and fluid inside the head, through the auditory organs, and that this change inside the head induces the vibratory sensation. Because the magnitude of the pressure change induced inside the head is expected to be approximately proportional to the sound pressure level of a low-frequency noise stimulus, this speculation is consistent with our results that the vibratory sensation in the head correlated significantly with the sound pressure level. If we hypothesize that two factors (the mechanical vibration on the body surface and the pressure change inside the head) induce the vibratory sensation in the head interactively, then frequency-dependence in the vibratory sensation is expected to be more complicated than in the case where either factor dominates in the perception. This would be consistent with our results that the most apparent frequency-dependence was in the correlation between the vibratory sensation in the head and the VAL of the noise-induced vibration. Although the details remain to be investigated, the results of this study suggest that the mechanism to perceive vibration in the head differs from the mechanisms at work in the chest and abdomen. And it is considered that the auditory organs play more important roles in perceiving vibration in the head than in the other two parts.

Our results also suggested that people standing, rather than sitting, were more sensitive to changes in the magnitude of mechanical vibration induced on the abdomen (Figs. 4(c) and 5(c)). At the chest and the head, on the contrary, posture was not found to have any clear effect. Mechanical characteristics of the human body, such as skin tension, muscle tension and intra-abdominal pressure, are considered to be more posture-dependent in the abdomen than in the head or the chest. For example, the skin and muscle in the abdomen are relaxed when the subject is sitting and tightened when he is standing, whereas the mechanical characteristics of the head and the chest are hardly affected by bending at the hips. However, our results provide no suggestion on the relationship between posture and the sensitivity of mechanoreception in the abdomen. Further studies are needed to confirm our findings that the sensitivity of mechanoreception is dependent on posture.

This study yielded many interesting results, but some points remain to be improved and resolved in the future. First of all, the subjects did not wear hearing protectors. This prevented them from hearing their physiological noises abnormally loudly, and ensured that they would be exposed to low-frequency noise stimuli under normal conditions. As our results suggest, the contribution of the auditory stimulation to perception of vibration is not negligible. Secondly, vibration from noise was expected to be induced not only on the body surface but also in the inner body. The possibility is not excluded that the vibration induced in the inner body contributed to the perception of vibration. And the vibration induced at one location might contribute to the sensation of vibration at a different location. In addition, we used the noise stimuli in the limited range of sound pressure level and frequency, and the subjects rated their vibratory sensations according to three grades. Further studies are desirable to confirm the relationship between the noise-induced vibration and the vibratory sensation induced by low-frequency noise.

ACKNOWLEDGMENTS

This study was supported by a fund from the Environment Agency of Japan .

REFERENCES

1. Berglund B, Hassmén P and Job R.F.S., (1996) Sources and effects of low-frequency noise. *J. Acoust Soc, Am.*, Vol. 99, No. 5, pp. 2985-3002.

2 Pawlaczyk Luszczyńska M., (1998) Occupational exposure to infrasonic noise in Poland. *J. Low Freq., Noise Vib. & Active Control*, Vol. 17, No. 2, pp. 71-83.

3 Yeowart N,S, and Evans M.J. (1974) Thresholds of audibility for very low-frequency pure tones. *J. Acoust. Soc. Am.*, Vol. 65, No. 4, pp. 814-818.

4 Watanabe T and Møller H., (1990) Low frequency hearing thresholds in pressure field and in free field. *J. Low-Freq Noise Vib.*, Vol. 9, No. 3, pp. 106-115.

5 Persson K and Rylander R (1988) Disturbance from low-frequency noise in the environment: A survey among the local environmental health authorities in Sweden. *J. Sound Vib.*, Vol. 121, No. 2, pp. 339-345.

6 Mirowska M.J (1998) An investigation and assessment of annoyance of low frequency noise in dwellings. *J. Low Freq. Noise, Vib. & Active Control*, Vol. 17, No. 3, pp. 119-126.

7 Moller H.Ø (1984) Physiological and psychological effects of infrasound on humans. *J. Low Freq. Noise & Vib.*, Vol. 3, No. 1, pp. 1-17.

8 Inukai Y., Taya H., Miyano and H. and Kuriyama H., (1986) A multidimensional evaluation method for the psychological effects of pure tones at low and infrasonic frequencies. *J. Low Freq, Noise & Vib.*, Vol. 5, No. 3, pp. 104-112.

9 Nakamura S and Tokita Y. (1981) Frequency characteristics of subjective responses to low frequency sound. *Proc Inter-noise 81*, pp. 735-738.

10 Takahashi., Yonekawa Y., Kanada K and Maeda S (1999) A pilot study on the human body vibration induced by low frequency noise. *Ind. Health,* Vol. 37, No. 1, pp. 28-35.

11 Smith S.D., (2002) Characterizing the effects of airborne vibration on human body vibration response. *Aviat. Space. Environ. Med.*, Vol. 73, No. 1, pp. 36-45.

12 Takahashi Y., Yonekawa Y., Kanada K, and Maeda S.,(1997) An infrasound experiment system for industrial hygiene. *Ind. Health,* Vol. 35, No. 4, pp. 480-488.

13 Kitazaki S and Griffin M,J, (1995) A data correction method for surface measurement of vibration on the human body. *J. Biomech.*, Vol. 28, No. 7, pp. 885-890.

14 International Organization for Standardization (1990) Human response to vibration—Measuring instrumentation. ISO 8041.

15 Takahashi Y., Kanada K. and Yonekawa Y. (2002) Some characteristics of human body surface vibration induced by low frequency noise, *J. Low Freq., Noise, Vib & Active Control*, Vol. 21, No. 1, pp. 9-19.

16 Yamada S., Ikuji M., Fujikata S., Watanabe T and Kosaka T (1983) Body sensation of low frequency noise of ordinary persons and profoundly deaf persons. *J. Low Freq, Noise & Vib.*, Vol. 2, No. 3, pp. 32-36.

17 Bolanowski S.J Jr., Gescheider G. A., Verrillo R.T and Checkosky C.M., (1988) Four channels mediate the mechanical aspects of touch. *J. Acoust. Soc. Am.*, Vol. 84, No. 5, pp. 1680-1694.

18 Bolanowski S.J., Gescheider G,A and Verrillo R.T., (1994) Hairy skin: Pshychophysical channels and their physiological substrates. *Somatosens Mot. Res.*, Vol. 11, No. 3, pp. 279-290.

Measurement of human body surface vibrations induced by complex low-frequency noise composed of two pure tones

Yukio Takahashi[1,2] and Setsuo Maeda[1]
[1]Department of Human Engineering, National Institute of Industrial Health, 6-21-1, Nagao, Tama-ku, Kawasaki 214-8585, Japan
[2]E-mail: takahay@niih.go.jp

ABSTRACT
To clarify the mechanical responses of the human body to airborne vibrations, six male subjects were exposed to eight kinds of low-frequency noise stimuli: airborne white noise, two pure tones (31.5 and 50 Hz), and five complex noises composed of pure tones. The vibrations induced on the body surface were measured at five locations: the forehead, the right and left anterior chest, and the right and left anterior abdomen. It was found that the vibration acceleration levels of both the 31.5- and 50-Hz components in the chest vibration increased as an approximately linear function of the sound pressure levels of each corresponding frequency component in the noise stimulus. No clear interference was found between the 31.5- and 50-Hz components in the chest vibration. Similar characteristics were also found in the vibrations induced at the forehead and abdomen. These findings suggest that within the limited range of frequency and sound pressure level used here, the human body acts as a mechanically linear system in response to airborne vibrations induced by complex low-frequency noise.

1. INTRODUCTION
Low-frequency noise, which is noise in the frequency range below 100 Hz, is commonly generated in living and working environments [1]. In particular, in the working environment, various machines such as blowers, air compressors, large engines and the like generate high-level low-frequency noise, the sound pressure levels of which occasionally exceed 100 dB(SPL).

Mechanical vibrations are induced in the human body when a person is exposed to high-level, low-frequency noise [2, 3]. These vibrations (noise-induced vibrations) are an interesting subject for researchers who study the mechanical responses of the human body to vibration. If a person is exposed to vertical vibrations in one direction, the whole body is uniformly vibrated in the same direction as the stimulating vibration. In the case of noise-induced vibrations, however, it is expected that the human body is exposed to approximately isotropic pressure changes because the wavelength of low-frequency noise (approximately 6.8 m for 50-Hz noise, for example) is longer than the human being. Thus, it is speculated that his abdomen and back, for example, are forced to move in opposite directions. This isotropic characteristic distinguishes noise-induced vibrations from other vibrations in which the human body is excited in only one direction. To the authors' knowledge, however, there have been few studies to investigate in detail the characteristics of noise-induced vibrations.

In our previous study [2], we used pure tonal stimuli in the 20- to 50-Hz range and measured noise-induced vibrations on the body surface. We found that the increment in the vibration acceleration levels of the noise-induced vibrations agreed well with the increment in the sound pressure levels of the noise stimuli, implying that the human body is a mechanical system that responds linearly to airborne vibrations generated by pure tones in the low-frequency range. However, low-frequency noises generated in real environments are not pure tones but complex noises whose frequency spectra are spread over a wide range. If the human body responds linearly to complex low-frequency noise, it is expected that (1) the vibrations at different frequencies are induced independently of each other, and (2) the vibrations are induced as a linear function of the magnitude of the corresponding frequency component in the noise stimulus, regardless of the magnitude of the other noise component.

The aim of the present study was to clarify the mechanical responses of the human body to airborne vibrations generated by complex low-frequency noise. We measured noise-induced vibrations on the body surface using eight kinds of low- frequency noise stimuli.

2. METHODS

Six healthy male subjects whose ages ranged from 19 to 25 yr (mean = 22.8, SD = 2.1) participated in the experiment. Their heights and weights (mean ± SD) were 174.3 ± 4.8 cm and 71.0 ± 8.3 kg, respectively.

The measurements were carried out in a soundproof test chamber with a capacity of approximately 25 m^3 (2.85 m (W) × 3.16 m (D) × 2.80 m (H)) [4] in winter (a dry season in Japan). The temperature in the test chamber was set at 25∞C, and the humidity in the chamber was maintained at 40% by a humidifier throughout the measurement period.

We used eight types of low-frequency noise stimuli (Table I). One type was white noise with an approximately flat spectrum within the 4- to 100-Hz range. Two were pure tones with frequencies of 31.5 and 50 Hz, respectively. These pure tones were selected so that we could compare the results of the present study with our previous results [2]. The other five stimuli were complex noises composed of the two pure tones. The source of the noise stimuli was WAV-type data generated at a sampling rate of 48 kHz on a PC. The five complex noises were generated by synthesizing the two source data for the pure tones at a phase difference of zero degrees. All of the source data were D/A converted by an audio data interface (AD216, Nittobo Acoustic Engineering, Japan).

Table I. The low-frequency noise stimuli used in the present study

Number	Combinations and sound pressure levels (dB(SPL))
1	White noise (100 dB)
2	31.5-Hz pure tone (100 dB)
3	50-Hz pure tone (100 dB)
4	Complex noise composed of 31.5-Hz (100 dB) and 50-Hz (100 dB) tones
5	Complex noise composed of 31.5-Hz (100 dB) and 50-Hz (95 dB) tones
6	Complex noise composed of 31.5-Hz (100 dB) and 50-Hz (90 dB) tones
7	Complex noise composed of 31.5-Hz (95 dB) and 50-Hz (100 dB) tones
8	Complex noise composed of 31.5-Hz (90 dB) and 50-Hz (100 dB) tones

It has previously been shown that the frequency response of the test chamber is not flat over the entire frequency range [4]. In our previous study [2], we did not apply any compensation for the frequency response because we used only pure tonal stimuli at frequencies equal to or below 50 Hz, a range in which sound pressure levels have proven to be closely uniform in the test chamber [4]. In the present study, however, we used several methods to compensate for the frequency response of the test chamber so as to reproduce the complex noise stimuli as precisely as possible. The frequency spectrum of the source data was modified through one of two digital filters generated by a digital audio convolution processor (CP4, Lake Technology, Australia). We prepared five kinds of digital filters, each of which worked appropriately at 110, 120, 130, 140, or 150 cm high, at the center of the test chamber. In the measurement, the height of the measuring location was adjusted at any one of these five heights by setting an appropriate support under a subject. The other digital filter, which was set subsequently to the first filter, was used to eliminate electrical noise at frequencies above 100 Hz. After the compensation for and amplification of the source data, the noise stimulus was reproduced by 12 loudspeakers (TL-1801, Pioneer, Japan) which were installed in one wall of the test chamber.

Figure 1 shows examples of the frequency spectra of the noise stimuli, as measured at the center of the test chamber (120 cm in height). Our results indicate that the desired frequency spectra were obtained for both the white noise stimulus and the complex noise stimulus. It should be noted that the sound pressure levels of both the 31.5- and 50-Hz components in the white noise stimulus were approximately 80 dB(SPL), as the overall sound pressure level was adjusted to 100 dB(SPL). It should also be noted that the sound pressure levels for the white noise stimulus were not uniform along the whole body surface of a subject because the sound pressure levels in the test chamber, as mentioned earlier, lacked uniformity along the vertical direction at frequencies higher than 50 Hz [4].

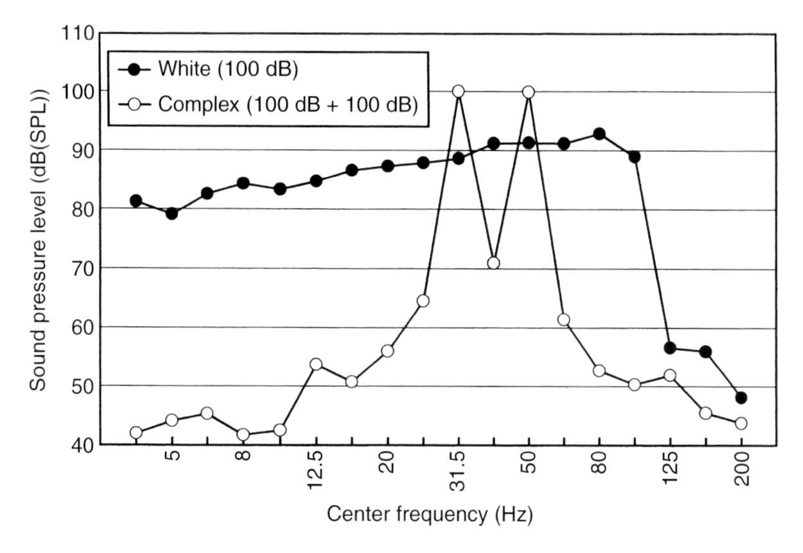

Figure 1. An example of the frequency spectra of the noise stimuli. The spectra shown are for the white noise stimulus and for the complex noise stimulus composed of a 31.5-Hz, 100-dB tone and a 50-Hz, 100-dB tone. These were measured at the center of the test chamber (120 cm in height).

To measure noise-induced vibrations, we used a small (3.56 mm × 6.86 mm × 3.56 mm) and lightweight (0.5 g) accelerometer (EGA-125-10D, Entran, USA). Prior to the present experiment, the accelerometer was calibrated on a vibrating table (AST-11V, Akashi, Japan) the vibrations of which were monitored with a set of a calibrated acceleration pickups (PV-85, Rion, Japan) and a vibration meter (VM-80, Rion, Japan). In the calibration, we investigated the frequency response of the accelerometer in the frequency range below 100 Hz and confirmed that the accelerometer responded linearly to the magnitude of the acceleration of the vibrating table.

Noise-induced vibrations were measured at five locations on the body surface: the forehead (2 cm above the level of the eyebrows and on the midline), the right anterior chest (2 cm above the right nipple), the left anterior chest (2 cm above the left nipple), the right anterior abdomen (5 cm below the pit of the stomach and 5 cm to the right of the midline), and the left anterior abdomen (5 cm below the pit of the stomach and 5 cm to the left of the midline). Five sets of an accelerometer and a strain amplifier (6M92, NEC-Sanei Instruments, Japan), each of which corresponded to one measuring location, were arranged. With the accelerometer attached to each measuring location using double-sided adhesive tape and no other supporting material, we detected vibrations perpendicular to the body surface. The detected vibrations were amplified by the strain amplifier and recorded on DAT with a multi-channel data recorder (PC216Ax, Sony Precision Technology, Japan).

Analysis by an FFT analyzer (HP3566A, Hewlett Packard, USA) yielded the power spectrum of the noise-induced vibration. The spectral components at 31.5 and 50 Hz were then transformed to vibration acceleration levels (VALs) defined as:

$$\text{Vibration acceleration level (VAL)} = 20 \times \log_{10} (a_{meas}/a_{ref}) \text{ [dB]},$$

where a_{meas} was a measured acceleration (m/s^2(r.m.s.)) and a_{ref} was the reference acceleration equal to 10^{-6} m/s^2. To eliminate any effects of transient vibrations corresponding to the beginning and end of each 1-min noise exposure, the first and last 10 seconds of each data recording were disregarded and only the remaining 40 seconds were analyzed. It was expected that the measured VAL would be contaminated by inherent vibrations originating in vital body activities [5]. In the above transformation, however, we did not separate the inherent vibrations from the total vibrations measured because the phase relationship between the inherent and the true noise-induced vibrations was unknown.

The measurements were conducted in three sessions (Fig. 2). In the first session, the subject stood at the center of the test chamber, and noise-induced vibrations were measured at the right and left chest. In the second session, noise-induced vibrations were measured at the right and left abdomen of the subject standing at the same place. In the last session, noise-induced vibrations were measured at the forehead of the subject who sat on a stool at the center of the chamber. At the beginning of each session, inherent vibrations were recorded (1 min) with no noise stimulus. Eight kinds of noise stimuli were then presented in random order for every session, and noise-induced vibrations were recorded during exposure to each 1 min noise stimulus. Between any two recordings, a 1-minute-long rest period with no noise stimulus was assigned. Throughout the measurements, the subject, who wore no clothes on the upper half of the body to allow the accelerometers to be attached, faced the wall in which loudspeakers were installed. The subjects wore no hearing protectors. We instructed each subject to keep his upper body in a relaxed erect position during the exposure period.

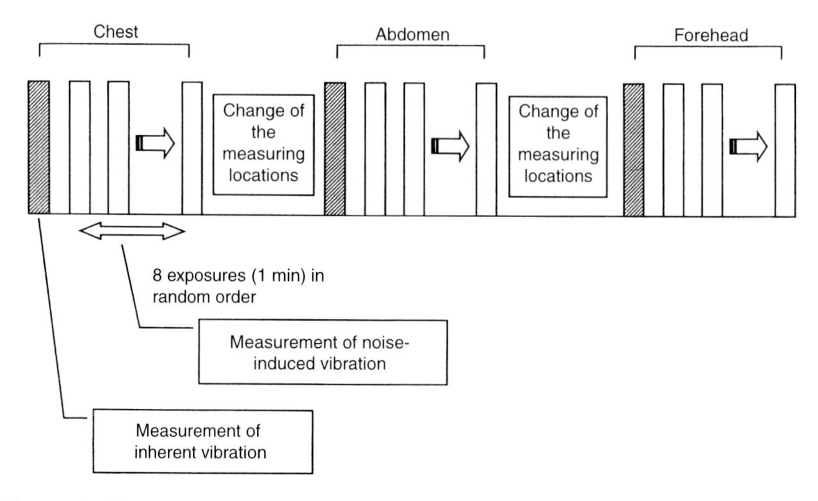

Figure 2. The procedure of the measurement. In the first session, noise-induced vibrations were measured at the chest. In the second and last sessions, noise-induced vibrations were measured at the abdomen and at the forehead, in that order.

To examine the differences between the VALs of the 31.5-Hz vibration measured under nine noise exposure conditions, including the inherent vibration measured with no noise stimulus, first the Kruskal-Wallis test then the Mann-Whitney test was performed using a statistics software package (SPSS for Windows 10.0J, SPSS Japan, Japan). This test was carried out for every measuring location. The same test was also performed on the VALs of the 50-Hz vibration measured. For these tests, a p-value lower than 0.05 (two-sided) was considered to be statistically significant.

The experiment was approved by the ethics committee of the National Institute of Industrial Health, Japan, and informed consent was obtained from each subject before the measurements were taken.

3. RESULTS

Figure 3 shows the VALs (means ± SD) of the 31.5-Hz component in the noise-induced vibration measured for no noise stimulus, the white noise stimulus, the 31.5-Hz pure tonal stimulus, and the three complex noise stimuli in which the sound pressure level of the 31.5-Hz noise component was 100 dB(SPL). Figure 4 shows the VALs (means ± SD) of the 50-Hz component in the noise-induced vibration measured for no

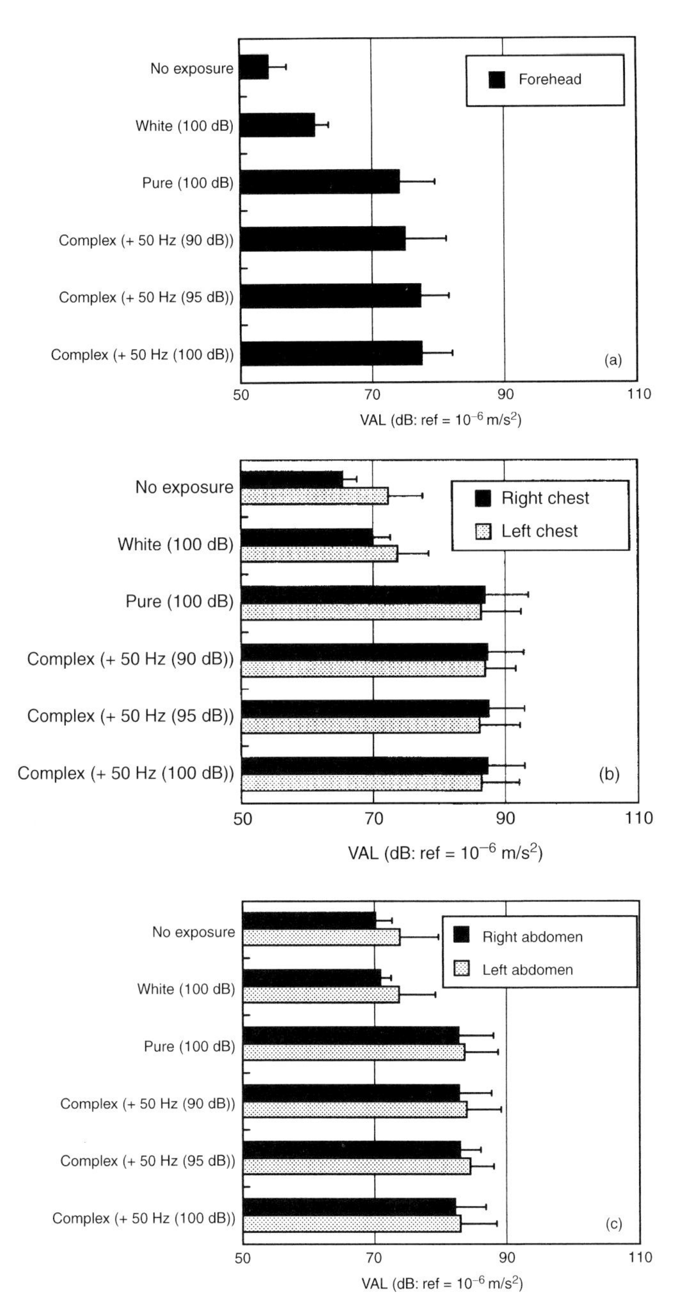

Figure 3. The vibration acceleration levels (means ± SD) of the 31.5-Hz component in the noise-induced vibration measured for no noise stimulus, the white noise stimulus, the 31.5-Hz pure tonal stimulus, and the three complex noise stimuli in which the 31.5-Hz noise component was set at 100 dB(SPL). They were measured (a) at the forehead, (b) at the chest, and (c) at the abdomen.

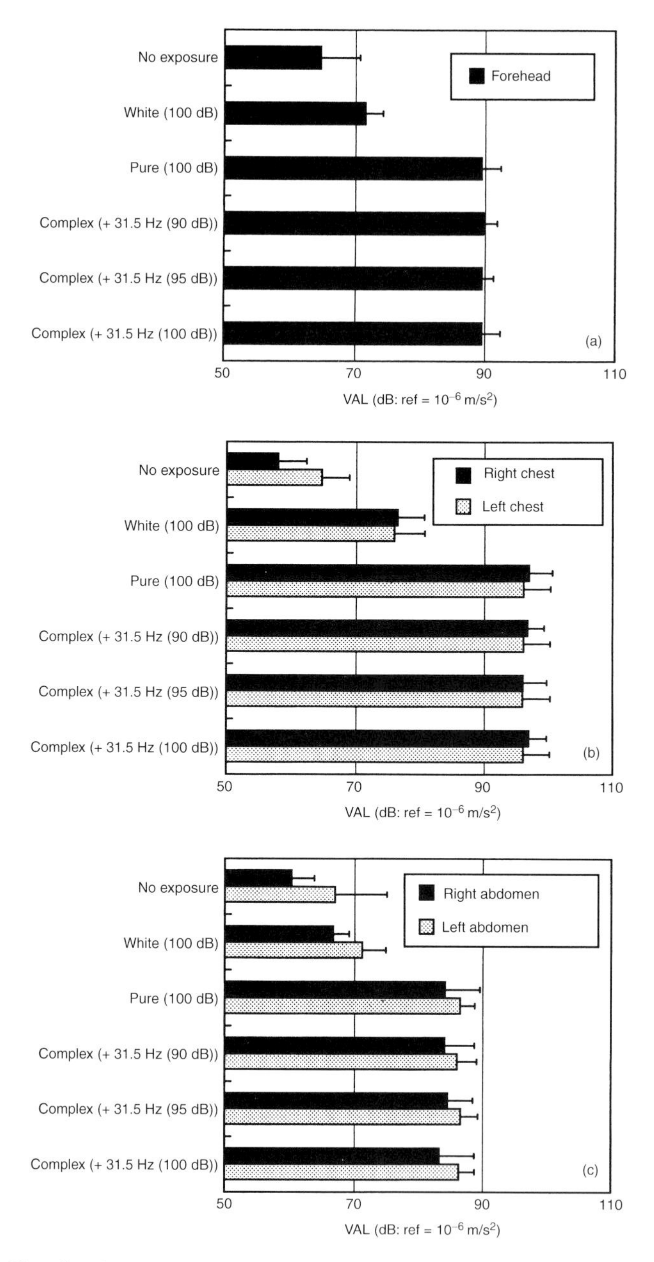

Figure 4. The vibration acceleration levels (means ± SD) of the 50-Hz component in the noise-induced vibration measured for no noise stimulus, the white noise stimulus, the 50-Hz pure tonal stimulus, and the three complex noise stimuli in which the 50-Hz noise component was set at 100 dB(SPL). They were measured (a) at the forehead, (b) at the chest, and (c) at the abdomen.

noise stimulus, the white noise stimulus, the 50-Hz pure tonal stimulus, and the three complex noise stimuli in which the sound pressure level of the 50-Hz noise component was 100 dB(SPL). The VALs for the white noise stimulus were measured to be lower than the VALs for the other noise stimuli because the sound pressure levels of the 31.5- and 50-Hz components in the white noise stimulus were approximately 80 dB(SPL). Half the VALs measured for the white noise stimulus (31.5-Hz vibrations measured at the left chest, the right abdomen, and the left abdomen, and 50-Hz vibrations measured at the forehead and the left abdomen) were found not to be significantly different from the VALs of the inherent vibrations measured for no noise stimulus. In contrast, the VALs measured for the pure tonal stimuli and the complex noise stimuli were found to be significantly different from the VALs of the inherent vibration, except for one case (31.5-Hz vibration measured for a complex noise stimulus composed of a 31.5-Hz, 100-dB(SPL) tone and a 50-Hz, 100-dB(SPL) tone. The VALs of the 50-Hz vibration induced by the 50-Hz pure tonal stimulus were found to be higher than those of the 31.5-Hz vibration induced by the 31.5-Hz pure tonal stimulus, a finding which was consistent with our previous results [2, 5].

At all the measuring locations, no statistically significant difference was found between the VALs of the 31.5-Hz vibration measured for four noise stimuli in which the 31.5-Hz noise component was set at 100 dB(SPL). Similarly, with respect to the 50-Hz vibration, no statistically significant difference was found between the VALs measured for four noise stimuli in which the 50-Hz noise component was set at 100 dB(SPL). Thus, no clear interference was found between the 31.5- and 50-Hz vibrations induced by complex low-frequency noise stimuli, which suggested that noise-induced vibrations at two frequencies were induced independently of each other.

Figure 5 shows the VALs (means ± SD) of the 31.5-Hz vibration measured for three complex low-frequency noise stimuli in which the sound pressure level of the conjugate (50-Hz) noise component was 100 dB(SPL). In the figure; the VALs of the 31.5-Hz vibration are depicted by black circles and plotted as a function of the sound pressure level of the corresponding (31.5-Hz) noise component. The solid line and expression incorporated in the figure represent a regression line calculated with these three VALs. The value attached to the expression (r^2) is the coefficient of determination obtained in the calculation. In Figure 5, the VAL (mean ± SD) of the 31.5-Hz vibration measured for the white noise stimulus is also depicted by a white square on the assumption that the sound pressure level of the 31.5-Hz noise component in the white noise stimulus is accurately equal to 80 dB(SPL). In addition, the mean VAL of the inherent vibration is shown by a horizontal dashed line. Figure 6 provides a similar treatment for the VALs of the 50-Hz vibration.

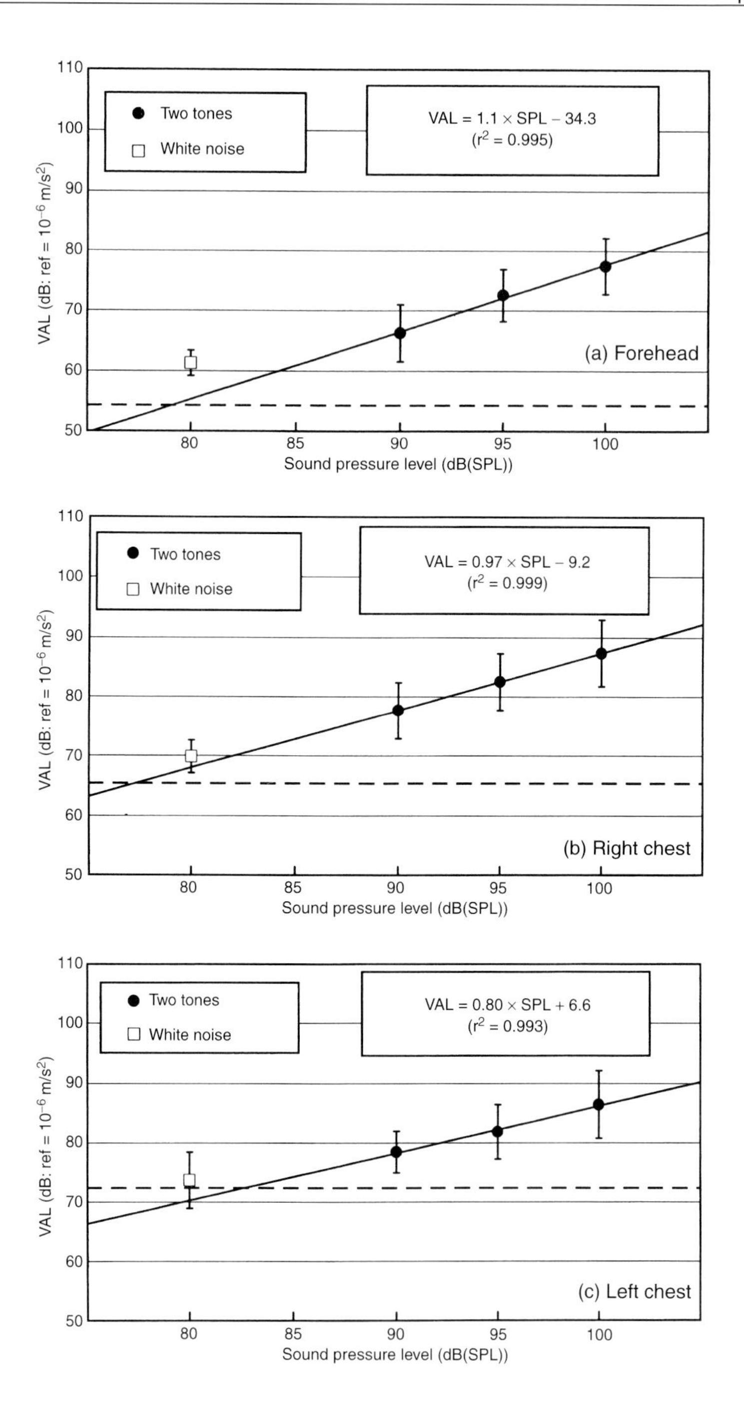

Figure 5. *continued opposite*

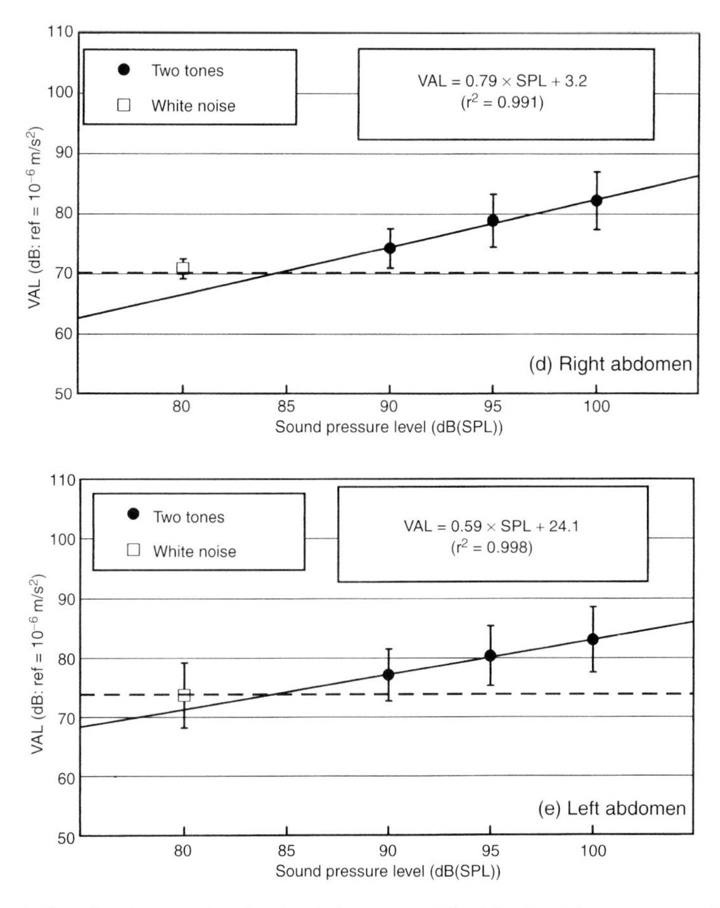

Figure 5. The vibration acceleration levels (means ± SD) of the 31.5-Hz component in the noise-induced vibration measured for the three complex noise stimuli in which the 50-Hz noise component was set at 100 dB(SPL). They (black circles) were measured (a) at the forehead, (b) at the right chest, (c) at the left chest, (d) at the right abdomen, and (e) at the left abdomen. Please refer to the Results section for more details.

Although the regression was not highly significant because of the small number of VALs used in the calculation, the coefficients of determination were very close to 1 at all the measuring locations and at both frequencies. At the forehead and chest, the slopes of the regression lines were nearly equal to 1, indicating that the increment in the VALs of the noise-induced vibration was in good agreement with the increment in the sound pressure levels of the corresponding noise component in the complex noise stimulus. These results suggest that the body surface vibration is induced as a linear function of the sound pressure level of the corresponding noise component, regardless of the sound pressure level of

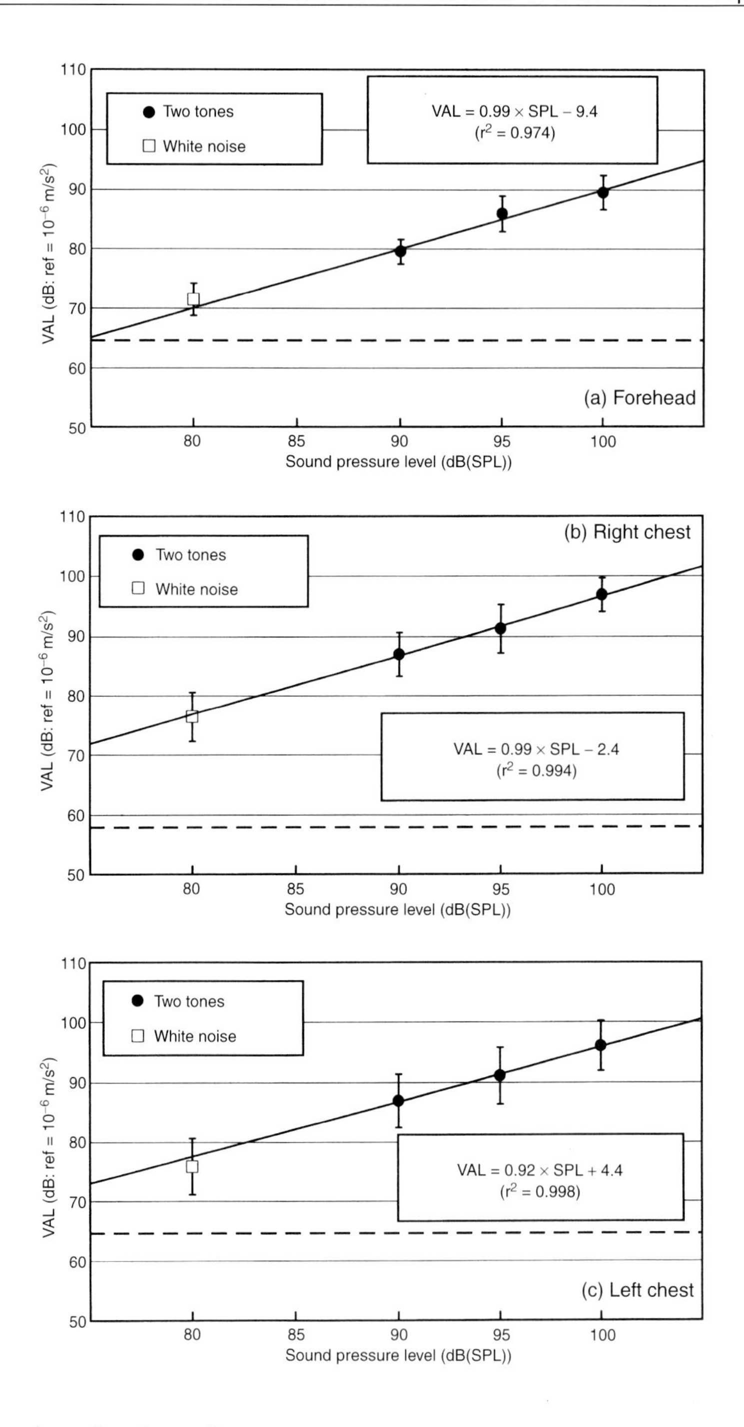

Figure 6. *continued opposite*

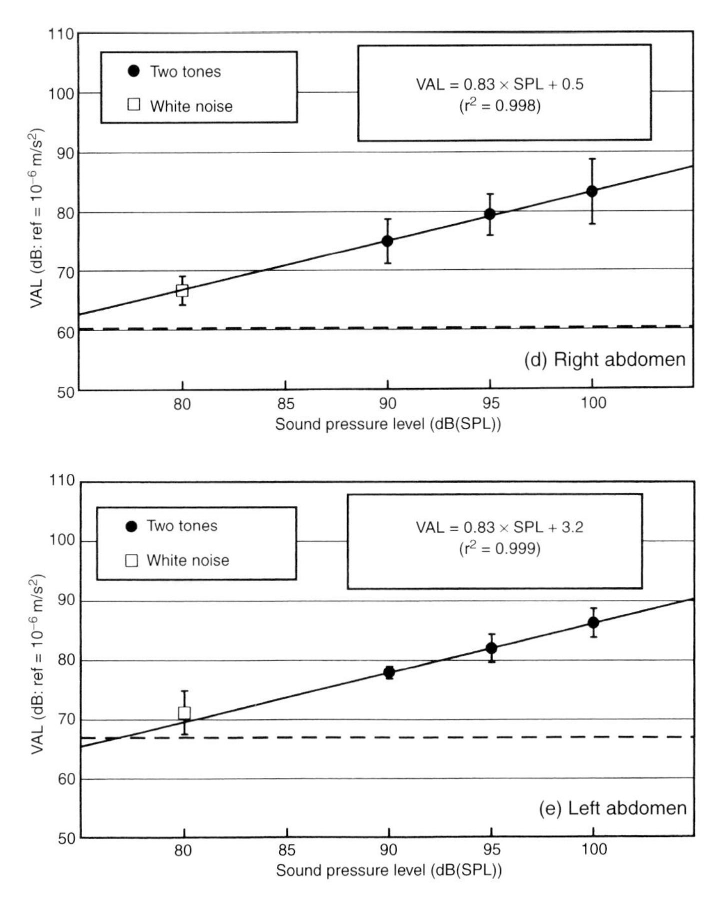

Figure 6. The vibration acceleration levels (means ± SD) of the 50-Hz component in the noise-induced vibration measured for the three complex noise stimuli in which the 31.5-Hz noise component was set at 100 dB(SPL). They (black circles) were measured (a) at the forehead, (b) at the right chest, (c) at the left chest, (d) at the right abdomen, and (e) at the left abdomen. Please refer to the Results section for more details.

the conjugate noise component. In contrast, at the abdomen, the slopes of the regression lines were found to be smaller than 1 at both frequencies.

Figures 5 and 6 show that if we estimate the VAL induced by a corresponding noise component at 80 dB(SPL) by extrapolating the VALs measured for the noise stimuli at higher sound pressure levels, the estimated VAL coincides with the VAL measured for the white noise stimulus, except for the one case shown in Figure 5 (a). At sound pressure levels equal to or below 80 dB(SPL), the influence of the inherent vibration on measurement of noise-induced vibrations is considered to be significant. Taking the contamination by the inherent vibration into account, it is considered that the VAL corresponding to a noise at 80 dB(SPL) is well estimated by the above extrapolation.

4. DISCUSSION

In the present study the VAL of the 31.5-Hz vibration induced by the 31.5-Hz, 100-dB(SPL) pure tone was measured to be 87.0 ± 6.6 dB at the right chest, while the corresponding VAL was measured to be 87.6 ± 2.4 dB in our previous study [2]. With respect to the 50-Hz vibration induced by the 50-Hz, 100-dB(SPL) pure tone, the VAL was measured to be 97.0 ± 3.7 dB in the present study and 96.9 ± 3.1 dB in the previous study. At the other measuring locations, the VALs measured under the same exposure conditions were consistent across the two studies. These results indicate that the sound pressure levels of the noise stimuli reproduced in the test chamber were not influenced by the compensation we introduced in the present study.

At all the measuring locations, the 31.5- and 50-Hz components in the noise-induced vibrations appeared to be induced independently of each other. The independence in the induction of noise-induced vibrations at the two frequencies lends support to the idea that the human body acts as a mechanically linear system in response to airborne vibrations generated by complex low-frequency noise. Moreover, we found that the VALs of both the 31.5- and 50-Hz components increased approximately as a linear function of the sound pressure level of the corresponding noise component, regardless of the sound pressure level of the conjugate noise component. We also found that the VALs induced by the noise at 80 dB(SPL) was well-estimated by extrapolating the VALs measured by the noise stimuli at higher sound pressure levels. These results provide additional support for the idea that the human body responds linearly to airborne vibrations generated by complex low-frequency noise.

At the abdomen, the gradients of the VALs of noise-induced vibrations with the sound pressure levels were found to be smaller than 1 (Figs. 5 and 6). It is considered likely that this result is primarily due to contamination caused by the vibrations inherent in the human body. As shown in Figures 5 (d) and (e), the VALs of the 31.5-Hz vibration measured for the white noise stimulus were at levels equal to or just above those of the inherent vibrations. Because the inherent vibrations were not eliminated from the total noise-induced vibration measured, the VALs measured for the noise stimuli at lower sound pressure levels were possibly higher than they were naturally. This undesirable effect is expected to be greater at lower frequencies and at lower sound pressure levels for noise stimuli [5], which results in gentler gradients of the VAL with sound pressure levels. However, the 50-Hz vibrations at the abdomen were found to be higher than the 31.5-Hz vibrations, and the slopes of the regression lines for the 50-Hz vibrations were closer to 1 than those for the 31.5-Hz vibrations (Figs. 5 and 6). These results imply that if the inherent vibrations are eliminated, the increment in the 31.5-Hz noise-induced vibration is in good agreement with the increment in the sound pressure levels of the 31.5-Hz noise component.

Because we measured noise-induced vibrations under limited experimental conditions, it is unclear whether or not the human body responds linearly to airborne vibrations in a wide range of frequency and sound pressure levels. With respect to vibrations induced in the head, for example, Håkansson et al. [6] have reported that the human skull responds linearly to vibration stimuli at frequencies above 100 Hz. Their results suggest that the human head responds linearly to airborne vibrations generated by low-frequency noise at frequencies higher than 100 Hz. In their study, however, they excited the head with a miniature vibrator attached near the right ear and detected the induced vibrations with a miniature accelerometer attached near the left ear. In contrast, we excited the whole body by approximately isotropic pressure changes induced by low-frequency noise. To clarify the effective range of frequency and sound pressure levels where the human head responds linearly to airborne vibrations induced by low-frequency noise, further studies are needed using noise stimuli in a wider range of frequency and sound pressure levels. Similarly, further investigations are needed for vibrations induced at the chest and abdomen.

As for the experimental conditions used in the present study, there are some other points to be discussed. First, each subject was exposed to each noise stimulus for only 1 min. It is considered that the extent of human psychological responses to noise would be influenced by the duration of the exposure. However, the magnitude of the noise-induced vibration was expected to be independent of the duration of the exposure, because the noise-induced vibration is a mechanical response of the human body to low-frequency noise stimuli. This is the reason why we set the duration of each exposure at 1 min only and did not use longer durations. Secondly, we used only young male subjects in the present study. Due to the difference in the amount of body fat, for example, it is considered that the physical characteristics of the female body are different from those of the male body. This may possibly be a confounding factor for our experiments. Similarly, the difference between the physical characteristics of older persons and those of younger persons are expected to be another confounding factor. To minimize the effects of these possible confounding factors, we did not use female or older subjects. However, it is important to verify whether or not the characteristics of noise-induced vibration depend on sex difference and age. It is desirable to conduct further studies under a variety of experimental conditions, and on a variety of experimental subjects.

We did not measure the transfer function from the low-frequency noise stimulus to the noise-induced vibration, as it was considered that the signal-to-noise ratio in the measured noise-induced vibration was poor due to contamination by inherent vibrations, especially at 31.5 Hz. Wodicka and Shannon [7] introduced a white noise stimulus into the mouth with an approximately flat spectrum in the 100- to 1000-Hz range and measured the amplitude of sound transmitted from the mouth to two sites on the posterior chest wall. Their results showed the transfer function of the transmission as having a single peak around 150 Hz and as decreasing with increasing frequencies. Although their results should not be compared directly with our present results due to differences in measuring methods, they suggest that the amplitude of noise-induced vibrations at the chest decreases at higher frequencies. It is not clear whether or not noise-induced vibrations at the head and abdomen behave in a similar manner. Because the transfer function is an important quantity in studying the mechanical response of the human body, further studies are needed to measure the transfer function for noise-induced vibrations.

In addition, we did not measure the phase relationship between the 31.5- and 50-Hz components in the noise-induced vibration. Provided that the noise-induced vibration is induced depending only on the atmospheric pressure change at the body surface, the phase relationship between two vibration components should be approximately the same as the relationship in the noise stimuli, except for the effect of the acoustical properties of the test chamber. However, if the pressure change in the inner body interactively contributes to the induction of noise-induced vibrations, the phase relationship between two vibration components on the body surface is possibly more complicated. Lu et al. [8] injected a broadband noise stimulus in the 300- to 1600-Hz range into the mouth and measured the vibrations induced on the chest wall. They estimated the phase-delay of the transmitted sound and reported a tendency for sounds at higher frequencies to reach the chest wall faster than that at lower frequencies. If lung pressure changes due to sound being transmitted through the airways, thus influencing the induction of noise-induced vibrations, some discrepancies may occur between the phases of different frequency components in the noise-induced vibration measured at the chest. This point also remains to be investigated in the future.

Peters et al. [9] exposed sheep to 100-dB airborne broadband noise and measured the inntraabdominal sound pressure by a hydrophone located within the abdomen, revealing that the sound pressure of the noise within the abdomen was less attenuated at frequencies below 100 Hz. Similar frequency-dependent transmission of intraabdominal vibration has been reported in other studies in which the abdomens of sheep were exposed to be mechanical vibrations by a shaker, not to airborne vibration by airborne sound [10-12]. These results suggested that transmission of vibrations in the inner body is frequency-dependent and the vibration is less attenuated at lower frequencies. In the present study we measured noise-induced vibrations in a limited range of frequency and only on the body surface. Although the effects of noise-induced vibration on vibrations within the body remain unclear, there is a good possibility that the noise-induced vibration on the body surface is transmitted deeply into the inner body.

Provided that detailed characteristics of the transmission of the noise-induced vibration within the human body are clarified, these findings may be helpful in the study of adverse health effects caused by low-frequency noise. Castelo Branco et al. [13-15] have reported that long-term exposure to high-level, low-frequency noise causes vibroacoustic disease involving some extra-aural pathologies such as pericardial thickening, pulmonary fibrosis, and so on. Although they presume that noise-induced vibrations are associated with vibroacoustic disease, the mechanisms by which vibroacoustic disease develops have not yet been investigated in detail. To assess the effects of low-frequency noise from a medical viewpoint, noise-induced vibrations should be measured not only on the body surface but also in the inner body, and the characteristics of noise-induced vibration must be quantitatively related to the prevalence and/or stages of adverse health effects such as vibro-acoustic disease. Already, some frequency-weighting curves, such as the LF-weighting curve [16] and the G-weighting curve [17], have been proposed. Because they have been designed on the basis of the human psychological and perceptual responses to low-frequency noise, they are considered to be useful and effective in assessing perceptions of noise. However, it is difficult to use these frequency-weighting curves to assess in an appropriate fashion the possible physical effects caused by low-frequency noise. A quantity related to noise-induced vibrations could be useful in establishing a new evaluating method for the adverse health effects caused by low-frequency noise.

CONCLUSIONS

The results of the present study suggest that the human body acts as a mechanically linear system in response to airborne vibrations generated by complex low- frequency noise. Nonetheless, some points remain to be investigated: i.e., the effective range of frequency and sound pressure levels at which this linear response occurs, the transfer function between noise stimuli and noise-induced vibrations, and so on. To clarify the details of mechanical responses of the human body to complex low-frequency noise, further studies should be conducted using noise stimuli within a wider range of frequency and sound pressure levels.

ACKNOWLEDGMENTS

The present study was supported by a Grant-in-Aid for Encouragement of Young Scientists (13780699), founded by the Japan Society for the Promotion of Science.

The authors would also like to thank Dr. Y. Matsumoto and Mr. K. Yamada, Saitama University, for their useful suggestions.

REFERENCES

1. Berglund, B., Hassmén, P. and Job, R.F.S., Sources and effects of low-frequency noise, *Journal of the Acoustical Society of America,* 1996, 99(5), 2985-3002.

2. Takahashi, Y., Kanada, K. and Yonekawa, Y., Some characteristics of human body surface vibration induced by low frequency noise, *Journal of Low Frequency Noise, Vibration and Active Control,* 2002, 21(1), 9-20.

3. Smith, S.D., Characterizing the effects of airborne vibrations on human body vibration response, *Aviation, Space, and Environmental Medicine,* 2002, 73(1), 36-45.

4. Takahashi, Y., Yonekawa, Y., Kanada, K. and Maeda, S., An infrasound experiment system for industrial hygiene, *Industrial Health,* 1997, 35(4), 480-488.

5. Takahashi, Y., Yonekawa, Y., Kanada, K. and Maeda, S., A pilot study on the human body vibration induced by low frequency noise, *Industrial Health,* 1999, 37(1), 28-35.

6. Håkansson, B., Carlsson, P., Brandt, A. and Stenfelt, S., Linearity of sound transmission through the human skull in vivo, *Journal of the Acoustical Society of America,* 1996, 99(4), 2239-2243.

7. Wodicka, G.R. and Shannon, D.C., Transfer function of sound transmission in subglottal human respiratory system at low frequencies, *Journal of Applied Physiology,* 1990, 69(6), 2126-2130.

8. Lu, S., Doerschuk, P.C. and Wodicka G.R., Parametric phase-delay estimation of sound transmitted through intact human lung, *Medical & Biological Engineering & Computing,* 1995, 33(3), 293-298.

9. Peters, A.J.M., Gerhardt, K.J., Abrams, R.M. and Longmate, J.A., Three-dimensional intraabdominal sound pressures in sheep produced by airborne stimuli, *American Journal of Obstetrics and Gynecology,* 1993, 169(5), 1304-1315.

10. Peters, A.J.M., Abrams, R.M., Gerhardt, K.J., and Longmate, J.A., Three dimensional sound and vibration frequency responses of the sheep abdomen, *Journal of Low Frequency Noise and Vibration,* 1991, 10(4), 100-111.

11. Graham, E.M., Peters, A.J.M., Abrams, R.M., Gerhardt, K.J., Burchfield, D.J., Intraabdominal sound levels during vibroacoustic stimulation, *American Journal of Obstetrics and Gynecology,* 1991, 164(4), 1140-1144.

12. Peters, A.J.M., Abrams, R.M., Gerhardt, K.J., and Wasserman, D.E., Acceleration of fetal head induced by vibration of maternal abdominal wall in sheep, *American Journal of Obstetrics and Gynecology,* 1996, 174(2), 552-556.

13. Castelo Branco, N.A.A. and Rodriguez, E., The vibroacoustic disease — An emerging pathology, *Aviation, Space, and Environmental Medicine,* 1999, 70(3 Pt 2), A1-A6.

14. Alves-Pereira, M., Noise-induced extra-aural pathology: A review and commentary, *Aviation, Space, and Environmental Medicine,* 1999, 70(3 Pt 2), A7-A21.

15. Castelo Branco, N.A.A., The clinical stages of vibroacoustic disease, *Aviation, Space, and Environmental Medicine,* 1999, 70(3 Pt 2), A32-A39.

16. Inukai, Y., Taya, H., Nagamura, N., Kuriyama, H., An evaluation method of combined effects of infrasound and audible noise, *Journal of Low Frequency Noise and Vibration,* 1987, 6(3), 119-125.

17. International Organization for Standardization, Acoustics — Frequency-weighting characteristic for infrasound measurements, *ISO 7196,* 1995.

An investigation on the physiological and psychological effects of infrasound on persons

Chen Yuan Huang Qibai and Hanmin Shi
School of Mechanical Science & Engineering, Huazhong University of Science and Technology, Wuhan, 430074, P .R. of China

INTRODUCTION

In order to study the physiological and psychological effects of infrasound on person, we have measured the changes of blood pressure and heart rate and also investigated subjective feelings of subjects exposed to infrasound. Hood et al and Evans and Tempest the subjective reaction and the hearing threshold etc Mfller reported the physiological and psychological effects of infrasound (with frequency bandwidth of from 6.3 Hz to 31.5 Hz) on persons. Considering that the main resonant frequency of person's internal organs is below 5 Hz, the study has adopted two different infrasonic conditions, about 2 Hz 110 dB (A) and 4 Hz 120 dB (A). Two groups of subjects were used in the experiment—Group A and Group B. Group A were exposed to infrasound with about 2 Hz 110 dB for 1 hour Group B were exposed to infrasound with about 4 Hz 120 dB for 1 hour. During the experiment, noise with frequency of over 20 Hz was minimized. The physiological and psychological effects of infrasound on persons have been objectively demonstrated.

METHODS

The experiment was arranged in a large room (L × W × H = 7.3 × 6.9 × 3.6 m^3). To acquire experimental data, a B&K4155 microphone and B&K2231 sound level meter were used. we also used a frequency spectrum analyser to analyze infrasonic conditions and a blood-pressure meter and cardiotachometer to measure the changes of the subjects blood pressure and heart rat. At the same time we provided the subjects with pure water and requested them to remain calm. By a previous experiment, we found that the sound level was approximately uniform in this room. The microphone and sound level meter are fixed on a tripod and located on the centre of the room as shown in Figure 1. Figures 2 and 3 show the frequency spectrum characteristic of the two different infrasound signals—2.14 Hz 110 dB (A) and 4.10 Hz 120 dB (A).

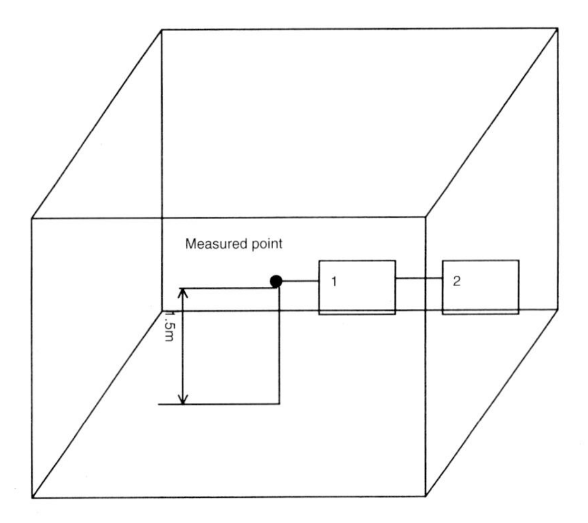

Figure 1. Infrasound measuring and signal analysis system 1. B&K4155 microphone and B&K2231 sound level meter 2. Signal analysis system.

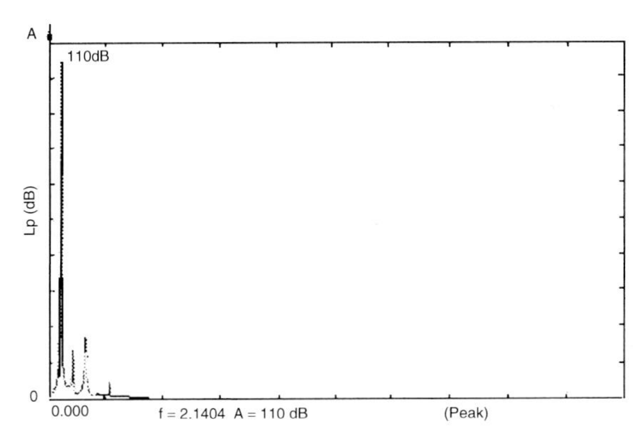

Figure 2. Frequency Spectrum of 2.14 Hz signal.

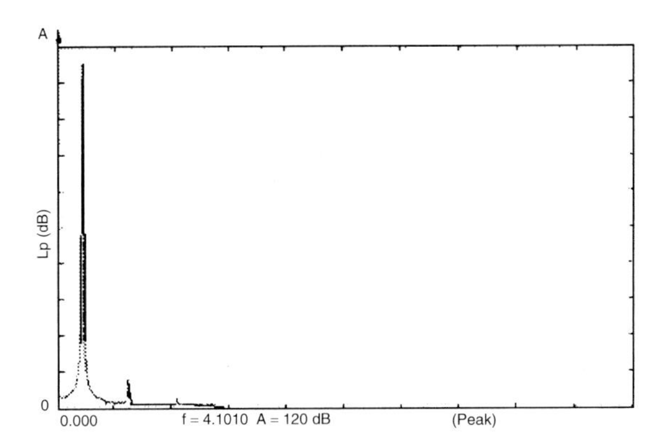

Figure 3. Frequency Spectrum of 4.10 Hz signal.

A total of 6 male and 4 female university students were used as subjects, all in good health, between 22 and 28 years. They were divided into 2 groups–Group A and Group B. Each group includes 3 males and 2 females. The temperature was 25±2°C, the degree of humidity of the air was 70±5% and the noise level in the experimental room was below 60 dB (A).

Physiological effects were examined by measuring the changes of blood pressure and heart rate and each subject recorded their subjective feelings. Before the experiment, we measured the normal blood pressure and heart rate. After being exposed to infrasound for one hour, blood pressure and heart rate were measured again.

In the normal condition, measurements were performed three times at 3-minute intervals and the average of the three values was regarded as the initial value before starting the exposure. The two groups' blood pressure and heart rate in the normal condition were shown in Table I and Table II (the unit of blood pressure: mmHg and the unit of heart rate: beats/min).

Group A was exposed to infrasound at 4.10 Hz, 120 dB and Group B was exposed to infrasound at 2.14 Hz 110 dB. After one hour, we measured the blood pressure and heart rate 3 times at 2-minute intervals and the average of 3 measurements was regarded as the final value of the experiment. The final values of blood pressure and heart rate are shown in Tables III and IV.

In order to investigate the subjective reactions of the subjects, we have designed a questionnaire of subjective feeling. After the experiment, the subjects were requested to answer questions. Everyone must answer the following questions by himself.

1. Do you feel windy?
2. Do you feel pressure in your ears?
3. Do you feel nauseated?
4. Do you have headache?
5. Do you feel fretful?
6. Do you feel tired?
7. When you feel uncomfortable during the experiment?
8. Do you feel trouble in this room?
9. Where have you ever experienced this feeling?
10. Have you experienced carsickness?
11. Do you have carsickness today?

RESULTS

The physiological and psychological effects of infrasound appeared as changes in heart rate, blood pressure and subjective reactions. Among physiological effects, heart-rate ratio and blood pressure ratio were expressed as the ratio of each final value of the experiment to the initial

Table I. Group A Blood pressure and Heart rate before exposure

		The members of Group A														
		A1			A2			A3			A4			A5		
		SP	DP	R	SP	DP	R	SP	DP	R	SP	DP	R	SP	DP	R
Measuring	1	106	64	67	113	62	73	107	60	76	128	79	76	115	76	62
Times	2	109	58	71	111	61	67	106	61	77	126	76	74	115	75	68
	3	101	58	73	114	59	68	105	58	76	123	81	80	108	67	61
	Average	105	60	70	113	60	69	106	60	76	126	79	77	113	73	64

SP: Systolic Pressure (mmHg)
DP: Diastolic Pressure (mmHg)
R: Heart Rate (times/min)

Table II. Group B Blood pressure and Heart rate before exposure

		The members of Group B														
		B1			B2			B3			B4			B5		
		SP	DP	R	SP	DP	R	SP	DP	R	SP	DP	R	SP	DP	R
Measuring	1	105	62	72	107	68	65	107	64	65	104	60	72	94	54	64
Times	2	105	57	71	105	67	69	107	59	58	102	56	73	94	60	60
	3	100	57	76	107	68	69	106	58	60	104	56	70	89	61	59
	Average	103	59	73	106	68	68	107	60	61	103	57	72	92	58	61

SP: Systolic Pressure (mmHg)
DP: Diastolic Pressure (mmHg)
R: Heart Rate (times/min)

Table III. Group A. Blood pressure and heart rate after symptoms of 4.10Hz 120dB

		The members of Group A														
		A1			A2			A3			A4			A5		
		SP	**DP**	**R**	**SP**	**DP**	**R**	**SP**	**DP**	**R**	**SP**	**DP**	**R**	**SP**	**DP**	**R**
Measuring	1	115	73	68	130	72	78	119	81	82	129	84	84	113	66	74
Times	2	118	74	75	127	75	89	118	81	85	136	89	86	120	68	78
	3	121	71	68	129	72	73	113	75	86	137	85	87	113	64	76
	Average	118	73	70	129	73	80	117	79	84	134	86	86	115	66	76

SP: Systolic Pressure (mmHg)
DP: Diastolic Pressure (mmHg)
R: Heart Rate (times/min)

Table IV. Blood B. Blood pressure and heart rate after exposure at 2.14Hz 110dB

		The members of Group B														
		B1			B2			B3			B4			B5		
		SP	**DP**	**R**	**SP**	**DP**	**R**	**SP**	**DP**	**R**	**SP**	**DP**	**R**	**SP**	**DP**	**R**
Measuring	1	110	74	76	112	75	60	124	77	73	111	73	77	116	72	70
Times	2	107	67	79	105	67	60	117	68	64	113	70	77	107	68	62
	3	113	72	82	107	72	61	118	65	69	107	67	77	105	72	65
	Average	110	71	79	108	71	60	120	70	69	110	70	77	109	71	66

SP: Systolic Pressure (mmHg)
DP: Diastolic Pressure (mmHg)
R: Heart Rate (times/min)

Table V. Group A. Heart-rate and blood pressure ratio

	The members of group A														
	A1			A2			A3			A4			A5		
	SP	DP	R	SP	DP	R	SP	DP	R	SP	DP	R	SP	DP	R
Initial Value	105	60	70	113	60	69	106	60	76	126	79	77	113	73	64
Final Value	118	73	70	129	73	80	117	79	84	134	86	86	115	66	76
Changes	13	13	0	16	13	11	11	19	8	8	7	9	2	-7	12
Ratio (%) (Increase)	12.4	21.7	0	12.4	21.7	15.9	10.4	31.7	10.5	6.3	8.9	11.7	1.8	-9.5	18.8

SP: Systolic Pressure (mmHg)

DP: Diastolic Pressure (mmHg)

R: Heart Rate (beats/min)R: Heart Rate (times/min)

Table VI. Group B. Heart-rate and blood pressure ratios

	The members of Group B														
	B1			B2			B3			B4			B5		
	SP	DP	R	SP	DP	R	SP	DP	R	SP	DP	R	SP	DP	R
Initial Value	103	59	73	106	68	68	107	60	61	103	57	72	92	58	61
Final Value	110	71	79	108	71	60	120	70	69	110	70	77	109	71	66
Changes	7	12	6	2	3	-8	13	10	8	7	13	5	17	13	5
Ratio (%) (Increase)	6.8	20.3	8.2	1.9	4.4	-11.7	12	16.7	13.1	6.8	22.8	6.9	18.5	22.4	8.2

SP: Systolic Pressure (mmHg)

DP: Diastolic Pressure (mmHg)

R: Heart Rate (times/min)

value of the experiment before starting the exposure. Psychological effects were expressed as the subjective reactions of the subjects.

Psychological Effects of Infrasound

By analyzing their answers we found: All the subjects felt uncomfortable and 8 persons said they had experienced the same feelings as travelling on vehicles or trains, two of them said they had experienced this feeling elsewhere. No one felt nauseated or carsick. Nine of them felt pressure in their ears. Six persons felt headachy and fretful. Five people felt tired and troubled in this room.

Physiological Effects of Infrasound

The changes of blood pressure and heart rate were shown in Table V and Tab1e VI. Table V is the heart-rate ratio and blood pressure ratio of group A after being exposed to infrasound (4.10 Hz, 120 dB) for over 1 hour. Table VI is the heart-rate ratio and blood pressure ratio of group B after being exposed to infrasound (2.14 Hz, 110 dB) for over 1 hour.

The data in Table V shows: After Group A was exposed to infrasound (4.10Hz 120 dB) for over 1 hour we found the changes of systolic pressure, diastolic pressure and heart rate. For each person, at least one index among systolic pressure, diastolic pressure and heart rate changes by more than 10 percent. Some changes exceed 30 percent, for example A3's change ratio of diastolic pressure.

The data in Table VI shows: After Group B was exposed to infrasound (2.14 Hz, 110 dB) for over 1 hour, for each person, at least one index among systolic pressure, diastolic pressure and B4's change ratio of distolic pressure.

In addition, the change ratio of A5's diastolic pressure was –9.5 percent. This indicates that A5:s diastolic pressure fell after A5 being exposed to infrasound (4.10 Hz, 120 dB) for over 1 hour.

CONCLUSIONS

The physiological and psychological effects of infrasound (2.14 Hz 110 dB and 4.10Hz 1200 dB) on persons are summarized as follows:

1. Being exposed to infrasound, a person feels headachy, fretful and tired.
2. Infrasound can cause the changes of blood pressure and heart rate.
3. In the infrasound condition with 4.10 Hz and 120 dB for over 1 hour, systolic pressure and heart rate of most subjects rose with the exception that the diastolic pressure of some subjects fell.
4. In the infrasound condition with 2.14 Hz, 110 dB for over 1 hour, systolic pressure and diastolic pressure of subjects rose. But heart rate of some subjects rose and others fell.

5. Different individuals have different responses to infrasound and the change ratio of blood pressure and heart rate are also different.
6. By comparing physiological and psychological effects of infrasound on persons in two different infrasound conditions, we find that there are not obvious differences.
7. Studying the relationships between the physiological and psychological effects of infrasound and the frequency and pressure level of infrasound is very necessary.

ACKNOWLEDGEMENT
This work is supported by national nature science foundation of China (Grant No. 50075029).

REFERENCES

1. H. Møller. Physiological and psychological effects of infrasound on humans. *Journal of Low Frequency Noise and Vibration.* Vol. 3, No. 1, 1984.

2. Hood, R. A, Leventhall, H. G, Kyriakides, K. Some subjective effects of infrasound, *British Acoustical Society of Meeting On Infrasound and Low Frequency Vibration*, University of Salford Nov., 1997.

3. Evans, Margaret, J., Tempest, W. Some effects of infrasonic noise in Transportation, *Journal of Sound and Vibration*, Vol. 22, 1972.

Effects on spatial skills after exposure to low frequency noise

Jessica Ljungberg[1, 2] Gregory Neely[1] and Ronnie Lundström[2]
[1]*National Institute for Working Life North, Umeå, Sweden*
[2]*Department of Public Health and Clinical Medicine, Occupational Medicine, Umeå University, Umeå, Sweden*

ABSTRACT
A study of spatial skills was conducted with 27 male and 27 female participants. The aim of the study was to examine the post-exposure effect of a complex low frequency noise (21 Hz) on a mental rotation task. It was hypothesised that reaction time and number of errors would increase after 20 minutes exposure to noise exposure compared to performance after a control condition and that groups exposed to higher intensity noise would exhibit greater impairment. Three groups of participants were exposed to a control condition and a noise condition (either 77, 81 or 86 dB (A)). After each exposure, subjects completed a mental rotation task where the st imulus consisted of one of three letters presented in five different rotations, showed either normally or mirrored. The participants were asked to respond as quickly and accurately as possible, affirmatively if the letter presented wasn't mirrored and negatively if mirrored. Statistical analysis revealed that the medium intensity level generated significant post-exposure effects while no effects were seen at the low or high intensity levels.

1. INTRODUCTION

Many people are exposed daily to unwanted noise in their working environment. One of the primary effects of noise is loss of attention during a cognitive activity, such as disruptions during reading or writing. Sounds often seem to influence our awareness. Research in this area has been focused on effects during exposure, both in terms of physiological and cognitive functioning.

For instance, studies have demonstrated the negative influence of noise on focused attention tasks and reduced hit-rate on detection tasks (Smith, 1988, 1991), but also on memory performance during exposure to low frequency noise (Gomes et al, 1999). A speed to accuracy trade – off effect was also detected in a selective attention task. Subjects worked faster during exposure to noise but with lower accuracy compared to a silent control (Hygge, Boman & Enmarker, 2003). Other researchers have studied interactive effects of multiple stressors on cognitive performance. For example, students who have to cope with demands of exams and papers at the end of their semester had greater psychophysiological stress and significant slower reaction time when they executed a dual task during a noise condition (Evans et al, 1996).

However, less is understood about post-exposure effects generated by noise. Glass and Singer (1972) were among the first to demonstrate post-exposure effects on mental performances. They found that predictability and the individual's possibility to control the noise reduces these effects and other researchers have also found similar results (Bullinger et al, 1999; Evans and Johnson, 2000).

Cohen et al, (1980) investigated these effects in school children, and the results showed that children from noisy schools are more likely to fail on cognitive tasks and appear to give up before they complete their task than children from quiet schools. Haines and colleagues (Haines et al, 2001a, 2001b) revealed the same effects in studies on school children who had been exposed to aircraft noise from a local airport. Aircraft noise had an impairing effect on reading comprehension, and generated higher levels of annoyance and perceived stress. Another study conducted in the field by Lindström and Mäntysalo (1981) indicated similar results when measuring post-exposure effects from industrial noise, before, in the middle and after a work-shift on a reaction time task. Results indicated a trend towards decrements in reaction time after being exposed to noise during the work-shift.

Post-exposure effects generated from noise on cognitive functioning need to be more closely examined. Large numbers of people regularly work in environments with intermittent exposures to noise of varying intensity and duration. Thus it is important to understand not only the effects during exposure, but also following or in between. The aim of the study was to examine the post-exposure effect of a low frequency noise on spatial skills.

We hypothesized that reaction time and number of errors would increase after exposure to noise compared to a control condition, and that participants exposed to more intense noise would exhibit greater impairment.

2. METHOD

2.1 Subjects

Fifty-four participants, (27 men and 27 women) with a mean age of 25 years (ranging from 19 to 30), participated in the study. They were tested individually and were reimbursed 300 Swedish crowns (approximately 30 USD) for their participation. All subjects reported good physical health and were tested for normal hearing, <20 dB HL. The study was reviewed by the ethics committee at Umeå University.

2.2 Physical stimuli

The noise condition consisted of a sound from a helicopter played at 21 Hz and was emitted from a loudspeaker positioned 60 cm behind the participants. The noise was registered with an integrating sound level meter (Brüel & Kjær 2237). No experimental noise was used during the control condition which consisted simply of the background noise in the laboratory, which remained steady at 60 dB(A).

2.3 Procedure and task

The spatial test was a part of a larger data collection material (Ljungberg, Neely and Lundstrom, 2004), but the following presentation focuses on post-exposure effects from noise on a mental rotation task.

The subjects received both written and verbal instructions about the tests and procedures and written consent was collected. The study applied a mixed model design where the participants were randomly assigned to one of three groups: low-intensity exposure (77 dB (A)), medium-intensity exposure (81 dB (A)), and high-intensity exposure (86 dB (A)). All participants were tested in both the noise and control condition, see Table 1 for a design overview.

Table 1. Environmental stimulus levels for the three intensity groups.

Intensity levels

Environmental exposures	Low group n = 17	Medium group n = 19	High group n = 18
Noise	77 dB(A)	81 dB(A)	86 dB(A)
Control		-	-

The participants were instructed to sit in an upright position. A familiarizing phase started the session. The mental rotation task was introduced and all subjects were allowed to practice the test until they had successfully completed 10 trials without error. Feedback was provided on the screen during the practice session indicating if the participant had responded correctly or not. The participants were also briefly presented to the noise stimulus. The experimental test session started with exposure from noise or a quiet condition for twenty minutes, the order of which was randomized over participants. During the exposure they completed an unrelated short-term memory task, and after each exposure the mental rotation task was conducted for five minutes. Between the noise and control exposures, there was a five minutes break.

The spatial orientation task consisted of three letters that were presented either normally or mirrored and rotated at one of the positions (0°, 60°, 120°, 180°, 240° or 300°) on a monitor. The participants had two hand-held, thumb-operated response buttons, one marked YES and one NO. The subject's task was to respond as quickly and accurately as possible pushing the "YES" marked response button if the letters were normal and rotated and the "NO" marked response button if the letters were mirrored and rotated. Half of the participants had the YES button in their significant dominant hand and the other half in the non-dominant. Reaction time and numbers of errors were measured as dependent variables.

3. RESULTS

An initial inspection of data showed that the reaction time data was skewed negatively, and was therefore logarithmically transformed. An analysis of the reaction time during the control condition revealed that the three intensity level groups differed in their baseline performance, see Table 2. Direct comparisons between the groups where therefore difficult. Reaction time was then recalculated for each individual into difference scores by dividing the reaction time during the noise exposure with the silent control. A value of 1 would represent the case where there was no change in the reaction time during the noise condition compared to the control condition. For the statistical tests, an alpha level of .05 was used.

Table 2. Means and standard deviations for reaction time in the noise and control condition for the intensity groups.

	Control		Noise	
Intensity levels	**Mean**	**SD**	**Mean**	**SD**
Low intensity	732,0	153,8	718,7	167,2
Medium intensity	729,8	148,1	777,8	170,7
High intensity	776,3	156,9	770,1	158,1

Reaction time data were analyzed for significant differences between the control and noise condition by testing each group's difference ratios with the hypothesized result of 1 using one-sample $\underline{t}$-test. The results showed that there were significantly slower reaction time in the noise condition for the participants in the medium intensity group ($\underline{t}$ = 2,151, df = 18, $\underline{p} < .05$). Neither of the other intensity levels generated any significant effect, when comparing noise and quite condition, see Figure 1. Analyses of the number of errors committed in each condition revealed no statistically significant effects.

4. DISCUSSION

The results from this experiment showed a post-exposure effect generated from noise at the medium intensity level, but not at the low or high intensity levels. These results indicate that at the intensity levels investigated here, there is not a direct relationship between intensity and

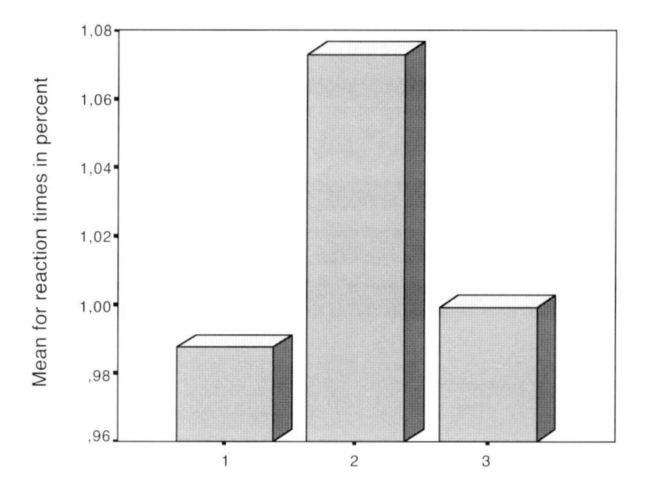

Figure 1. Mean reaction time ratios for the mental rotation task obtained at different exposure levels. For more information, see text.

post-exposure performance in a mental rotation task. The fact that there was no significant differences in the number of errors made between a control condition and the noise condition for any of the groups indicate that the observed differences in reaction time wasn't a result of different strategies in speed/accuracy tradeoff. While gender and age were controlled in this experiment, preference and noise sensitivity was not. Weinstein (1978) has observed that people differ in their sensitivity to noise intensity. It may be the case that the middle intensity group tested here contained in overrepresentation of individuals who are more sensitive for noise in general. However, preliminary data from our lab indicate there are no effects of noise sensitivity on performance in cognitive tasks.

Anyway, the predictability of the noise and individuals possibility to control the noise has in earlier experiments shown to be variables that influence post-exposure effects generated from noise (Glass & Singer, 1972). Noise with a more informative character and more complex cognitive tasks has also shown to have stronger predictability. Further investigations where these factors are included in the design might help to better understand the post-exposure effects of noise exposure.

REFERENCES

1. Bullinger, M., Hygge, S., Evans, G. W., Meis, M., & von-Mackensen, S. *The psychological cost of aircraft noise for children*. Zbl Hyg Umweltmed. 1999, 202, pp. 127-138.

2. Cohen, S., Evans, G. W., Krantz, D. S., & Stokols, D. Physiological, motivational and cognitive effects of aircraft noise on children. *Am Psychol.*, 1980, Vol. 35, No. 3, pp. 231-243.

3. Evans, G. W., Allen, K. M., Tafalla, R., & O´meara, T. Multiple stressors: performance, psychophysiological and affective responses. *J Environ Psychol.*, 1996, Vol. 16, pp. 147-154.

4. Evans, G. W., & Johnson, D. Stress and open-office noise. *J Appl Psychol.*, 2000, Vol. 85, No. 5, pp. 779-783.

5. Glass, D. C., & Singer, J. E. *Urban stress: Experiments on noise and social stressors.* 1972, New York: Academic Press.

6. Gomes, L. M. P., Martinho Pimenta, A. J. F., & Castelo Branco, N. A. A. Effects of occupational exposure to low frequency noise on cognition. *Aviat Space Environ Med.* 1999, Vol. 70, No. 3, pp. 115-118.

7. Haines, M. M.,Stansfeld, S. A., Job, R. F., Berglund, B., & Head, J. A follow-up study of effects of chronic aircraft noise exposure on child stress responses and cognition. *Int J Epidemiol.* 2001, Vol. 30, No. 4, pp. 839-845.

8. Haines, M. M.,Stansfeld, S. A., Job, R. F., Berglund, B., & Head, J. Chronic aircraft noise exposure, stress responses, mental health and cognitive performance in school children. *Psychol Med.*, 2001, Vol. 31, No. 2, pp. 265-277.

9. Hygge, S., Boman, E., & Enmarker, I. The effect of road traffic noise and meaningful irrelevant speech on different memory systems. *Scand J Psychol.*, 2003, Vol. 44, No.1, pp. 13-21

10. Lindström, K., Mäntysalo, S. Attentive behavior after exposure to continuous industrial noise. In G. Salvendy, & M. J. Smith (Ed.), *Mach pac Occup Stress.* London: Taylor & Francis. (pp. 91-96), 1981.

11. Ljungberg, J., Neely, G., & Lundstrom, R. Cognitive performance and subjective experience during exposure to whole-body vibration and noise. *Int Arch Occup Environ Health.*, 2004, Vol. 77, pp. 217-221.

12. Smith, A. P. Acute effects of noise exposure: an experimental investigation of the effects of noise and task parameters on cognitive vigilance tasks. *Int Arch Occup Environ Health.*, 1988, Vol. 60, pp. 307-310.

13. Smith, A. P. Noise and aspects of attention. *Brit J Psychol.*, 1991, Vol. 82, pp. 313-324.

14. Weinstein, N. D. Individual differences in reactions to noise: a longitudinal study in a college dormitory. *J Appl Psychol.*, 1978, Vol. 63, No. 4, pp. 458-466.

Does low frequency noise at moderate levels influence human mental performance?

Małgorzata Pawlaczyk-Łuszczyńska[1], Adam Dudarewicz[1], Małgorzata Waszkowska[2], Wiesław Szymczak[3], Maria Kameduła[1], Mariola Śliwińska-Kowalska[1]

Nofer Institute of Occupational Medicine, [1]Department of Physical Hazards, [2]Department of Work Physiology, [3]Department of Environmental Epidemiology, 8, Sw. Teresy Str., 90-950 Lodz, Poland

e-mail address: mpawlusz@imp.lodz.pl

ABSTRACT

The aim of this study was to assess the influence of low frequency noise (LFN) at levels normally occurring in industrial control rooms on human mental performance (attention, visual perception and logical reasoning) and subjective well-being. Subjects were 191 male volunteers categorised in terms of subjective sensitivity to noise in general. They performed standardised tests: the Signal Detection Test (test I), the Stroop Colour-Word Test (test II), and two sub-tests of the General Aptitude Test Battery, i.e. the Math Reasoning Test (test III) and the Comparing of Names Test (test IV). Three different acoustic conditions were used in the between-subjects design: the background laboratory noise of about 30 dB(A), LFN, and a broadband noise without dominant low frequency components (reference noise) at 50 dB(A). Each subject was tested only once in randomly-assigned exposure conditions. Generally, no significant differences in performance related to exposure conditions were noted. Some of the results from test I and test II were influenced by sensitivity to noise. However, there were no significant differences between high- and low-sensitive subjects during exposure to LFN. The annoyance of LFN and reference noise was rated higher than that of the background noise. Subjects highly-sensitive to noise reported higher annoyance due to LFN in comparison with low-sensitives. No significant differences related to noise sensitivity in annoyance assessment of background and reference noises were noted. In conclusion, no effects due to LFN on mental performance compared to background and reference noises were found.

1. INTRODUCTION

There is a growing body of data showing that low frequency noise (LFN) differs in its nature from other noises at comparable levels[1, 2, 3, 4]. Low frequency noise is not only ubiquitous in the general environment but also in the occupational environment (e.g. in industrial control rooms, office-like area etc.). Ventilation systems, pumps, compressors, diesel engines, gas turbine power stations, means of transport, etc., may be quoted as some examples of common sources of LFN. Its prevalence in offices and control rooms is mainly due to indoor network installations, ventilation, heating and air conditioning systems as well as from outdoor sources of noise and poor attenuation of low frequency components by the walls, floors and ceilings[1, 5]. However, LFN effects are less well recognised compared to the effects of noise at higher frequencies and the specific regulations on the hygienic control in the occupational environment are unsatisfactory.

Annoyance seems to be the primary and the most frequent effect of LFN exposure. However, differences in responses seem to exist between exposure to low and higher frequency noises. The annoyance experienced from LFN is higher than from noise without dominant low frequency components. It is frequently suffered at relatively low sound pressure levels and subjects sensitive to this type of noise (LFN) were not necessarily sensitive to noise in general[1, 2]. Furthermore, some symptoms related to LFN annoyance, especially fatigue, concentration problems, headache and irritation could reduce working capacity[6, 7, 8].

Over the years, a great deal of research has been carried out to evaluate adverse effects on performance from different kind of noises, but most of them have been based on noise at rather high levels. Considerably fewer studies were concerned with noise at moderate levels, including moderate levels of low frequency noise. Moreover, their results are rather inconsistent, probably due to considerable differences in individual sensitivity to noise[9, 10].

All in all, a few previous studies indicated that LFN might reduce performance at levels that could occur in the occupational environment[11,12,13]. While recent investigations showed that LFN at relatively low A-weighted sound pressure levels (about 40-45 dB) could be perceived as annoying and adversely affecting the performance, particularly when more demanding tasks were executed. Moreover, persons classified as sensitive to LFN may be at the highest risk [14, 15, 16]. Thus, LFN could possibly influence the working capacity of personnel in control rooms and offices, particularly when the job involves an element of considerable unpredictability, or requires selective attention and/or processing a high load of information.

The aim of the study was to investigate the influence of LFN on human mental performance. An attempt was made to answer the following questions:

- Can LFN at levels normally occurring in the industrial control rooms affect attention, visual perception, logical reasoning and subjective well-being?
- Does a relationship exist between sensitivity to noise and noise effects?

2. MATERIAL AND METHODS

2.1 Study Design

Subjects performed a series of standardised psychological tests designed for the assessment of attention, visual perception and logical reasoning (figure 1). Three different acoustic conditions were used in the experiment: a background laboratory noise at about 30 dB(A), LFN, and a reference noise at the same equivalent-continuous A-weighted sound pressure level (SPL) of approx. 50 dB.

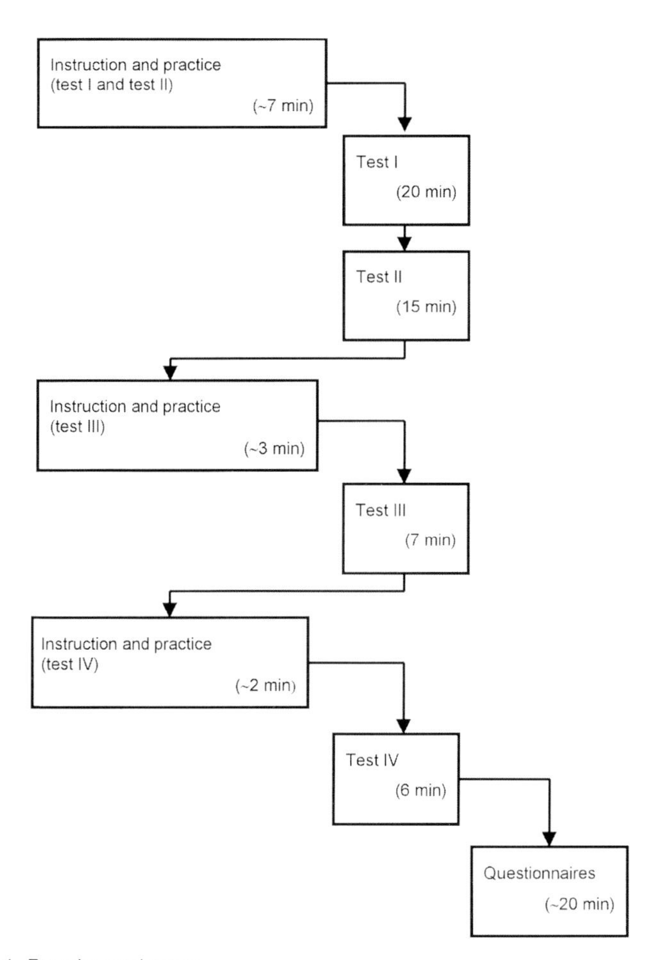

Figure 1. Experimental setup.

Each subject took part in only one test session. Exposure conditions were changed in the following order: background noise, LFN, reference noise, background noise, etc. Subjects were assigned to noise conditions in order of application for the experiment.

After the test session, persons completed questionnaires aimed at:

- subjective rating of annoyance and effort put into performing tasks,
- symptoms experienced during the tests conditions such as a headache, a feeling of pressure on the eardrum, nausea, dizziness and concentration difficulties, etc.,
- self-assessment of hearing status,
- evaluation of individual sensitivity to noise,
- temperament assessment.

A 100-score graphical rating scale was used for the annoyance and effort assessment.

In order to evaluate subjects' sensitivity to noise the Weinstein noise sensitivity evaluation questionnaire[17], consisting of 21 statements with proposed degrees of agreement (from "do not agree at all" to "agree completely"), graded from 1 to 5, was adopted. The questionnaire had a total of 105 points, the higher the score, the higher the sensitivity to noise. Thus, persons who obtained more than median score were categorised as highly sensitive (high-sensitive) to noise in general (NG+). The others were classified as less sensitive (low-sensitive) (NG-).

The Strelau Temperament Inventory (STI) was applied for temperament assessment[18]. Generally, the STI is an instrument designed to measure broad characteristics of the central nervous system. It consists of six sub-scales: spryness, perseverance, sensory sensitivity, reactivity, resilience and activity.

The local ethics committee approved the study.

2.2 Study Population

Subjects of the study comprised 191 non-preselected male volunteers, with an average age 35.3 years (SD = 13.7), not occupationally exposed to noise (table I). The majority of them were high school graduates. No persons reported any hearing problems. They were recruited by advertisement and received financial compensation for their participation in the experiment.

2.3 Exposure Conditions

The experiment was performed in a special chamber for psychological tests (6,8 m^2 area) furnished as an office environment. The noise was generated from a set of loudspeakers placed in the corners of the room.

LFN was of a tonal character with dominant components centred at 1/3-octave bands of 25, 31.5, 80 and 100 Hz (figure 2). The reference noise was a broadband noise without dominant low frequency components of a predominantly flat frequency character. Both noises were of an artificial origin and rather steady-state character. The background noise consisted of noise accompanying computer and air conditioning operation. Noise exposure parameters were monitored during the test session (table II).

Table I. Study group characteristics

	Total	Type of exposure: Background noise	Low frequency noise	Reference noise
Number of subjects	191	64	62	65
High-sensitive to noise	95	31	33	31
Low-sensitive to noise	96	33	29	34
	Mean ± SD			
Age, in years	35.3 ± 13.7	35.9 ± 13.7	36.0 ± 13.1	34.0 ± 14.4
Sensitivity to noise*	68.0 ± 12.3	68.5 ± 14.0	68.7 ± 11.0	66.8 ± 11.6
Temperament, in sten scale*				
Spryness	6.5 ± 2.2	6.5 ± 2.2	6.5 ± 2.3	6.5 ± 2.1
Perseverance	5.6 ± 2.3	5.4 ± 2.1	5.3 ± 2.1	6.0 ± 2.5
Sensory sensitivity	6.1 ± 1.8	6.5 ± 1.7**	5.7 ± 1.7**	5.9 ± 1.9**
Reactivity	5.1 ± 2.0	5.3 ± 2.0	5.0 ± 1.9	5.1 ± 2.1
Resilience	5.5 ± 2.0	5.6 ± 2.2	5.2 ± 2.1	5.6 ± 1.9
Activity	5.4 ± 2.0	5.1 ± 2.0	5.6 ± 2.1	5.6 ± 2.0

* Score in a questionnaire

** Significant differences ($p<0.05$)

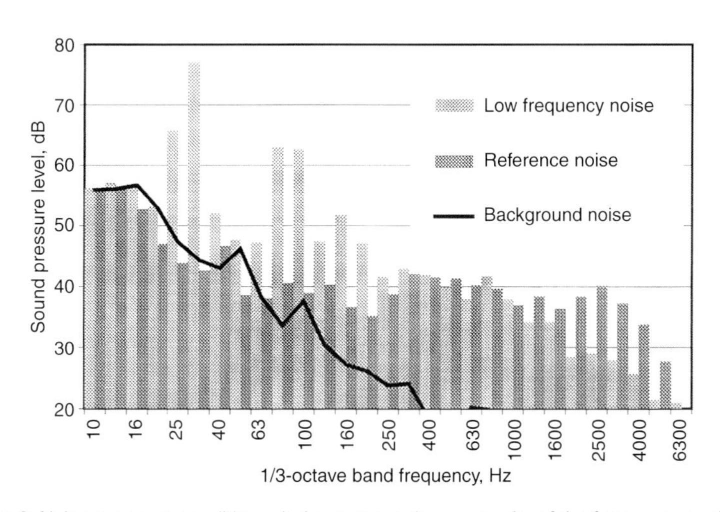

Figure 2. Noise exposure conditions during test sessions – results of the frequency analysis.

Table II. Noise exposure parameters during test sessions

		Type of exposure		
		Background noise	Low frequency noise	Reference noise
Noise parameters	Mean ± SD			
Equivalent-continuous A-weighted sound pressure level, $L_{A\,eq\,T}$, in dB	40.7 ± 3.7*	29.9 ± 1.4**	50.9 ± 1.7*	50.6 ± 1.2*
Equivalent-continuous C-weighted sound pressure level, $L_{C\,eq\,T}$, in dB	55.8 ± 2.2*	53.6 ± 1.8**	74.4 ± 0.8*	56.9 ± 1.4*
$L_{C\,eq\,T} - L_{A\,eq\,T}$, in dB	16.0 ± 7.1	23.5 ± 2.0	23.5 ± 1.6	6.2 ± 1.3

* Measurements at the subject's ear during performing test I and test II

** Measurements at the subject's head in empty chamber

2.4 Performance Tasks

Subjects performed four standardised tests, i.e.: the Signal Detection Test (test I), the Stroop Colour-Word Test (test II), Math Reasoning Test (test III) and the Comparing of Names Test (test IV).

Tests I and II involved working with a computer, but tests III and IV – with pen and paper. Before the test session the subjects were informed how to perform the first two tests. Instructions concerning test III and IV took place just before performing them. The subjects were instructed to work as accurately and quickly as possible.

The Signal Detection Test is a computerised test applied to measure the ability of visual differentiation. The screen is covered with dots, and then, one after another, they are faded out apparently by pure chance and are substituted by new ones. Subjects are expected to detect cases when four dots represent the shape of square. The main variables include the number of correct and delayed reactions as a measure for reliability of the detection process, and the mean detection time as a measure of the speed of the detection process[19, 20].

The Stroop Colour-Word Test is a computerised realisation of the Colour-Word interference paradigm by Stroop[19, 21]. It is based on the assumption that the reading speed of a colour-word is slower, if the word is written in a differently coloured font. There is always a delay in naming the colour of this word, if colour and colour-word do not match.

This test is used for registration of the colour-word interference tendency, i.e. impairment of the reading speed or colour recognition due to interfering information. Therefore, it is useful in determining the individual susceptibility to stimulus disturbing mental processes.

The test consists of four parts:

- the first – in which the names of colours (RED, GREEN, YELLOW or BLUE) are exposed in grey on the screen and subject is expected to push the button corresponding to the name – "reading in the baseline conditions";
- the second – in which colour rectangles are shown and subject is asked to press the button in the same colour – "naming in the baseline conditions";
- the third – in which the names of colours are presented in different colours (e.g. name "GREEN" is written in red, blue or yellow) and subject is expected to push the button corresponding to the name – "reading in the interference conditions";
- the fourth – in which names of colours are shown in similar way as in a preceding part, but person is told to respond to the colour of fonts – "naming in the interference conditions".

The main evaluated variables are:

- the reading interference, i.e. the difference between the median reaction times of reading in the interference and baseline conditions;
- the naming interference, i.e. the difference between the median reaction times of naming in the interference and baseline conditions;
- median reaction times and the number of incorrect answers for each individual test part[19].

The Math Reasoning Test (test III) is a sub-test of the General Aptitude Test Battery (GATB) adapted to Polish population[22]. It consists of 25 mathematical tasks and is designed to measure of skills in the four basic arithmetic operations and ability to perform them quickly and accurately. The number of correct and erroneous answers given within a 7-minute period are the main test results.

The Comparing of Names Test (test IV) is a second sub-test of the GATB[22]. It consists of two columns of words (names). Respondent decides whether couples of words (names) in both columns are exactly the same. This test is desired to measure the ability to see pertinent detail in verbal material. Test results are the number of correct and incorrect answers given within a 6 minutes period.

The test session lasted in total about 60 minutes.

2.5 Statistical Analysis

The influence of noise exposure and subjective sensitivity on the different performance tests and subjective annoyance ratings were analysed using covariance analysis, ANCOVA.

In the first stage of ANCOVA two main effects, i.e. noise exposure (3 noise conditions) and sensitivity to noise (2 sensitivity sub-groups) were analysed with age, education and temperament features (scores in the STI sub-scales) as covariates. These covariates were introduced to the model to avoid their possible influence on test results and subjective ratings.

In the second stage of ANCOVA, each group performing tasks in different noise conditions was considered separately and only the main effect of sensitivity to noise was analysed, while the covariates were unchanged.

The relationships between subjective annoyance rating and symptoms reported during the test session were analysed using Pearson's correlation coefficient (r). However, the differences in rates of registered sensations and complaints due to various noise conditions were evaluated using the Fisher test.

All statistical tests were done with an assumed significance level $p<0.05$. The statistical analysis employed SPSS software for Windows (Chicago, IL) and Statistica 5.1 (StatSoft).

3 RESULTS

3.1 Performance Tests

Results from the Signal Detection Test are shown in table III. No significant main effects of exposure conditions and sensitivity to noise on the test results were found. However, in the reference noise case differences related to noise sensitivity were observed ($p=0.049$). Subjects classified as high-sensitive to noise achieved longer mean detection times compared to low-sensitive (figure 3). Whereas, in the background noise and LFN conditions there were no differences between subjects with different sensitivity to noise.

Results from the Stroop Colour-Word Test are given in table IVa and IVb). There were no differences between the results obtained in various exposure conditions.

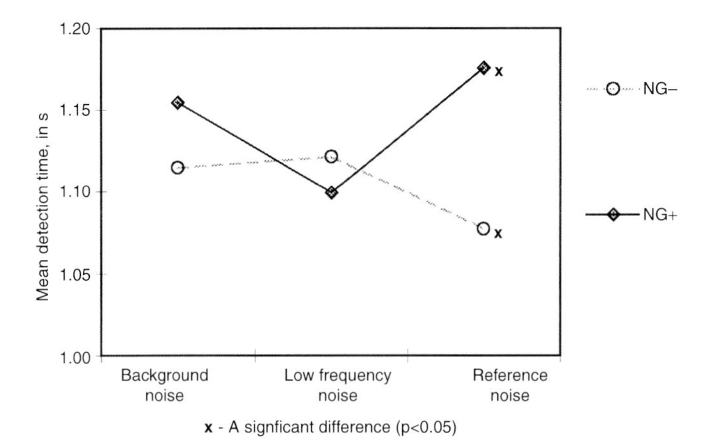

Figure 3. Mean detection time in the Signal Detection Test – the results obtained in various noise exposure conditions by persons with various sensitivity to noise – mean values adjusted for age, education and temperament features.

However, a significant main effect of sensitivity to noise was found in case of the median reaction time of reading in the interference conditions (p=0.023) as well as in case of the reading interference (p<0.001). Regardless of the exposure conditions, subjects classified as high-sensitive to noise had a higher value of reading interference and longer median reaction time compared to the low-sensitives.

Table III. The results from the Signal Detection Test

Test parameters	Study group	Total	Type of exposure: Background noise	Low frequency noise	Reference noise
		Mean ± SD (Mean adjusted for covariates)			
Number of correct reactions	All subjects	50 ± 5	50 ± 5 (50.3)	50 ± 5 (50.0)	51 ± 5 (51.0)
	NG-	51 ± 4 (50.5)	51 ± 4 (50.8)	50 ± 4 (49.2)	52 ± 4 (51.6)
	NG+	50 ± 5 (50.3)	49 ± 6 (49.7)	50 ± 5 (50.9)	50 ± 5 (50.4)
Number of delayed reaction	All subjects	0.68 ± 0.86	0.73 ± 0.80 (0.724)	0.63 ± 0.86 (0.618)	0.68 ± 0.93 (0.715)
	NG-	0.55 ± 0.74 (0.635)	0.64 ± 0.74 (0.672)	0.58 ± 0.72 (0.656)	0.42 ± 0.75 (0.576)
	NG+	0.82 ± 0.96 (0.736)	0.84 ± 0.86 (0.775)	0.67 ± 0.99 (0.579)	0.97 ± 1.03 (0.855)
Number of omitted stimuli	All subjects	8.9 ± 4.9	9.0 ± 5.1 (9.09)	9.5 ± 4.6 (9.48)	8.3 ± 4.8 (8.30)
	NG-	8.4 ± 4.3 (8.84)	8.2 ± 4.3 (8.58)	9.6 ± 4.5 (10.12)	7.5 ± 4.1 (7.82)
	NG+	9.5 ± 5.3 (9.07)	9.9 ± 5.9 (9.60)	9.5 ± 4.8 (8.84)	9.2 ± 5.4 (8.78)
Number of incorrect reactions	All subjects	1.7 ± 1.9	1.6 ± 1.6 (1.54)	1.7 ± 2.2 (1.68)	1.7 ± 1.8 (1.75)
	NG-	1.6 ± 1.4 (1.57)	1.8 ± 1.6 (1.69)	1.4 ± 1.4 (1.30)	1.7 ± 1.2 (1.70)
	NG+	1.7 ± 2.3 (1.75)	1.3 ± 1.6 (1.39)	2.0 ± 2.8 (2.06)	1.8 ± 2.3 (1.79)
Mean detection time, in s	All subjects	1.12 ± 0.27	1.13 ± 0.26 (1.135)	1.11 ± 0.31 (1.110)	1.12 ± 0.25 (1.126)
	NG-	1.05 ± 0.26 (1.104)	1.07 ± 0.27 (1.115)	1.06 ± 0.30 (1.121)	1.00 ± 0.22* (1.077)
	NG+	1.20 ± 0.26 (1.143)	1.19 ± 0.24 (1.155)	1.16 ± 0.33 (1.099)	1.25 ± 0.20* (1.176)

* A significant difference between sub-groups of various sensitivity to noise ($p < 0.05$)

During exposure to LFN, there were no differences in the values of the reading interference between subjects with different noise sensitivities (table IVa figure 4). Whereas, in the background and reference noises, persons classified as NG+ achieved higher values of the reading interference than NG- (p=0.001, p=0.033). On the other hand, in the case of median reaction time (in the interference conditions) differences related to noise sensitivity were observed only in the background noise conditions (p=0.005). As can be seen in figure 5, subjects NG+ had longer median reaction time than NG- persons

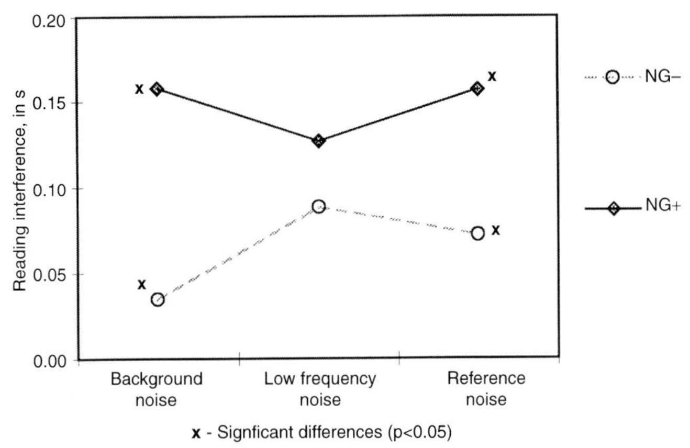

Figure 4. Reading interference in the Stroop Colour-Word Test – the results obtained in various noise exposure conditions by persons with different sensitivity to noise – mean values adjusted for age, education and temperament features.

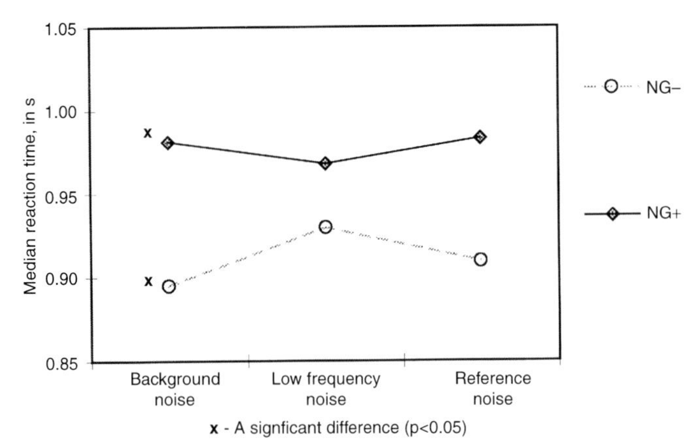

Figure 5. Median reaction time of reading (intereference conditions) in the Stroop Colour–Word Test – the results obtained in various noise exposure conditions by persons with different sensitivity to noise – mean values adjusted for age, education and temperament features.

Table IVa. The results from the Stroop Colour-Word Test – part I

Test parameters	Study group	Total	Background noise	Low frequency noise	Reference noise
			Type of exposure		
		Mean ± SD (Mean adjusted for covariates)			
Reading interference, in s	All subjects*	0.10 ± 0.12	0.09 ± 0.14 (0.096)	0.11 ± 0.11 (0.107)	0.12 ± 0.11 (0.114)
	NG-	0.08 ± 0.12 (0.065)	0.04 ± 0. 13** (0.035)	0.10 ± 0.14 (0.088)	0.09 ± 0.09** (0.072)
	NG+	0.14 ± 0.11 (0.147)	0.14+0. 13** (0.158)	0.12+0.08 (0.127)	0.15+0.13** (0.157)
Naming Interference, in s	All subjects	0.09 ± 0.20	0.12 ± 0.19 (0.121)	0.05 ± 0.18 (0.054)	0.09 ± 0.22 (0.092)
	NG-	0.10 ± 0.19 (0.108)	0.11 ± 0.16 (0.105)	0.06 ± 0.15 (0.073)	0.13 ± 0.24 (0.147)
	NG+	0.08 ± 0.22 (0.069)	0.14 ± 0.23 (0.136)	0.04 ± 0.21 (0.034)	0.05 ± 0.20 (0.038)
Median reaction time of reading (interference conditions), in s	All subjects*	0.94 ± 0.17	0.93 ± 0.15 (0.938)	0.95 ± 0.19 (0.949)	0.95 ± 0.17 (0.946)
	NG-	0.90 ± 0.14 (0.911)	0.88 ± 0.12** (0.895)	0.92 ± 0.14 (0.929)	0.89 ± 0.15 (0.909)
	NG+	0.99 ± 0.19 (0.977)	0.98 ± 0.17** (0.981)	0.99 ± 0.23 (0.968)	1.01 ± 0.17 (0.983)
Median reaction time of regarding (baseline conditions), in s	All subjects	0.84 ± 0.15	0.84 ± 0.14 (0.841)	0.84 ± 0.18 (0.842)	0.83 ± 0.14 (0.832)
	NG-	0.82 ± 0.15 (0.845)	0.84 ± 0.16 (0.859)	0.82 ± 0.15 (0.841)	0.81 ± 0.15 (0.836)
	NG+	0.86 ± 0.15 (0.831)	0.84 ± 0.12 (0.824)	8.87 ± 0.21 (0.842)	0.86 ± 0.12 (0.828)
Median reaction time of naming (interference conditions), in s	All subjects	0.88 ± 0.21	0.90 ± 0.24 (0.903)	0.84 ± 0.14 (0.839)	0.89 ± 0.22 (0.889)
	NG-	0.85 ± 0.19 (0.879)	0.88 ± 0.18 (0.889)	0.81 ± 0.13 (0.834)	0.87 ± 0.25 (0.931)
	NG+	0.90 ± 0.22 (0.876)	0.93 ± 0.30 (0.917)	0.87 ± 0.14 (0.845)	0.91 ± 0.20 (0.866)
Median reaction time of naming (baseline conditions), in s	All subjects	0.79 ± 0.16	0.78 ± 0.11 (0.782)	0.79 ± 0.20 (0.786)	0.80 ± 0.15 (0.798)
	NG-	0.75 ± 0.12 (0.772)	0.77 ± 0.11 (0.784)	0.74 ± 0.13 (0.762)	0.75 ± 0.13 (0.770)
	NG+	0.83 ± 0.18 (0.805)	0.79 ± 0.10 (0.779)	0.83 ± 0.25 (0.811)	0.86 ± 0.15 (0.826)

* A significant main effect of sensitivity to noise ($p < 0.05$)

** A significant difference between sub-groups of various sensitivity to noise ($p < 0.05$)

Moreover, in the background noise conditions, there were significant differences between subjects categorised as high- and low-sensitive to noise in the number of errors of naming in the baseline as well as in the interference conditions (p=0.031, p=0.041). In both cases subjects NG- made more errors than NG+ (table IVb)

Results from the Math Reasoning Test are shown in table V. No influence of exposure conditions, subjective sensitivity to noise and their interaction on test results was found. Similar relations were noted in case of the Comparing of Name Test (table VI).

Table IVb. The results from the Stroop Colour-Word Test - part II

Number of incorrect reactions	Study group	Total	Type of exposure: Background noise	Low frequency noise	Reference noise
			Mean ± SD (Mean adjusted for covariates)		
Reading (interference conditions)	All subjects*	2.8 ± 9.3	4.1 ± 15.6 (4.10)	1.5 ± 1.9 (1.61)	2.7 ± 3.5 (2.68)
	NG-	2.4 ± 2.6 (2.26)	2.4 ± 3.2 (2.04)	2.0 ± 2.2 (2.01)	2.8 ± 2.4 (2.74)
	NG+	3.1 ± 13.0 (3.33)	6.0 ± 22.1 (2.68)	1.0 ± 1.4 (1.20)	2.5 ± 4.4 (2.63)
Reading (baseline (conditions)	All subjects	1.0 ± 1.4	1.0 ± 1.3 (0.88)	0.9 ± 1.5 (0.97)	1.2 ± 1.4 (1.20)
	NG-	1.3 ± 1.6 (1.15)	1.2 ± 1.4 (0.92)	1.2 ± 1.9 (1.14)	1.5 ± 1.6 (1.40)
	NG+	0.7 ± 1.0 (0.89)	0.8 ± 1.1 (0.85)	0.5 ± 0.8 (0.81)	0.9 ± 1.1 (1.01)
Naming (interference conditions)	All subjects	2.3 ± 3.9	2.4 ± 2.2 (2.28)	1.7 ± 1.9 (1.82)	2.9 ± 6.0 (2.88)
	NG-	2.9 ± 4.9 (2.82)	3.0 ± 2.6* (2.77)	2.2 ± 1.9 (2.29)	3.4 ± 7.9 (3.41)
	NG+	1.8 ± 2.1 (1.83)	1.8 ± 1.5* (1.78)	1.1 ± 1.7 (1.36)	2.4 ± 2.7 (2.35)
Naming (baseline conditions)	All subjects	0.9 ± 1.3	1.0 ± 1.3 (0.95)	0.5 ± 1.0 (0.54)	1.1 ± 1.5 (1.05)
	NG-	1.1 ± 1.5 (1.06)	1.4 ± 1.4* (1.28)	0.7 ± 1.1 (0.62)	1.3 ± 1.7 (1.28)
	NG+	0.6 ± 1.0 (0.63)	0.6 ± 1.0* (0.62)	0.3 ± 0.8 (0.45)	0.8 ± 1.0 (0.82)

*A significant difference between sub-groups of various sensitivity to noise ($p < 0.05$)

Table V. The results from the Math Reasoning Test

			Type of exposure		
		Total	Background noise	Low frequency noise	Reference noise
Test parameters	**Study group**	Mean ± SD (Mean adjusted for covariates)			
Number of correct answers	All subjects	10 ± 3	10 ± 3 (9.8)	10 ± 2 (9.6)	10 ± 3 (10.1)
	NG-	10 ± 3 (9.9)	10 ± 3 (9.8)	10 ± 3 (9.6)	10 ± 3 (10.4)
	NG+	10 ± 2 (9.7)	10 ± 2 (9.8)	9 ± 2 (9.6)	10 ± 3 (9.7)
Number of Incorrect Answers	All subjects	2.5 ± 2.0	2.7 ± 2.2 (2.73)	2.7 + 2.0 (2.63)	2.2+1.7 (2.13)
	NG-	2.5 ± 2.1 (2.64)	2.9 ± 2.4 (3.15)	2.6 ± 2.3 (2.72)	2.0 ± 1.5 (2.06)
	NG+	2.5 ± 1.9 (2.35)	2.4 ± 2.0 (2.32)	2.7 ± 1.8 (2.54)	2.3 ± 1.9 (2.20)

Table VI. The results from the Comparing of Names Test

			Type of exposure		
		Total	Background noise	Low frequency noise	Reference noise
Test parameters	**Study group**	Mean ± SD (Mean adjusted for covariates)			
Number of correct answers	All subjects	53 ± 16	52 ± 16 (52.4)	53 ± 16 (53.2)	54 ± 16 (53.3)
	NG-	56 ± 15 (53.4)	55 ± 15 (53.3)	56 ± 15 (52.8)	56 ± 16 (54.2)
	NG+	50 ± 16 (52.5)	50 ± 16 (51.5)	51 ± 16 (53.6)	51 ± 15 (52.4)
Number of incorrect answers	All subjects	2.7 ± 3.0	3.3 ± 3.5 (3.30)	2.6 ± 3.1 (2.56)	2.1 ± 2.2 (2.19)
	NG-	2.6 ± 2.7 (2.67)	2.8 ± 3.2 (2.90)	2.6 ± 2.9 (2.77)	2.3 ± 1.9 (2.34)
	NG+	2.8 ± 3.3 (2.69)	3.8 ± 3.8 (3.70)	2.6 ± 3.4 (2.35)	2.0 ± 2.6 (2.03)

3.2 Subjective Ratings

Subjective assessments of annoyance and effort put into performing tasks are given in table VII.

Table VII. The subjective ratings of annoyance related to exposure conditions and efforts put into performing tests

			Type of exposure		
		Total	Background noise	Low Frequency noise	Reference noise
Subjective rating	**Study group**	Mean ± SD (Mean adjusted for covariates)			
Annoyance	All subjects*	22 ± 20	8 ± 12 (8.5)	29 ± 22 (28.8)	29 ± 17 (28.7)
	NG-	18 ± 18 (18.0)	7 ± 14 (7.4)	22 ± 20** (21.7)	25 ± 13 (24.8)
	NG+	26 ± 21 (26.1)	10 ± 10 (9.6)	36 ± 22** (35.9)	32 ± 19 (32.7)
Effort	All subjects	29 ± 18	26 ± 19 (26.6)	29 ± 19 (27.9)	32 ± 17 (31.8)
	NG-	26 ± 17 (27.8)	23 ± 18 (26.2)	23 ± 18 (24.8)	31 ± 16 (32.4)
	NG+	32 ± 19 (29.8)	28 ± 20 (27.1)	34 ± 18 (31.0)	33 ± 18 (31.2)

* Significant main effects of exposure conditions and noise sensitivity ($p < 0.001$)

** A significant difference between sub-groups of various sensitivity to noise ($p = 0.001$)

No influence of exposure conditions and sensitivity to noise on effort rating was noted. The significant main effects of both noise exposure ($p<0.001$) and subjective sensitivity to noise ($p<0.001$) on annoyance rating were found. Despite the noise sensitivity, the background noise annoyance was the lowest assessed (table VII). On the other hand, generally subjects categorised as high-sensitive to noise assessed annoyance related to exposure conditions higher than low-sensitives.

There were no significant differences in annoyance ratings of the background noise and reference noise conditions among persons of different noise sensitivity. The annoyance due to LFN was higher rated by subjects high-sensitive to noise than low-sensitive ($p=0.001$) (figure 6).

Regardless of exposure conditions, a considerable fraction of subjects (from 46.9 to 64.1%) reported no complaints during the test session (table VIII). In the background noise, 57.8% of subjects did not report any sensations, whereas during exposure to LFN and reference noise – only 12.5% and 9.4%, respectively (significant differences between various exposure conditions). Noise present in the room was perceived by nearly all the subjects exposed to reference noise and by over two-thirds of those exposed to LFN. On the other hand, in the background noise conditions, only 17.2% persons perceived the sounds accompanying computer and air conditioning operation.

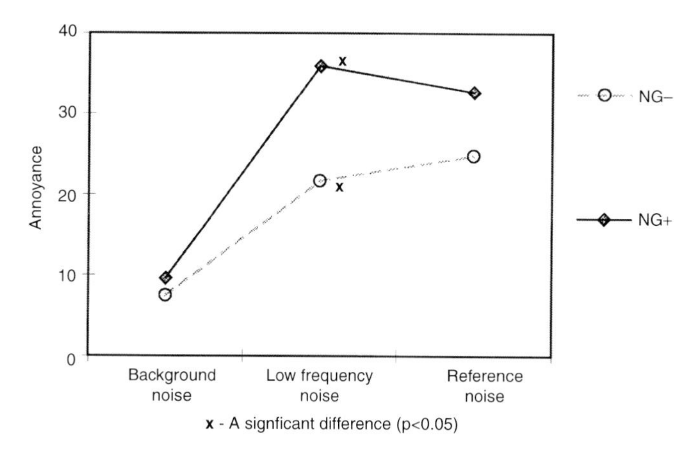

Figure 6. Annoyance assessments of various exposure conditions during test session by persons with different sensitivity to noise – mean values adjusted for age, education and temperament features.

During exposure to the reference noise, subjects most frequently reported problems with concentration (32.8%). Moreover, this rate of answers was significantly higher in comparison with the other exposure conditions. On the other hand, the LFN subjects most frequently reported fatigue (26.6%) and drowsiness (18.8%), but the complaints were not significantly more frequent than during other noise conditions (table VIII).

Generally, the annoyance rating on the graphical scale was significantly correlated with the number of reported sensations ($r=0.49$ $p<0.001$) and complaints ($r=0.36$ $p<0.001$) related to exposure condition during performing the tasks (table IX).

4. DISCUSSION

The study was designed to investigate whether exposure to moderate levels of LFN can influence human mental performance and subjective well-being. A further objective was to analyse the relation between sensitivity to noise and LFN effects on both performance and subjective annoyance rating. Four standardised psychological tests were applied for that purpose. Those tests are usually used as a measure of ability of visual differentiation and attention (test I), perceptiveness and concentration (test IV), susceptibility to stimulus disturbing mental processing (test II) and ability of logical (arithmetic) reasoning (test III). A high workload was generated by instructing the subjects to work as quickly and accurately as possible.

Table VIII. The subjective sensations and complaints during test session

	Background noise	**Low frequency noise**	**Reference noise**
		Rates of answers in %	
	Sensations		
No sensations	57.8*	12.5*	9.4*
I heard sounds (noise)	17.2*	68.8*	82.8*
I felt pressure in ears	10.9	15.6	21.9
I felt pressure in head	12.5*	15.6*	3.1*
I felt vibrations in room	3.1	6.3	6.3
I felt vibrations in part of body	6.3	1.6	3.1
I felt discomfort	17.2	23.4	21.9
Other	26.6*	10.9*	32.8*
	Complaints		
No complaints	64.1	50.0	46.9
Headache	1.6	7.8	1.6
Problems with concentrations	12.5*	14.1*	32.8*
Dizziness	1.6	4.7	1.6
Drowsiness	9.4	18.8	15.6
Fatigue	17.2	26.6	15.6
Other	15.6	4.7	14.1

*Significant differences between groups in various exposure conditions($p < 0.05$)

Table IX. Correlation coefficients between subjective annoyance rating of acoustic conditions during test session and reported sensations and complaints

	Correlation coefficient r			
Subjective annoyance assessment		**Type of exposure**		
	Total	**Background noise**	**Low frequency noise**	**Reference noise**
Number of reported sensations	$r = 0,49$ $p < 0.001$	$r = 0,46$ $p < 0.001$	$r = 0,51$ $p < 0.001$	$r = 0,40$ $p < 0.001$
Number of reported complaints	$r = 0,36$ $p < 0.001$	$r = 0,30$ $p = 0,017$	$r = 0,32$ $p = 0,010$	$r = 0,52$ $p < 0.001$

A subject's sensitivity to noise was rated on the basis of the score in the Weinstein noise sensitivity evaluation questionnaire[17]. This questionnaire was widely used in other studies[15, 25, 26]. Because people recognised as more sensitive to noise might demonstrate higher arousal levels and are likely to react with larger changes in sympathetic activity in noisy conditions, it was suggested that nervous system features had a mediating role in the effects of noise on mental performance. Therefore, the subjective sensitivity to noise was complemented by temperament assessment using the Strelau Temperament Inventory[18].

The psychological tests were carried out in three different acoustic conditions, including the background laboratory noise, and low frequency and reference noises at the same equivalent continuous A-weighted sound levels of 50 dB. An earlier study showed that sound pressure levels normally occurring in industrial control rooms remained within the range of 48-66 dB(A)[23]. Thus, an SPL of 50 dB(A) corresponded with the lower limit of the measured levels. Moreover, it was 15 dB lower than the currently admissible level established in Poland to ensure suitable working conditions for operators of control equipment in control booths, remote control rooms, etc.[24]

The low frequency noise was of an artificial origin, thus its frequency spectrum differed from typical spectra of LFN in control rooms[23]. It had a tonal character with dominant components centred within 1/3-octave bands of 25, 31.5, 80 and 100 Hz. It is worth noting that the background noise also contained low frequency components from 10 Hz to 100 Hz.

It has been usually assumed that the difference between C- and A-weighted sound pressure levels exceeding 15 dB indicates the occurrence of LFN[2]. This condition was satisfied in both low frequency and background noise conditions.

To avoid learning effects each subject performed tests in only one exposure condition. Although non-preselected male volunteers took part in the experiment, three sub-groups working in various exposure conditions did not differ according to age, education and most of temperament features. There were only differences in sensory sensitivity. Subjects from the sub-group performing tasks in the background noise conditions achieved a higher score in the questionnaire, i.e. showed higher capability to react to faint stimuli than others.

Generally, no differences related to exposure conditions in test results were noted. Thus, no effects of LFN on performance compared to background and reference noises were found. However, some of the results from the Stroop Colour-Word Test and the Signal Detection Test were influenced by noise sensitivity.

A significant main effect of sensitivity to noise was found for median reaction time of reading in the interference conditions and reading interference in the Stroop Colour-Word Test. Regardless of the exposure conditions, subjects classified as highly-sensitive to noise had higher values of reading interference and longer median reaction times in comparison with the low-sensitives.

In general, no differences related to noise sensitivity were found in the LFN. Whereas, during exposure to the reference noise poorer performance in subjects highly-sensitive to noise compared to low-sensitives were observed in the case of mean detection time (in the Signal Detection Test) and reading interference (in the Stroop Colour-Word Test). On the other hand, in the background noise conditions differences related to noise sensitivity were found for some results from the Stroop Colour-Word Test (i.e. reading interference, median reaction time of reading in the interference conditions, number of errors of naming in the baseline and in the interference conditions).

The influence of noise sensitivity on performance during exposure to noise was shown in earlier studies. For example, Jelnikova[25] found that persons recognised as sensitive to noise had a reduced working ability and attention during exposure to recorded traffic noise at equivalent continuous A-weighted sound pressure level of 75 dB in comparison with persons tolerant to noise. Similarly, Bolejovic et al.[26] not only confirmed the influence of noise sensitivity on performance during exposure to traffic noise at 55 and 75 dB(A), but also found a relationship between noise sensitivity and subjective assessment of noise annoyance.

Generally, the research data concerning the influence of LFN at dB(A) levels normally occurring in control rooms and office areas are sparse. For example, Persson et al.[14] found in a pilot study that 42 dB(A) LFN from a ventilation system could increase the response time in a verbal grammatical reasoning task in comparison with a ventilation noise at the same level, but without the low frequency components.

The next laboratory study, a continuation of the work performed in the pilot study quoted above, confirmed that LFN at relatively low A-weighted sound pressure levels (about 40 dB) could be perceived as annoying and adversely affecting performance, particularly when mentally demanding tasks were executed, while the effects on the routine tasks were less clear. Moreover, persons classified as sensitive to LFN may be at the highest risk[15].

In the quoted study, subjects categorised in terms of sensitivity to noise in general and to LFN in particular, performed a series of tasks involving different levels of mental processing (i.e. simple reaction-time task, short-term memory task and bulb-task, proof-reading task and verbal grammatical reasoning task) during exposure to ventilation noise of a low frequency character or a flat frequency (reference) noise, both at the same level of 40 dB(A). All performance tasks were carried out twice in each test sessions, once in phase A and once in phase B. Thus, the experiment had 2 noises × 2 phases × 2 sensitivity groups, but sensitivity to noise in general and to LFN in particular, was considered separately. The results showed that there was a large improvement of response time over time, during work with a verbal grammatical reasoning task in the reference noise, indicating a better learning effect in this noise condition in comparison with LFN. The results also showed that LFN interfered with a proof-reading task by lowering the number of marks made per line read. The persons reported a higher degree of annoyance and impaired working capacity during exposure to LFN. The effects were more pronounced for subjects classified as sensitive to LFN, whereas somewhat different results were found in subjects rated as sensitive to noise in general[15].

In another study aimed at evaluating effects of moderate levels of LFN on attention, tiredness and motivation in a low demanding work situations, only subjects categorised as highly-sensitive to LFN were enrolled. As previously, two ventilation noises at the same A-weighted sound pressure level of 45 dB were used, one of predominantly low frequency content and one with flat frequency spectrum. Subjects worked with six performance tasks. Most of them were of a monotonous and routine type. The major finding in that study was that LFN adversely affected performance in two tasks sensitive to reduced attention and in a proof-reading task. Performances of tasks aimed at evaluating motivation were not significantly influenced. Moreover, no significant difference between noise conditions was found in annoyance assessment[16].

In this study, noise annoyance assessment was related both to exposure conditions and subjective sensitivity to noise. It is not surprising that, regardless of the noise sensitivity, the background noise annoyance was rated lowest. However, there was no difference between the annoyance assessment of low frequency and reference noises. The LFN annoyance was rated higher by subjects categorised as highly-sensitive to noise than by low-sensitive subjects. Whereas no significant differences related to noise sensitivity were noted in annoyance assessment in the background noise and reference noise conditions.

Regardless of the exposure conditions, the annoyance rating was correlated with the number of reported sensations and complaints related to conditions during the test session. It was not surprising that the majority of subjects did not report any sensations and complaints in the background noise conditions. But it is worth noting that during exposure to LFN, subjects complained most frequently of fatigue and drowsiness, especially as there are indications that LFN may intensify tiredness more easily than noise of higher frequencies[27, 28].

To sum up, no effects due to LFN on mental performance compared to reference and background noise were found. Moreover, during exposure to LFN no significant differences in performance between higher and lower sensitive to noise subjects were noted. However, annoyance caused by LFN was rated higher by subjects high-sensitive to noise.

As the experiment was carried out under laboratory conditions and between- subject design was chosen, the relevance of the results for normal working conditions must be evaluated with care. Nevertheless, the study does not support the hypothesis that LFN at levels normally occurring in the control rooms (at about 50 dB(A)) may adversely influence human mental performance and lead to work impairment and hence is not in agreement with some previous studies. However, findings presented here do not exclude that such a noise might be perceived as annoying, especially by people particularly sensitive to noise.

Thus, it seems that studies on LFN effects on human mental performance should be continued, but more attention should be paid to the evaluation of subjective sensitivity to this type of noise.

ACKNOWLEDGEMENT

This study is supported by the Polish State Committee for Scientific Research (Grant no. IMP 18.1/2002).

REFERENCES

1. Berglund B., Hassmen P., Job R. F., Sources and Effects of Low-Frequency Noise, *Journal of Acoustical Society of America,* 1996, Vol. 99, No. 5, pp. 2985–3002.

2. Persson Waye K., *On the Effects of Environmental Low Frequency Noise.* PhD Thesis, Gothenburg University, 1995.

3. Leventhall G., Pelmear P., Benton S., *A Review of Published Research on Low Frequency Noise and Its Effects.* Department for Environment, Food and Rural Affairs, Crown copyright, 2003 (website, *www.derfra.gov.uk*).

4. Bengtsson J., Low Frequency Noise During Work – Effects on Performance and Annoyance, PhD Thesis, Gothenburg University, 2003.

5. Pawlaczyk-Luszczyńska M., Occupational Exposure to Infrasonic Noise in Poland, *Journal of Low Frequency Noise, Vibration and Active Control,* 1998, Vol. 17, No. 2, pp. 71–84.

6. Tokita Y. Low Frequency Noise Pollution Problems in Japan, in: Moller H., Rubak P., eds, *Proceedings of Conference on Low Frequency Noise and Hearing,* Aalborg University Press, Aalborg, 1980, pp. 47–60.

7. Nagai N., Matsumoto M., Yamasumi Y., Shirahi T., Nishimura K., Matsumoto K., Miyashita K., Takeda S., Process and Emergence on Effects of Infrasonic and Low Frequency Noise on Inhabitants, *Journal of Low Frequency Noise and Vibration,* 1989, Vol. 8, No. 2, pp. 87–99.

8. Persson Waye K., Rylander R., The Prevalence of Annoyance and Effects after Long-term Exposure to Low Frequency-noise. *Journal of Sound and Vibration,* 2001, 240, pp. 483–97.

9. Smith A. P, Jones D. M., Noise and Performance, in: Jones D. M., Smith A. P., eds., *Handbook of Human Performance,* vol. 1: The Physical Environment, Academic Press, London, 1992, pp. 1–28.

10. Kjellberg A., Landström U., Noise in the Office: Part II- the Scientific Basis (Knowledge Base) for the Guide. *International Journal of Industrial Ergonomics* 1994, 14, pp. 93–118.

11. Benton S., Leventhall H., G. Experiments into the Impact of Low Level, Low Frequency Noise upon Human Behaviour. *Journal of Low Frequency Noise and Vibration,* 1986, Vol. 5, No. 4, pp. 143–62.

12. Benton S., Robinson G., The Effects of Noise on Text Problem Solving for Word Processor User (WPU), in Vallet M., ed., *Noise and Man '93. Noise as a Public Health Problem. Proceedings of the 6th International Congress,* Institut National de Recherche sur Le Transport et Luer Securite, Bron, Nice France, 1993, pp. 539–541.

13. Kjellberg A., Vide P., Effects of Simulated Ventilation Noise on Performance of a Grammatical Reasoning Task, in: Berglund B., Berglund U., Karlson J., Lindvall T., eds., *Proceedings of the 5th International Congress on Noise as a Public Health Problem,* Stockholm, 1988, pp. 31–36.

14. Persson Waye K., Rylander R., Benton S., Effects on Performance and Work Quality Due to Low Frequency Ventilation Noise, *Journal of Sound and Vibration,* 1997, 205, pp. 467–74.

15. Persson Waye K., Bengtsson J., Kjellberg A., Benton S., Low Frequency Noise Pollution Interferes with Work Performance, *Noise & Health,* 2001, 4, pp. 33–49.

16. Bengtsson J., Persson Waye K., Kjellberg A., Evaluation of Effects Due to Low Frequency Noise in a Low Demanding Work Situation (submitted for *Journal of Sound and Vibration*, 2003).

17. Weinstein N. D., Individual Differences in Reaction to Noise. A Longitudinal Study in a College Dormitory, *Journal of Applied Psychology,* 1978, 63, pp. 458–66.

18. Strelau J., Angleitner A., Bantelmann J., Ruch W., The Streulau Temperament Inventory- Revised (ST-R): Theoretical Consideration and Scale Development, *European Journal of Personality,* 1990, 4, pp. 209–235.

19. *www.alta.pl*

20. Luczak A., Sobolewski A., Psychometric Characteristic of Psychometric Tests, Central Institute for Labour Protection, Warsaw, 2002 (in Polish).

21. Stroop J. R., Studies of Interference in Serial Verbal Reactions. *Journal of Experimental Psychology,* 1935, 18, pp. 643–62.

22. Noworol C., Beauvale A., Laczala Z., Żarczyński Z., *General Aptitude Test Battery. Polish Adaptation of version B-10002B,* Ministry of Labour and Social Policy, Warsaw 1997 (in Polish).

23. Pawlaczyk-Luszczyńska M., Dudarewicz A., Waszkowska M., Annoyance of Low Frequency Noise in Control Rooms, in: Selamet A., Singh R., Maling G. C., eds., *Proceedings of the Inter-Noise 2002, the 2002 International Congress and Exposition on Noise Control Engineering,* Institute of Noise Control Engineering of USA, Inc., 2002, (CD-ROM, paper no N118).

24. PN-N-01307:1994 Noise. Permissible Values of Noisc in thc Workplace. Requirements Relating to Measurements. Polish Committee for Standardisation, 1995 (in Polish).

25. Jelnikova Z., Coping with Noise in Noise Sensitive Subjects, in: Berglund B., Berglund U., Karlsson J., Lindvall T., eds., *Proceedings of the 5th International Congress on Noise as a Public Health Problem,* Swedish Council for Building Research, Stockholm, 1988, pp. 27–30.

26. Belojevic G., Ohrstrom E., Rylander R., Effects of Noise on Mental Performance with Regard to Subjective Noise Sensitivity, *International Archives of Occupational and Environmental Health,* 1992, Vol. 64, No. 4, pp. 293–301.

27. Landstrom U., Bystrom M., Nordstrom B., Changes in Wakefulness during Exposure to Noise at 42 Hz, 1000 Hz and Individual EEG Frequencies, *Journal of Low Frequency Noise and Vibration,* 1984, 4, pp. 27–33.

28. Landstrom U., Noise and Fatigue in Working Environments, *Environment International,* 1990, 16, pp. 471–76.

Section 2.
Effects of Low Frequency Vibration on People

Chapter 4. Perception thresholds for low frequency vibration and the effect of low frequency vibration on people in terms of comfort and annoyance

The hearing thresholds at which test subjects first perceive low frequency vibration is discussed in the following two papers. The first paper also considers levels at which the subject becomes uncomfortable

The purpose of this study was to identify levels and frequencies of vibrations at which drivers of heavy articulated vehicles perceived the ride as uncomfortable. Whole body vibration measurements were taken in all three axes and pavement roughness was also measured. It was found that the vertical vibration axis was the most important and this was well correlated with pavement surface roughness.

187 healthy subjects were tested for tonal vibration perception in the range 63 Hz to 500 Hz and in the range 4 to 250 Hz. It was found that age weight and to a lesser degree height affected vibration perception thresholds for exposure of the digital pulp on their hand to a vibrating probe.

Perception of low frequency vibrations by heavy vehicle drivers

Rayya Hassan[1] and Kerry McManus[2]
[1]ARRB Group Ltd, 500 Burwood Highway, Vermont South VIC 3133, Australia.
[2]Faculty of Engineering and Industrial Sciences, Swinburne University of Technology, PO Box 218, Hawthorn, Vic. 3122, Australia.

ABSTRACT

A study was initiated to identify the levels and frequencies of heavy articulated vehicle body vibrations at which the drivers perceive road pavement rideability as uncomfortable. The study involved conducting a subjective assessment survey in which a panel of truck drivers were asked to rate the ride quality provided by a number of road sections with different surface roughness characteristics. The study's objective was achieved by correlating the mean panel ratings (MPRs) to road surface roughness contents in different one-third-octave bands of the roughness spectrum. The results showed that at 100km/hr, truck drivers object mainly to motions resulting from roughness excitations of the low frequency vibration modes of the truck body in the range 1.42-5.7Hz. These results were validated by correlating MPRs with the levels of whole body vibrations measured on the driver's seat in a representative vehicle while traversing some test sections. MPRs were found to correlate well with the measured overall vibration total values and the likely comfort reactions to various magnitudes of overall vibration total values given by ISO 2631-1. The influence on MPRs of vehicle and driver related factors were also investigated and commented upon.

1. INTRODUCTION

The ride environment of the truck driver is the product of applied excitation and response properties of the truck body. Road surface roughness is a major source of excitation in addition to the rotating tyre/wheel assemblies, the driveline and the engine (Gillespie 1985). When a vehicle traverses a rough road pavement, surface irregularities excite different vibration modes of the vehicle body at different frequencies. Past research has shown that heavy vehicle ride is most sensitive to excitations of the low frequency modes in the range 1-8Hz (Gillespie 1985). At these frequencies, modes such as body bounce, pitch, roll, frame bending and axle hop are excited. The resulting motions affect the ride quality perceived by the occupants and their comfort as at these frequencies humans are most sensitive to vertical (4-10Hz) and lateral and longitudinal (0.5-2Hz) vibrations (ISO-2631-1 1997).

The aim of this paper is to report on the results of a study conducted for the purpose of identifying the levels and frequencies of heavy articulated vehicle body vibrations, excited by road roughness, at which the drivers perceive the ride as uncomfortable. The study involved conducting a subjective assessment survey to collect drivers' ratings of ride quality provided by a number of road sections with varying roughness levels and spectral characteristics. Further, Whole Body Vibrations (WBV) transmitted through the driver's seat were measured at highway speed on a number of test sections.

Driver's perception of ride is influenced by factors other than the vibrations present in the vehicle's cabin. They include factors related to the road, the vehicle and the driver. Past research has shown that road roughness is the only road factor that has a significant effect on drivers' ratings (Cairney et al. 1989) however factors such as alignment, speed, pavement and surface types were considered and controlled in the selection of the test sections. Data relating to vehicles' properties and drivers' characteristics were collected during the survey to study their effects and establish if any of them have influenced drivers' judgements during the rating exercise.

2. DATA COLLECTION

2.1. Subjective Assessment Survey

The main objective of the study was to identify the range of roughness wavelengths/frequencies in the longitudinal profile of a road surface that influence heavy vehicle drivers' perceptions of ride quality. To achieve this objective, a subjective rating survey was conducted where a group of heavy vehicle drivers provided ratings of their perceptions of ride quality provided by 29 rating sections. They used the rating scale, shown in Figure 1, which ranges from 0 for extremely poor to 5 for perfect.

Figure 1. The rating scale

The rating sections were selected from two highways located in rural areas to the East and West of the State of Victoria/Australia. They varied in lengths, roughness levels and spectral characteristics however they had uniform roughness along their lengths. Roughness measure used in the selection process is the International Roughness Index (IRI). Each section was identified with a white sign at the start of it and a yellow sign at its end with the section number written on them. Details of the test sections are summarised in Table 1.

Table 1. Details of the test sections, the drivers and the test vehicles

Rating Sections

Number	Length Range	Roughness Range IRI (m/km)	Geometry	Pavement/ surface type	Ratings Range
29			Flat &	All flexible	
17 East	100-1000m	1.11- 3.48	straight	with chip	1.8-5
12 West	200-700m	1.68-4.24		or thin	1-4
				asphalt seal	

Vehicles

Number	Make	Cabin location	Age	Seat suspension	Drive axle suspension
28	15 Kenworth	24 above	1-14 years	Airbag for all	13 spring
	4 International	4 behind			15 airbag
	6 Volvo				
	3 Scania				

Rating Panel

Number	Age	Weight, kg	Experience	Sex	Familiarity on route
31	25-55	70-110	4-35 years	All males	29 experienced
16 East					2 new
15 West					

The survey was designed using the independent group method i.e. different panel of drivers rated the sections on each of the highways. The drivers had to do the rating exercise during their normal operations and recorded their ratings using voice-activated recorders. They were instructed to start concentrating on the ride quality when they see the start sign and to record their ratings after passing the end sign. They travelled at a speed of 100km/hr (normal speed on State Highways) in tractor-semi-trailer combination units but with different characteristics and loading conditions. The rating panels left their bases in Melbourne (Capital of Victoria) with fully loaded vehicles and returned the same day with empty vehicles. Eighteen sections were selected in the outbound directions (rated with loaded vehicles) and eleven in the inbound directions (rated with empty vehicles). The number of ratings collected for the different sections ranged between 10 and 16. Details of the rating panel and the vehicles are summarised in Table 1.

During the rating exercise, the drivers were also asked to describe the type(s) of vibration(s) that they felt while traversing each test section. The vertical vibrations were the most felt by majority of drivers. The second most felt type of vibration is sideway motion followed by fore and aft motion. In addition to that a number of drivers complained of vibrations in the steering wheel and one driver reported cab movement.

2.2. Measurement of WBV

Whole Body Vibrations (WBVs) transmitted through the driver's seat were measured in one of the test vehicles with fully loaded trailer. The loading condition of the test vehicle dictated the conduct of WBV measurements on the rating sections in the outbound directions only. The measurement was performed on 17 sections only as the signs for the eighteenth section were missing. The measurement was done using special vibration measuring equipment while travelling at a speed of 100km/hr. The acceleration spectrum for each rating section was measured in the three orthogonal directions x, y and z. The x-axis, which is in the direction of travel, the y-axis is transverse to it and the z-axis is in the vertical direction passing from the seat to the head of the driver. The measurement output included RMS accelerations (m/s^2) computed for consecutive bands of ~0.2Hz width covering frequencies from 0 to 87.7Hz.

2.2.1. The Test Vehicle

The vehicle used in WBV measurement is a one-year-old tractor-semi-trailer, Kenworth make, with cab over engine and an airbag suspended driver's seat. The vehicle had a tandem drive axle and a tandem trailer axle and both were fitted with airbag suspensions. There was no specific criterion for selecting this vehicle but voluntary contribution. However, it provides good representation of the vehicles used in the survey as the majority had similar configuration, cabin location and make. Past research has shown that cabin location has a significant effect on drivers' judgements whereas, the effects of other vehicle related factors including age and type of drive axle suspension were found to be not significant (Miller 1981).

2.2.2. The Test Equipment

The equipment used in measuring WBV transmitted through the driver's seat consists of a triaxial accelerometer pad, a four channel input module with four built in charge amplifiers and a vibration analyser (Svantek 2000). The measurement started a short distance (10-15m) before a test section and ended a short distance after its end.

3. ANALYSIS OF SUBJECTIVE AND ROUGHNESS DATA

3.1. Subjective Data

The subjective measure used in the analysis is the Mean Panel Rating (MPR). The MPR for each rating section was calculated by summing the ratings of the drivers who rated that section then dividing the sum by the number of drivers. The standard deviations of the ratings for the different sections ranged between 0.4 and 0.7. The reliability and agreement between the raters/drivers was tested using Ebel's intraclass correlation coefficient (ICC). The ICC is used to estimate the reliability of individual or average ratings of judges (Cramer 1998). The ICC was found to be equal to 0.91, which indicates that there is very good agreement between the different drivers.

3.2. Roughness Data

Road surface profile data is measured using a laser profiler system along the wheel tracks of a passenger car. Data analyses involved performing Power Spectral Density (PSD) analyses on road surface profile data of each of the rating sections. Then roughness spectrum of each section was divided into narrow frequency bands, namely, one-third-octave bands (OTOB). One-third-octave frequency bands were chosen to provide a clearer definition of the contribution of road profile to roughness across the frequency range of interest. The range of spatial frequencies considered is 0.011-2.83cycles/m, as recommended in ISO 8608 (1995). The PSD functions of the rating sections were then converted to a series of Root Mean Square (RMS) values of profile slope in each OTOB. The RMS value of the PSD function of road surface profile elevation or slope in any band represents roughness content in that band (Janoff et al. 1985).

3.3. Frequencies affecting drivers' perception

Roughness contents in different wavebands of the longitudinal surface profile for each section were correlated to the corresponding MPR values to identify roughness content(s) and band(s) that mainly influence drivers' perception of pavement rideability. The waveband that resulted in the highest correlation ($r = -0.88$) ranges between 4.88m and 19.51m (centre spatial frequencies 0.0513-0.205cycle/m). Spatial frequency is the inverse of the wavelength. When travelling at normal highway speeds

(60-100km/hr), roughness wavelengths within this band excite the low frequency vibration modes of the truck body in the range of 0.9-5.8Hz. The wavelength (length per cycle) equals the speed (length per second) divided by the frequency (cycle per second).

Since the IRI provided lower correlation with MPR (r = – 0.74), roughness content in this band was used to establish a new roughness index called the Profile Index for trucks (PI_t). The subscript (t) stands for truck and is added to differentiate this index from another index, also called Profile Index (PI) proposed by Janoff et al. (1985) which was developed to represent passenger car ride. The PI_t is defined as the RMS average of the RMS values of profile slope calculated for the outer (PI_{tOT}) and inner (PI_{tIN}) wheel paths in the waveband 4.88-19.5m and calculated using Eqn 1.

$$PI_t = \sqrt{\frac{PI^2_{tOT} + PI^2_{tIN}}{2}} \qquad \text{Equation 1}$$

The new index serves to predict drivers' perceptions of pavement rideability (r^2=0.85) better than the IRI (r^2=0.64) (Hassan 2002). Hassan (2002) developed a non-linear statistical transform called the Truck Ride Number (TRN) that can predict heavy vehicle drivers' ratings from PI_t with a margin of error of $\leq \pm 0.3$ of a scale interval at 95% confidence. TRN has a scale similar to the rating scale used in the survey. Using TRN transform and MPRs, a scale was developed for PI_t. A PI_t value of less than 0.5m/km represents a very good ride, 0.5-1.3 good ride, 1.3- 2.75 fair ride, 2.75-5.5 poor ride and PI_t greater than 5.5m/km represents a very poor ride.

4. ANALYSIS OF WBV DATA

4.1. Graphical analysis

Graphical analyses of measured accelerations in the different directions were performed to view their distributions over the frequency range of interest (0-20Hz). Vibration spectra measured on some of the test sections covering a range of PI_t values were plotted against the corresponding MPR values to identify the frequency bands and modes that contribute most to each type of vibration, vertical, longitudinal and lateral. Figure 2 shows clearly that the contribution to the vertical vibrations comes mainly from exciting the low frequency modes in the range 1-4.5Hz such as body bounce and pitch modes. In this range the human body has low sensitivity to this type of vibration. Excitation of body bounce by surface roughness is depicted in Figure 3.

The spectra of vibrations in the x direction (fore and aft motions) presented in Figure 4 also show that the highest peaks in the spectra

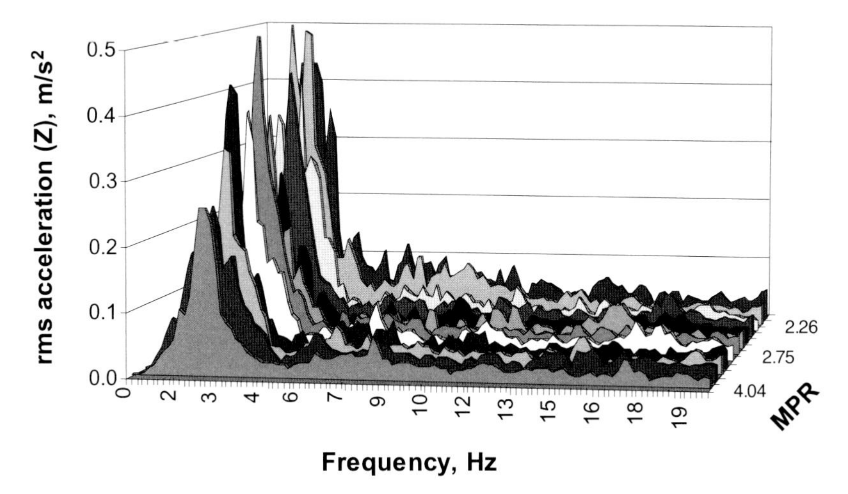

Figure 2. Acceleration spectra in the vertical direction (z) versus MPR

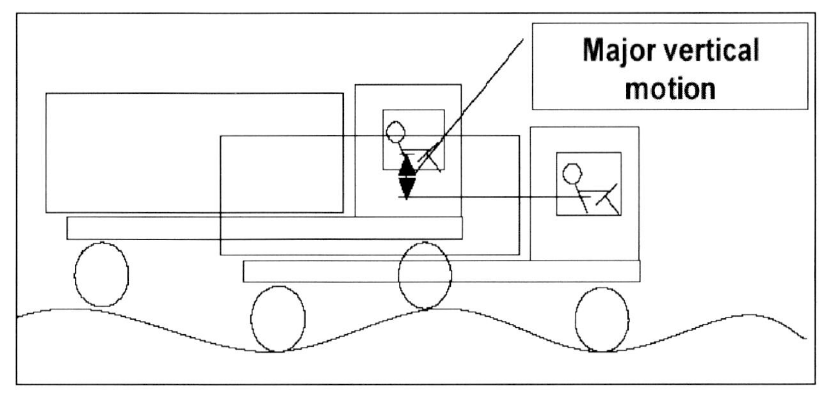

Figure 3. Excitation of truck body bounce motion by surface roughness (Granlund et al. 2000)

occur in the range 1-4Hz due to excitation of modes such as the body pitch. At these frequencies humans have high sensitivity to this kind of vibration. Excitation of body pitch by surface roughness is depicted in Figure 5. The figure also shows that there is also high contribution from higher frequency modes (such as unsprung mass pitch and frame bending) in the range 7-17Hz at which humans have low sensitivity to this kind of motion.

Figure 6 shows that there is a relatively good correlation between MPR and the spectra of sideway vibrations in the frequency band 1-4.5Hz. However, the low frequency modes such as body roll, contributes little to sideway vibrations. The highest levels of sideway vibrations are attributable to excitations of vibration modes in the range 7-9Hz and 16-18Hz. In this range human body is not sensitive to this type of vibration. The conclusion that could be drawn from these findings is that the seat and suspension systems of the test vehicle are providing good isolation of vertical and sideway vibrations and bad isolation of fore and aft vibrations.

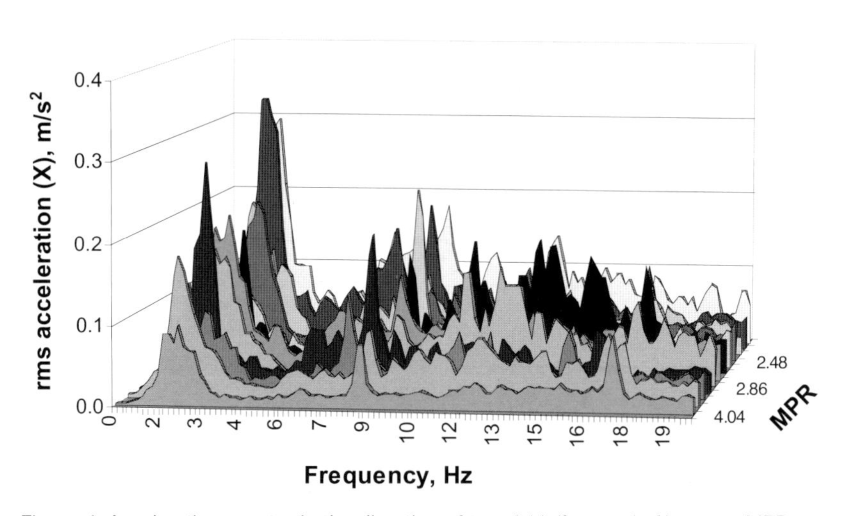

Figure 4. Acceleration spectra in the direction of travel (x) (fore and aft) versus MPR

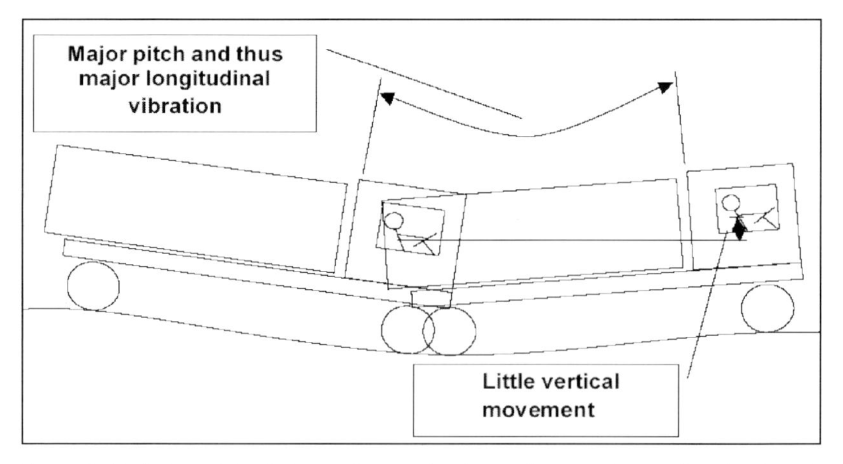

Figure 5. Excitation of truck body pitch motion by surface roughness (Granlund et al. 2000)

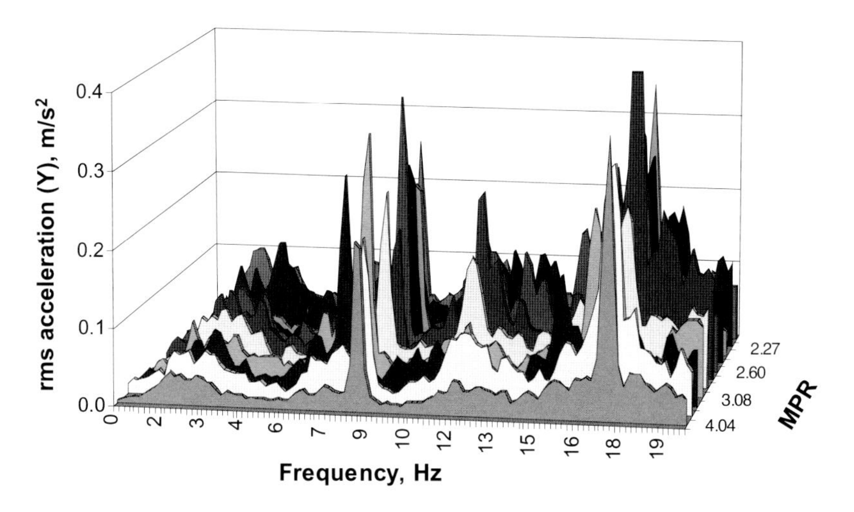

Figure 6. Acceleration spectra in the y direction (sideway) versus MPR

4.2. Measured Vibration levels

The rms accelerations were calculated in each OTOB with centre frequencies from 1-80Hz (ISO2631-1 1997). Knowing the bandwidth of each OTOB, the number of narrow bands (~0.2Hz) in each OTOB was calculated and their rms accelerations were summed (root sum of squares) to produce the rms acceleration in each OTOB. Then the weighted rms accelerations were calculated in each OTOB using frequency weightings provided in ISO-2631-1 (1997). The overall weighted accelerations were then determined for each axis and their vibration total values were calculated following the guidelines in ISO-2631-1 (1997).

Table 2 lists the values of overall raw (a_x, a_y, a_z) and weighted accelerations (a_{wx}, a_{wy}, a_{wz}) in each axis, the vibration total values for raw ($a_v$raw) and weighted acceleration (a_v) together with MPR, PI_t and IRI values for each of the 17 test sections. Weighted accelerations in the vertical direction are the highest followed by fore aft accelerations. Similar results were reported by Sweatman and McFarlane (2002). The backrest vibrations were also accounted for by increasing the multiplying factors for x and y axes (ISO-2631-1 1997). Applying these factors resulted in increased vibration total values of weighted rms accelerations (a_v-backrest) (see Table 2).

4.3. Analysis of subjective and WBV data

The correlation results between MPRs and the corresponding raw and weighted rms accelerations in each axis and their total values with and

Table 2. Raw and weighted rms acceleration values in different directions for each test section and their overall values with and without the backrest factors

Section	Accelerations, m/s^2									MPR	PI_t, m/km	IRI, m/km
No.	a_x	a_y	a_z	a_v, raw	a_{wx}	a_{wy}	a_{wz}	a_v	a_v-backrest			
1	0.73	0.89	0.78	1.39	0.27	0.20	0.54	0.64	0.72	3.16	1.18	1.74
2	0.93	1.17	1.10	1.86	0.43	0.25	0.72	0.88	1.00	2.75	1.46	2.29
3	0.76	1.01	0.69	1.44	0.27	0.22	0.48	0.59	0.68	3.36	1.19	2.08
4	0.81	0.95	0.93	1.56	0.45	0.23	0.59	0.78	0.92	3.08	1.39	1.82
5	0.76	1.04	0.87	1.55	0.27	0.20	0.58	0.67	0.75	3.57	0.89	1.66
6	0.65	0.92	0.65	1.30	0.20	0.15	0.45	0.52	0.57	4.04	0.56	1.16
7	1.08	1.41	0.79	1.94	0.34	0.28	0.55	0.70	0.83	2.60	2.02	3.48
8	0.94	1.30	0.95	1.87	0.40	0.27	0.65	0.81	0.94	2.48	1.53	2.55
9	0.87	1.18	0.73	1.64	0.34	0.22	0.50	0.65	0.76	3.19	0.88	2.07
10	1.06	1.62	0.88	2.13	0.38	0.30	0.60	0.77	0.91	2.86	0.95	2.42
11	0.89	1.23	0.72	1.68	0.26	0.20	0.54	0.57	0.71	3.35	0.82	1.78
12	1.10	1.10	1.36	2.06	0.67	0.29	0.88	1.15	1.35	2.26	2.12	2.82
13	0.84	1.22	0.66	1.62	0.25	0.18	0.47	0.56	0.64	2.56	1.68	3.12
14	0.86	1.23	1.04	1.82	0.43	0.29	0.69	0.86	1.00	2.27	2.44	4.24
15	0.81	0.92	0.75	1.44	0.32	0.18	0.53	0.65	0.75	2.99	1.18	1.98
16	1.20	1.59	1.34	2.40	0.60	0.36	0.88	1.13	1.32	2.24	2.38	3.73
17	0.84	1.01	0.96	1.63	0.38	0.23	0.65	0.79	0.90	3.01	1.47	2.35

without the backrest vibrations are presented in Table 3. It can be noticed from the table that there is good correlation between MPR and a_v and that allowing for the backrest vibrations does not affect the correlation.

Table 3. Correlation coefficients between different roughness indices and rms acceleration in different directions

	MPR	IRI	PIt
a_x	-0.76	0.65	0.62
a_y	-0.22	0.42	0.24
a_z	-0.68	0.50	0.69
$a_v$raw	-0.76	0.69	0.64
a_{wx}	-0.73	0.52	0.70
a_{wy}	-0.78	0.74	0.74
a_{wz}	-0.70	0.52	0.69
a_v	-0.74	0.56	0.72
$a_v$backrest	-0.75	0.56	0.72

MPR has good correlations (>0.70) with the weighted vibrations in the three directions. Norsworthy (1985) and Miller (1981) reported similar results for the vertical and fore and aft vibrations. Smith et al. (1976) also reported that lateral accelerations correlate well with drivers' ratings. The PI_t follows the same trend as the MPR, which is expected as it is based on the subjective responses of truck drivers. The IRI has poorer correlations with all weighted vibrations except for the sideway vibrations. The latter could be explained by the fact that the quarter car filter used for calculating the IRI is more sensitive to the shorter wavelengths of the roughness spectrum, which are responsible for exciting the high frequency sideway vibrations where the peak levels occur.

4.4. Assessment of perceptibility and comfort

The assessment of perceptibility of WBVs is made with respect to the highest weighted rms acceleration determined in any axis at any point of contact (seat surface for a sitting person) at any time (ISO 2631-1 1997). According to Table 2, the highest weighted rms accelerations occur in the vertical axis (shown in bold) with a minimum value of $0.45m/s^2$, which is much higher than the peak threshold of perceptibility, $0.02m/s^2$ (ISO 2631-1 1997). This implies that the levels of vibrations measured on all test sections would influence drivers' perceptions.

The perceptions (from the rating scale) that correspond to MPRs for the test sections and the measured weighted vibrations were compared to the likely comfort reactions to various magnitudes of overall vibration total values (a_v) in public transport given by ISO 2631-1 (1997). The comparison was based on the scale shown in Table 4 where a perception from the rating scale was assigned to each of the likely reactions given by ISO. The results of this comparison are shown in Table 5, which indicate that in the range of measured vibration levels, drivers' perceptions generally match the likely reactions proposed in ISO 2631. This implies that the rating scale used provides close representation of the vibrations felt by the drivers. For the sections, vehicles and operating conditions considered in this study, the table also indicates that generally the drivers would perceive the ride as fairly uncomfortable when the overall vibration total value exceeds $0.7 m/s^2$.

Table 4. Likely reactions to various magnitudes of overall vibration total values in public transport (ISO 2631, 1997) and the corresponding perceptions from the rating scale

WBV, m/s2	Likely Reaction, ISO	Perception
< 0.315	Not uncomfortable	Very good
0.315-0.63	A little uncomfortable	Good
0.5-1	Fairly uncomfortable	Fair
0.8-1.6	Uncomfortable	Poor
1.25-2.5	Very uncomfortable	Very poor
> 2	Extremely uncomfortable	Ext. poor

5. OTHER FACTORS AFFECTING RIDE PERCEPTION

The magnitudes of vibrations to which the driver is exposed in the cabin are influenced by vehicle properties. As mentioned earlier, the drivers travelled in vehicles of different properties and loading conditions. The vehicles varied in age, cabin location and type of drive axle suspension.

Using factorial analysis of variance (ANOVA), the statistical significance of the effects of these factors on drivers' perceptions of ride was tested at different PI_t levels. They were tested by including the ratings of the drivers who did the rating exercise under similar conditions other than the variable being studied and deleting the ratings of the others from the data set. Then the main effect of each factor and its interaction with PI_t was estimated by ANOVA using the ratings as the dependent variable. The tests' results showed that cabin location is the only factor that has a statistically significant effect on driver's perception of ride with cabins behind engines providing better ride (see Figure 7).

Factors relating to the driver including age, weight and years of driving experience may also influence driver's perception of ride. The effects of these factors were also studied and found to have no statistical significance. In all the tests performed, the roughness factor (PI_t) was found to have a significant effect at the 0.05a-level with a large effect size. More details on these tests could be found in Hassan and McManus (2001).

Table 5 Comparison between ISO likely reactions and drivers' perceptions

Section	MPR	a_v, m/s^2	ISO likely reactions	Perception	Reaction (Table 4)
1	3.16	0.64	A little to fairly uncomfortable	Good	A little uncomfortable
2	2.75	0.88	Fairly uncomfortable to uncomfortable	Fair	Fairly uncomfortable
3	3.36	0.59	A little to fairly uncomfortable	Good	A little uncomfortable
4	3.08	0.78	Fairly uncomfortable	Good	A little to fairly uncomfortable
5	3.57	0.67	Fairly uncomfortable	Good	A little uncomfortable
6	4.04	0.52	A little to fairly uncomfortable	V. Good	Not to little uncomfortable
7	2.60	0.70	Fairly uncomfortable	Fair	Fairly uncomfortable
8	2.48	0.81	Fairly uncomfortable to uncomfortable	Fair	Fairly uncomfortable
9	3.19	0.65	A little to Fairly uncomfortable	Good	A little uncomfortable
10	2.86	0.77	Fairly uncomfortable	Fair	Fairly uncomfortable
11	3.35	0.62	A little to fairly uncomfortable	Good	A little uncomfortable
12	2.26	1.15	Fairly uncomfortable to uncomfortable	Fair	Fairly uncomfortable
13	2.56	0.56	A little to fairly uncomfortable	Fair	Fairly uncomfortable
14	2.27	0.86	Fairly uncomfortable to uncomfortable	Fair	Fairly uncomfortable
15	2.99	0.65	Fairly uncomfortable	Good	A little to fairly uncomfortable
16	2.24	1.13	Fairly uncomfortable to uncomfortable	Fair	Fairly uncomfortable
17	3.01	0.79	Fairly uncomfortable to uncomfortable	Good	A little to fairly uncomfortable

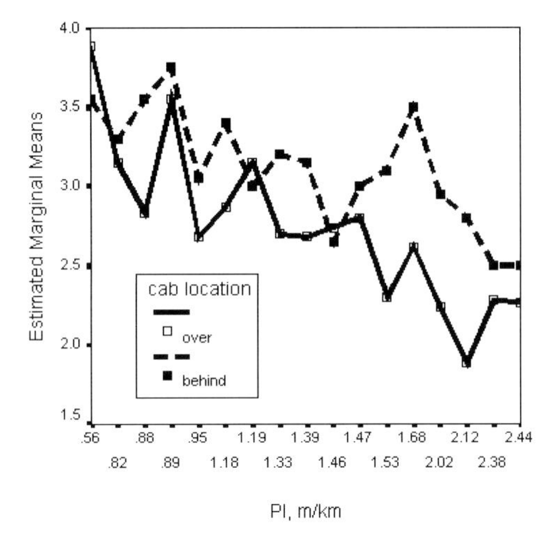

Figure 7. Effect of cabin location

6. DISCUSSION OF THE RESULTS

6.1. PI_t and IRI

Analyses of the subjective and roughness data have shown that excitation of the low frequency vibration modes of heavy articulated vehicle has greater impact on driver's perception than the high frequency ones. This was demonstrated by the high correlation between MPRs and roughness contents in the PI_t band. The lower correlation between MPR and IRI could be explained by the fact that the IRI is more sensitive at higher frequencies as it is based on a quarter car model. Passenger cars are more responsive at higher frequencies and the occupants were found to object to vibrations in the range of 10-51 Hz at 24.6m/sec (Janoff et al. 1985). Whereas heavy vehicle is generally more responsive at low frequencies as was shown in the graphical analyses.

6.2. PI_t and WBV

The PI_t/ MPR has good correlations with raw and weighted overall vertical and fore and aft vibrations and bad correlations with raw sideway vibrations. Roughness wavelengths, covered by the PI_t filter, excite the low frequency vibration modes in the range 1.42-5.7Hz (at 100km/hr). At these frequencies, the highest vertical and fore and aft vibrations occur and this explains the good correlation between raw (PI_t, 0.62 & 0.69) and weighted (PI_t, 0.70 & 0.69) accelerations in these directions with this index and MPRs. However, the correlation coefficients are slightly lower than with MPR. This is expected as the PI_t was developed from surface profile measurements in car wheel paths, whereas MPR represents driver's perception while travelling along truck wheel paths. Truck wheel paths are, in some cases, a bit rougher than the car wheel paths as they are closer to the unsealed road shoulder. Further, the PI_t represents surface roughness contribution to the vibrations present in the vehicle cabin and as mentioned earlier, there are other sources of vibrations that influence driver's perception.

The raw highest sideway vibrations occur at high frequencies i.e. not excited by wavelengths within the PI_t band. This explains the bad correlation with PI_t (0.24) and the better correlation with IRI (0.42). However, the correlation improves to (0.74) with applying the frequency weighting factors due to the reduced effects of high vibrations at high frequencies and emphasising the effects of low frequency vibrations (1-3Hz), which have good correlation with PI_t/MPR as can be seen in Figure 6. Although the levels of vibrations in this band are low, most of the drivers were able to feel this motion while traversing most of the test sections. This clearly indicates that road surface roughness contributes little to the excitation of sideway vibrations/body roll motion and that excitation of this motion can be attributed to other road characteristics. In this case it is believed that road cross fall and/or the differences in profile elevations of the two wheel tracks are the main excitation sources. Body roll excitation due to cross fall variation is depicted in Figure 8.

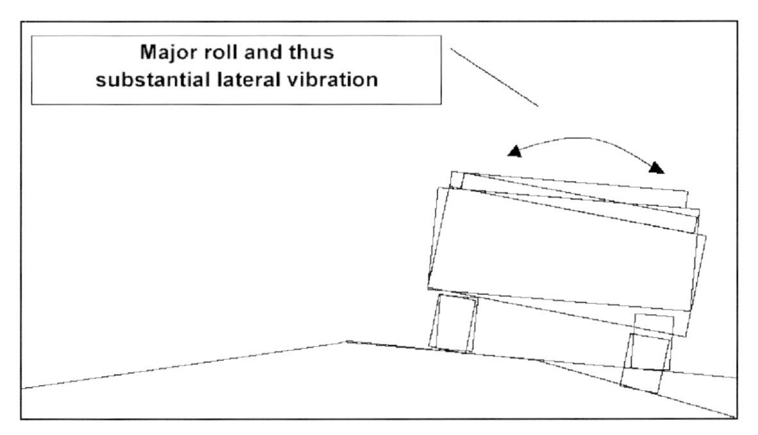

Figure 8. Excitation of truck body roll motion due to cross fall variation (Granlund et al. 2000)

Gillespie (1985) noted that the effect of body roll motion has little influence on the ride characteristics of trucks compared to the bounce motion. This is true when comparing the magnitude of low frequency sideway vibrations to the low frequency vertical accelerations. However, the findings discussed above, in addition to the fact that MPR/PI_t has high correlation with low frequency sideway vibrations, indicate that the roll motion significantly affects truck ride and ultimately influences the perception and comfort of heavy transport operators. Further, the results indicate that the drivers have low tolerance to this motion considering the fact that on most sections sideway accelerations are less than 0.30m/s^2. Using ISO likely reactions in table 5 as a guide, at this level of acceleration, the likely comfort reaction is considered 'not uncomfortable'.

The correlation coefficients between WBV with PI_t and/or MPR could be regarded as high considering the fact that driver's perception is influenced by vibrations transmitted through the floor and the steering wheel in addition to factors such as interior cab noise, seat design and its fit to the passenger, temperature, ventilation, interior space, hand holds (Gillespie 1984) and ease of steering (Miller 1981).

6.3. PI_t as a Measure of Heavy Vehicle Ride

The above discussion leads to the conclusion that the PI_t is a viable measure of heavy articulated vehicle ride. It explains 85% of the variation in MPR and has high correlations with the measured vibrations on driver's seat in a representative vehicle. Further, all the factors that may influence the perception of the driver were found to have no statistical significance except for the cabin location factor with cabins behind engines providing better ride. It could be argued that the low ratings of those who travelled in vehicles with cabins over engines may have influenced the correlation results. However, as the drivers who travelled in these conditions comprise the majority of the sample this makes the PI_t valid for the worst cases of vehicles and operating conditions considered in this study.

7. CONCLUSIONS

The results of this study showed that drivers of heavy articulated vehicles, namely semi-trailer combination units, mainly object to excitations caused by the roughness wavelengths in the range of 4.88-19.5m (PI_t band). At normal highway speeds (60-100 km/hr) these wavelengths excite the low frequency vibration modes of the vehicle body in the range of 0.9-5.8Hz. This implies that low frequency vibration modes of heavy vehicle body have greater influence on driver's perception of ride than the high frequency modes. For the sections, vehicles and operating conditions considered in this study, other conclusions that could be drawn from the findings are summarised below:

1. The new roughness index PI_t represents heavy vehicle ride better than the IRI.
2. The PI_t is a viable measure of heavy vehicle ride. The effect of road roughness in the PI_t band on drivers' ratings was found to be

significant with a large effect size. It serves to explain 85% of the variation in MPR for the sections considered in this study and provides good indications of the magnitudes of the different vibrations transmitted to the driver through the seat. Further the PI_t is valid for the worst cases of vehicles and operating conditions considered in this study.

3. Driver's perception of ride declines when PI_t exceeds 1.3 m/km or when WBV exceeds 0.7m/s^2.
4. Drivers' reactions at different levels of measured vibrations match those provided in ISO 2631-1. This implies that the rating scale used provides close representation of the vibrations felt by the drivers.
5. Road surface roughness has little contribution to exciting the low frequency roll motion of the vehicle body. However, sideway vibrations resulting from the roll motion, excited by road cross fall or differences in profiles elevations between the wheel paths, significantly affect heavy vehicle ride and ultimately influence the perceived comfort/ride of heavy transport operators.

8. REFERENCES

Cairney, P.J, Prem, H., McLean J.R. and Potter D.W. (1989), *A Literature Study of Pavement User Ratings*, Research Report ARR 161, Australian Road Research Board (ARRB).

Cramer, D. (1998), *Fundamental Statistics For Social Research*, step-by-step calculations and computer techniques using SPSS for Windows, Routledge, London/New York.

Gillespie, T.D. (1985), *Heavy Truck Ride*, Society of Automotive Engineers Inc, USA.

Gillespie T.D. (1984), *Tyre and Wheel Nonuniformities: Their Impact on Heavy Truck Ride*, Presented at a meeting of the Rubber Division, American Chemical Society, Denver, October 24-25.

Granlund, J.,Ahlin, K., and Lundström, R. (2000), *Whole Body Vibration When Riding on Rough Roads*, Vagverket Publication 2000:31E, Sweden.

Hassan, R. (2003), *Assessment of road roughness effects on heavy vehicles on State Highways in Victoria/Australia*, PhD Thesis, School of Engineering and Science, Swinburne University of Technology, Victoria/Australia.

Hassan, R.A, McManus, K., (2001), Factors affecting Truck Driver Comfort, *Proceedings of the SAE 2001 Noise and Vibration conference and*

Exhibition, paper No. 2001-01-1571, Society of Automotive Engineers, April 30-May 3, Traverse City, Michigan, USA.

International Standard ISO 2631-1(1997), *Mechanical vibration and shock Evaluation of human exposure to whole body vibration*, Part 1 General requirements, 2nd edition, International Organization for Standardisation.

International Standard ISO 8608, (1995), *Mechanical Vibration-Road Surface Profiles-Reporting of Measured Data,* International Organization for Standardisation.

Janoff, M.S., Nick J.B. and P.S. Davit Kerton Inc. and G.F. Hayhoe D&S Consulting, (1985), *Pavement Roughness and Rideability Field Evaluation*, National Cooperative Highway Research Program (NCHRP) Report 275, Transportation research board, Washington D.C.

Miller J.C., (1981), *A Subjective Assessment of Truck Ride Quality*, Society of Automotive Engineers, Inc., Technical Paper No. 810047.

Norsworthy T.H., (1985), "The Correlation of Objective Ride Measures to Subjective Jury Evaluation of Class 8 COE Vehicles", *Surface Vehicle Noise and Vibration Conference Proceedings*, SAE, Traverse City, Michigan, May 15-17.

Smith C.C. and others, (1976), *The Prediction of Passenger Riding Comfort from Acceleration Data*, Council for Advanced Transportation Studies, Univ. of Texas at Austin, Research Report, 107pp.

Sweatman, P., McFarlane, S. (2000), *Investigation into the Specification of Heavy Trucks and Consequent Effects on Truck Dynamics and Drivers*: Final Report, Federal Office of Road Safety, Australia.

Svantek, (2000), http://www.svanteck.com.

Vibration perception thresholds assessed by two different methods in healthy subjects

Ewa Zamyslowska-Szmytke, Wieslaw Szymczak and Mariola Sliwinska-Kowalska
Nofer Institute of Occupational Medicine, Lodz, Poland.
e-mail: zamysewa@imp.lopz.pl

ABSTRACT

The values of vibration perception threshold (VPT) should be related to the equipment and methods of measurement that differ between European countries. ISO Standard 13091-1-2001 specifies the general guidelines regarding device facilities and measurement methodology. This study was designed to compare VPTs in healthy subjects by using two different methods of measurement. The first one was a standard technique applied currently in Poland and the second one was based on the principles of the International Standard ISO 13091-1-2001. The study comprised 187 healthy subjects, aged 17–57 years, not occupationally exposed to vibration. Both measurements of VPTs were performed using the same equipment (P-8, EmsonMat). The main differences in methodology between the Polish standard method and the method of ISO included:

- different mode of stimuli presentation (ascending in standard vs. Bekesy in ISO method),
- frequency range applied (63–500 Hz vs. 4–125 Hz),
- probe contact force (larger in Polish method),
- probe diameter (12 mm vs. 5 mm) and its surface (plane vs. rough).

The results obtained found that the mean VPTs differed significantly at the overlapping frequencies depending on the method used, the differences varied from 2.2 to 6.4 dB at frequencies 125 and 250Hz, respectively. The correlation between the thresholds obtained by both methods at these frequencies was moderate (correlation coefficients about 0.6), although statistically significant. In the ISO method, the best VPTs were found at 4 and 125 Hz. There was a poor correlation between the thresholds at the extreme frequencies (i.e. 125 and 4 Hz) and rather high between adjacent frequencies (i.e. 25 and 32 Hz). Age, weight and height were significant covariates in both measurements. The model to calculate the normative values of VPTs adjusted for explanatory factors was set up in this study. It was concluded that different methods of VPT measurement are not comparable. The ISO method seems to supply more information about skin receptors and different kinds of tough sense than the Polish standard method. In order to perform international cohort studies there is a need to establish a unified standardised method of vibrotactile sense assessment.

1. INTRODUCTION

In different European countries vibration perception thresholds (VPTs) are determined using different devices. They include the Somedic vibrometer (Stockholm, Sweden), Bruel and Kjaer vibrometer (Naerum, Denmark), VibroMedic tactilometer (Malmo, Sweden), the biothesiometer of Biomedical Instruments (Newbury, Ohio, USA), Vibraton II (Clifton, New Jersey, USA), Optacon (Telesensory Systems, Palo Alto, California, USA) and P8 of EmsonMat (Cracow-Poland). It is known, that the VPTs can be influenced by the device parameters and the technique of measurement[1]. The diameter of the vibrating probe[2], the contact force between the probe and the finger[3], the frequency of vibration stimuli used[4], the method of stimuli presentation and the response acquisition[2] – all these are important in the VPT's assessment. Moreover, the sense of vibration can be influenced by individual factors, such as age[4,5,6], skin thickness, skin temperature, body weight and height[5,7] and habits such as smoking and alcohol consumption.

The aim of this study was to compare vibrotactile thresholds in healthy subjects using two different techniques of measurement. In both methods the same equipment (P8 of EmsonMat, Cracow, Poland) was used. The first method named as a standard one, is currently used in Poland, the second method was defined in the International Standard (ISO 13091-1-2001). The main differences between the measurements comprised: the mode of stimuli presentation, range of stimuli frequency, contact conditions between the finger and the probe, (including pressing force), probe diameter and type of surface. Moreover, we aimed to set up a model of calculating the normative data in both methods used.

2. SUBJECTS AND METHODS

2.1. Subjects

White-collar workers and students, with no previous occupational exposure to vibration were interviewed and examined by a physician following a written protocol. The protocol included questions on age, profession and work assignment, years of work and general state of health. Moreover, it also contained detailed questions as to sensorineural disturbances in hands, such as nocturnal numbness, trouble with dropping things easily, difficulties in buttoning, and its dependence on the season of the year. Physical features (height and weight), and habits (smoking, alcohol consumption) were also noted. Subjects reporting frequent and severe symptoms of numbness and tingling in the hands were excluded from the study. Eventually, 187 subjects aged 17 – 57 (mean ± SD: 27 ± 18) were selected as suitable for further examination. The characteristic of the study group is presented in table I.

Table I. Population characteristics

	No. of Subjects	Age (mean ± SD)	Weight (mean ± SD)	Height (mean ± SD)	Smokers (n)
Women	62	27.7 ± 7.8	59.9 ± 8.7	166.9 ± 5.9	8
Men	125	30.7 ± 10.8	80.6 ± 13.2	178.9 ± 6.7	36
Total	187	29.7 ± 9.9	73.7 ± 15.4	174.9 ± 8.5	44

2.2 Method

VPT measurement were conducted on the digital pulp of the right hand (53 subjects) or of both hands (134 subjects). Altogether 321 hands were examined.

The device used was the vibrometer P-8. This allows generating the vibration stimuli with increasing levels or the stimuli with levels automatically increasing and decreasing (Bekesy mode). Moreover, the device allows for control of vibration probe pressure on the skin and the use of two vibrating probes different in size and surface and two different probe contact forces.

The test was performed in a quiet laboratory room with a constant ambient temperature of 22–24°C. Before starting the test, the superficial skin temperature of the fingertips was measured. If the temperature was below 28°C, the hands were warmed up by warm air until the temperature reached 28°C. The subject was seated with the tested arm resting on a support and pressing the vibrating probe with constant force. One well-experienced examiner conducted the measurements for all subjects.

2.3 Testing procedures

2.3.1 Standard method

Stimulus frequencies used in this method were 63, 125, 250, 400 and 500 Hz. The vibrating probe, 12 mm in diameter, was pressed with force of 1.2 N. The test was started with a low intensity of the stimulus, then the magnitude of vibration was increased until a subject perceived the vibration. The VPT was defined as the lowest value of vibration stimulus that had been felt by the subject. VPTs were measured on two fingertips (2nd –index and 4th –ring finger). The measurement was done once for every tested finger and frequency. Technical parameters are shown in Table II.

2.3.2. ISO method

Stimulus frequencies used were 4, 25, 32, 125 and 250 Hz. This coincides with the Polish standard method at two frequencies only 125 and 250 Hz. The plane round probe tip (5 mm in diameter) was pressed with force equal to 0.1 N, a force lower than in the standard method. The intensity of stimuli was automatically increasing and decreasing according to subject signalling. The subjects were asked to press the button when they perceived the vibration stimulus and to release the button when not perceiving vibration stimulus (Bekesy mode). Several up and down values were averaged automatically.

In both methods the response data were estimated as root mean square acceleration values and expressed in decibels (rel. 10^{-6} m/s^2). Then, these values were transformed to velocity units decibels (rel. 5×10^{-8} m/s), as routinely used in Poland as a standard.

Table II. Technical parameters of measurements

Measurement technique parameters	**Standard method**	**Method according to ISO**
Stimulus frequency [Hz]	63, 125, 250, 400, 500	4, 25, 32, 125, 250,
Probe diameter [mm]	12	5
Contact force:		
–absolute values [N]	1.2	0.1
–relative values		
(in relation to probe area) [N/cm^2]	1.1	0.5
Acceleration changing rate [dB/s]	4	2 (initial magnitude increment)
		4 (test magnitude increment)
Probe surface	rough	plane

3. STATISTICAL ANALYSIS

Student's *t*-tests were used to compare the differences in mean vibration thresholds between 2nd and 4th fingers and between hands. Student's *t*-tests were used to compare mean values of VPT for overlapping frequencies (125 and 250 Hz). Pearson correlation coefficients were used to evaluate the linear relationship between the results of the Polish standard and ISO methods at these frequencies (125 and 250 Hz). In both methods the correlations between all frequencies were tested.

Other variables

Student's *t*-tests were used to estimate the relationships between the vibration perception thresholds and such variables as gender and smoking (binary yes-no variables). A linear regression model was used to evaluate the linear relationship between VPTs and the continuous variables (age, body height and weight). Fingertip temperature was not used in the final analysis because it was kept at a constant level (above 28°C).

Univariate initial analysis revealed no relationships between VPTs and gender or smoking habit. A linear regression model for other variables (age, body height and weight) was performed. Analysis of the linear regression showed statistically important relationships between VPTs and age, body weight and, at lesser degree, height (see Table III). These variables were included in the multivariate linear regression model for every frequency separately.

The significant differences between fingers' thresholds ($p<0.05$) were observed at 25, 32 and 125 Hz in the ISO method and at 63 and 125 Hz in the standard method. There were no differences in VPTs between left and right hand in any method used. Therefore, in further analysis the results of both hands were calculated together, but 2nd and 4th fingers' data were analyzed separately.

4. RESULTS

In both methods the VPTs differed depended on the stimulus frequency used (Fig 1). In the ISO method the lowest thresholds were obtained at 4 Hz among the low range of frequencies and at 125 Hz among the high frequency range. The mean thresholds ranged from 76.2 dB to 85.4 for the 2nd finger and from 78.1 to 87.9 for the 4th finger.

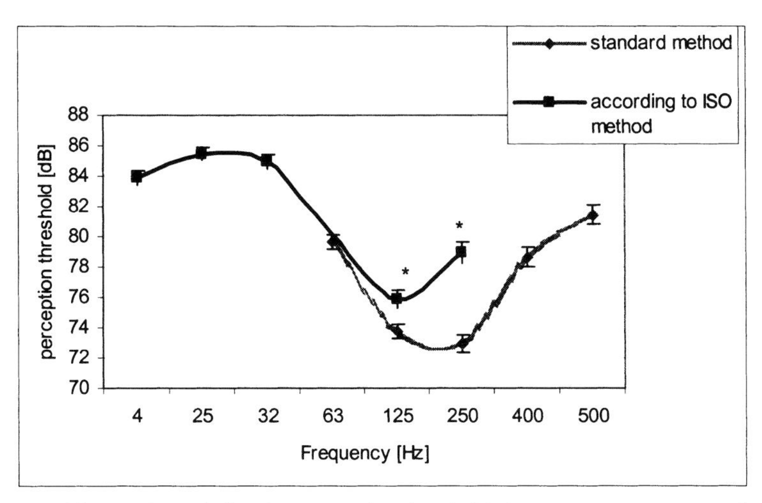

Figure 1. Mean values of vibration perception thresholds in two measurement methods (2nd finger)

The standard method included only the high-frequency range and the lowest threshold was noted at 250 Hz. The mean thresholds ranged from 72.9 to 81.4 dB for the 2nd finger and from 73.1.1 to 82.1 for the 4th finger.

All correlation coefficients between frequencies were statistically significant ($p<0.05$) in both methods. The highest correlation coefficients were found for adjacent frequency thresholds, while the lowest coefficients were found for the extreme frequencies (Table III).

The coincident stimulus frequencies for the standard and ISO methods were 125 and 250 Hz. The VPT's at these frequencies were significantly lower in the standard method as compared to the method according to ISO. The threshold differences varied from 2.2 to 3.2 dB at 125 dB Hz (2nd and 4th finger, respectively) and from 6.0 to 6.4 dB at 250 Hz (2nd and 4th finger, respectively). There was a significant correlation between VPTs obtained by both methods at these frequencies. Pearson correlation coefficients at 125 Hz were 0.56 ($p<0.05$) for both fingers and 0.61 and 0.56 at 250 Hz ($p<0.05$) for finger 2nd and 4th, respectively (fig. 2).

Table III. Linear correlation between vibration perception thresholds obtained with two methods, by finger and test frequency

Standard method

Frequency	2nd finger					4th finger				
	63 Hz	125 Hz	250 Hz	400 Hz	500 Hz	63 Hz	125 Hz	250 Hz	400 Hz	500 Hz
63 Hz	1					1				
125 Hz	0.58	1				0.71	1			
250 Hz	0.32	0.60	1			0.43	0.57	1		
400 Hz	0.27	0.50	0.78	1		0.34	0.51	0.81	1	
500 Hz	0.31	0.45	0.71	0.84	1	0.30	0.38	0.72	0.84	1

ISO Method

Frequency	2nd finger					4th finger				
	4 Hz	25 Hz	32 Hz	125 Hz	250 Hz	4 Hz	25 Hz	32 Hz	125 Hz	250 Hz
4 Hz	1					1				
25 Hz	0.58	1				0.60	1			
32 Hz	0.49	0.70	1			0.54	0.79	1		
125 Hz	0.36	0.44	0.55	1		0.43	0.58	0.66	1	
250 Hz	0.28	0.42	0.44	0.61	1	0.24	0.36	0.42	0.66	1

4.1 The influence of the other variables (age, weight and height)

Age was significantly associated with VPTs regardless of the method used. It concerned the frequencies 63–250 Hz in the standard method and all frequencies tested in the ISO method (Table III). VPTs were also significantly associated with weight. However at low frequencies positive correlation was found and at higher frequencies, negative correlation. Height was correlated with thresholds only at a few frequencies (Table IV).

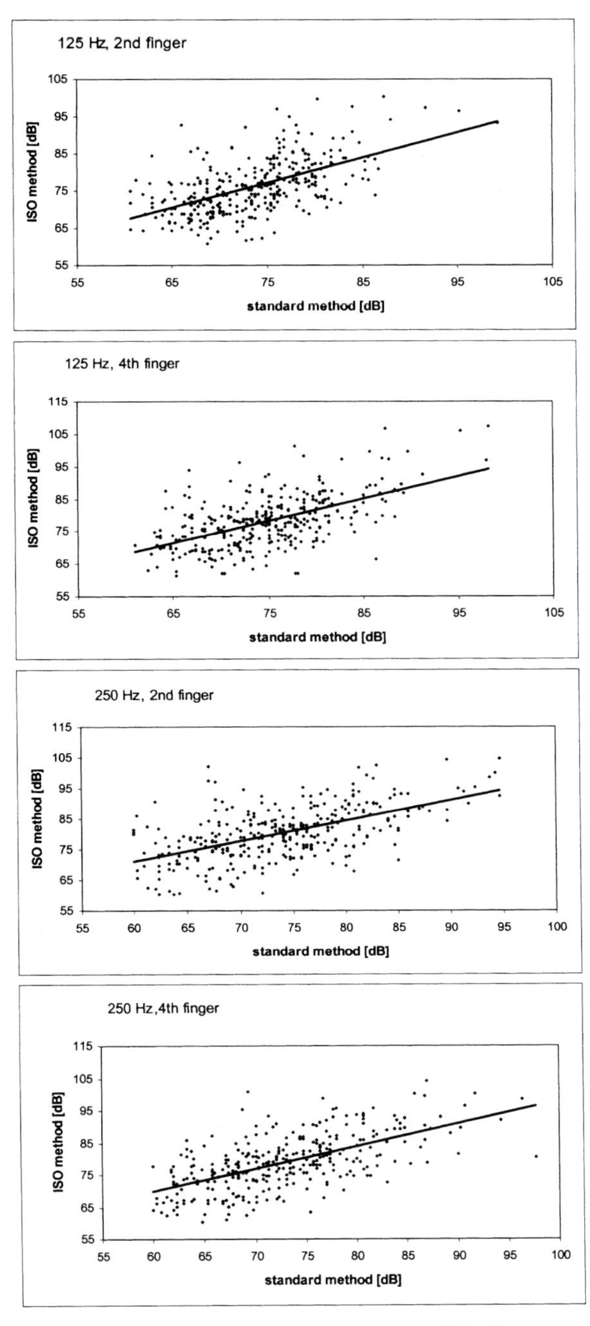

Figure 2. Scattergrams of VPTs for two methods and each of the fingers and frequencies
a) 125 Hz, finger 2nd b) 125 Hz finger 4th
c) 250 Hz, finger 2nd d) 250 Hz, finger 4th

Table IV. Linear regression analysis for explanatory variables

	Frequency [Hz]	Age		Weight		Height	
		Correlation coefficient	p	Correlation coefficient	p	Correlation coefficient	p
2nd finger							
Standard	63	**0.252**	<0.0005	**0.332**	<0.0005	**0.196**	0.0004
Method	125	**0.248**	<0.0005	**0.128**	0.0220	0.027	0.6280
	250	**0.143**	0.0100	**–0.116**	0.0380	–0.076	0.1760
	400	–0.004	0.9480	**–0.261**	<0.005	**–0.140**	0.0120
	500	0.012	0.8240	**–0.198**	0.0004	–0.072	0.2020
According to	4	**0.196**	0.0004	0.076	0.1750	–0.078	0.1650
ISO method	25	**0.172**	0.0020	**0.206**	0.0002	0.017	0.7580
	32	**0.137**	0.0140	**0.228**	<0.0005	0.064	0.2510
	125	**0.297**	<0.0005	**0.192**	0.0006	0.025	0.6580
	250	**0.164**	0.0032	–0.101	0.0706	**–0.181**	0.0011
4th finger							
Standard	63	**0.179**	0.0013	**0.280**	<0.0005	**0.181**	0.0011
Method	125	**0.195**	0.0005	**0.114**	0.0412	0.047	0.4070
	250	**0.134**	0.0163	**–0.151**	0.0069	–0.075	0.1800
	400	0.018	0.7530	**–0.221**	0.0001	**–0.128**	0.0224
	500	–0.004	0.9500	**–0.114**	0.0413	0.013	0.8200
According to	4	**0.175**	0.0017	**0.208**	0.0002	0.037	0.5100
ISO method	25	**0.197**	0.0004	**0.263**	<0.0005	0.050	0.3760
	32	**0.185**	0.0009	**0.322**	<0.0005	0.107	0.0563
	125	**0.189**	0.0007	0.174	0.0018	0.004	0.9370
	250	**0.144**	0.0097	–0.049	0.3800	**–0.129**	0.0205

The statistically significant correlation coefficients are marked in bold

4.2 Normative values of VPTs

Taking into account the explanatory variables, we set up formula (1) allowing the calculation of the VPT's range of normative values individually for a particular subject.

Range of normative values for every subject examined was evaluated with 95% confidence interval for single observation[8,9,10], according to equation (1):

$$\hat{y}_0 \pm {}_{t\alpha/2;(n-p-1)} \cdot s \cdot \sqrt{1+x'_0(X'X)^{-1}x_0} \qquad (1)$$

Where:

$\hat{y}_0$ – perception threshold value, calculated from linear multiple regression model (parameters are presented in Table V), ta/2;(n–p–1) – statistics of Student's *t*-test (equal to 1.9675 if a = 0.05, n = 320, p = 3),

x'_0 = (age_0, $weight_0$, $height_0$) vector of measured values of variables in individuals,

S – fit error for the model (s values are presented in Table V),

$(X'X)^{-1}$ – inverse product of design matrices; symmetrical matrix 3 x 3:

$$(X'X)^{-1} = \begin{bmatrix} 3{,}194076E-05 & -8{,}588293E-06 & -1{,}742191E-06 \\ -8{,}588293E-06 & 2{,}577145E-05 & -9{,}286776E-06 \\ -1{,}742191E-06 & -9{,}286776E-06 & 4{,}253026E-06 \end{bmatrix}$$

Regression coefficients used in the formula are presented in Table V.

Table V. Regression coefficients and fit error of the model to establish the confidence intervals for single observation

	Frequency	Multiple regression coefficients				
	[Hz]	age [years]	weight [kg]	height [cm]	constant	Fit error s
2nd finger						
Standard	63	0.1077	0.0945	0.0501	60.9167	5.170
Method	125	0.1469	0.0189	0.0216	64.3153	6.007
	250	0.1571	–0.1276	0.0817	63.2878	6.972
	400	0.0792	–0.1982	0.07512	77.5582	7.532
	500	0.0988	–0.1992	0.1354	69.3133	8.238
According to	4	0.0831	0.0492	–0.0873	93.2576	5.576
ISO method	25	0.0421	0.1083	–0.0877	91.6788	5.215
	32	0.0273	0.1226	–0.0700	87.4892	5.762
	125	0.1852	0.0770	–0.0274	70.1864	7.036
	250	0.1433	–0.0515	–0.1047	96.7914	8.082
4th finger						
Standard	63	0.0775	0.0863	0.0522	63.4850	5.670
Method	125	0.1267	0.0133	0.0404	63.1674	6.516
	250	0.1714	–0.1664	0.1201	59.8298	7.088
	400	0.0831	–0.1692	0.0590	78.6205	7.597
	500	0.0730	–0.1560	0.1728	61.0482	8.363
According to	4	0.0516	0.1023	–0.0674	87.9314	5.436
ISO method	25	0.0495	0.1382	–0.0934	91.8244	5.432
	32	0.0404	0.1791	–0.0880	88.2226	5.997
	125	0.0931	0.1265	–0.1048	84.5767	7.889
	250	0.1232	–0.0211	–0.0939	94.5689	8.626

5. DISCUSSION

The International Standard ISO 13091-1-2001 Standard specifies vibrotactile perception threshold measurement conditions and measuring device properties. However this standard does not specify or indicate any particular method, thus leaving a wide choice of options for the examiner. According to the document stimulation at a minimum of three frequencies (4, 25 and 125 Hz) is required to establish the acuity of the three subpopulations of mechanoreceptors primarily involved in the sense of touch (slow-adapting SAI, fast-adapting FAI, and FAII receptors). In this study, apart from the recommended frequencies, two other frequencies were included, namely: 32 Hz (related to FAI receptors) and 250 Hz (related to FAII receptors). According to ISO, stimulation above 160 Hz does not improve the test reliability. However the frequency of 250 Hz was included in the ISO method for allowing the comparison between the ISO method and the standard method used currently in Poland. The previous study has shown this frequency to be the most sensitive to vibration stimuli[11a]. The whole frequency range of measurement used in both tests i.e. from 4Hz to 640 Hz is much more time consuming and strenuous for a patient.

Two different algorithms of stimuli presentation in the ISO standard are proposed. The first algorithm require intermittent stimulation, which consists of repeated tone-bursts, the second one – continuous sinusoidal stimulation with time-varying magnitude (von Bekesy method). According to Morioka and Griffin[2] changing the stimulus mode from tone-bursts to a continuous stimulus can affect perception thresholds in the range from 3 to 6 dB. In this study the Bekesy method has been chosen because of its greater similarity to the standard method used so far in Poland. The differences in perception thresholds that may occur between ascending and von Bekesy method would be a result of presentation mode (one appearance threshold vs. mean value of several appearance and disappearance thresholds). In our earlier study we did not reveal statistical differences in perception thresholds using both algorithms, as long as the other measurement conditions were unchanged[11a]. The differences between thresholds actually noticed (in the range from 2 to 6 dB) may be a result of different contact force and the probe diameter, similarly to Harada et al[11b].

As mentioned above, several receptor populations are involved in the vibration sensation. Pacinian corpuscles (functionally FAII) are deep dermal and subcutaneous sensory receptors, responsible for two-point discrimination and the sensation of vibration above 60 Hz. Meissner corpuscles are localized in the dermal papillary ridge. These cells are mostly stimulated by fingertip stroking, moving two-point esthesiometry and vibration below 60 Hz. Merkel cells (SAII) are less important perception than other receptors. They are located in the basal layer of epidermis. These receptors respond to vertical skin displacement and classic two-point discrimination, fingertip touch and pressure sensation. The differences between receptors in their localization and adaptation to stimuli may result in different perception thresholds when two levels of contact force were used. The contact force required by the ISO method is half that of the Polish standard method. It would result in significantly higher perception thresholds in the ISO method at high frequencies, and to lesser stimulation of the receptors located deeper in the skin. Only moderately significant correlations between tests (coefficients: 0,56 and 0,61 at 125 and 250 Hz respectively) suggest the different character of stimulation i.e. in the standard method the deeper skin structures (as Ruffini corpuscles) may be stimulated besides PC corpuscles or even more PC receptors' activity is included. Test-retest correlations (coefficients in the range 0.71–0.87)[1,13,14] and our actual correlation coefficients between the adjacent frequencies (about 0.8) were markedly higher. Additionally the differences at the high frequency range may be involved with large intra-individual variability in VPT measurements (27% at 2 Hz vs. 67% at 200 Hz[16]). VPTs are hardly influenced by the changes of contact force in the low frequency range[11a].

The high and low frequency stimuli used in the ISO method employ different receptor populations. Thus, very low values of correlation coefficients between the thresholds at the highest and the lowest frequencies compared to relatively high correlations between adjacent frequencies might be explained by the variability of stimulated receptors.

Individual factors known to affect the peripheral sensory function are age, skin temperature and body height. In addition to these variables some others are also believed to be of interest in relation to sensory function, e.g. weight, body mass index, tobacco smoking, ethanol consumption and occupational exposure to solvents and vibration.

Age is the most important factor among all mentioned above. Hilz et al.[6] tested vibration perception at 120 Hz in 530 children, juveniles and adults aged 3–79 years. They noticed that VPTs slightly increased with age in adults, subjects above 40 years old revealed higher thresholds than younger persons. Skov et al.[5] examined a large group of 1663 subjects (stimulus 120 Hz) and demonstrated J-shaped increase in finger threshold value (in log scale) with age (no increase up to the age of 35 and a linear increase by 0.025 units per year thereafter). On the contrary, Torgen and Swerup 7 (484 middle-aged subjects) found the increase of VPTs with age, but only on the toes of the foot, not hands. Lindsell and Griffin[16] found that age was positively correlated with vibrotactile thresholds measured only on the little fingers ($p<0.08$), but not on index or middle fingertips. However the authors examined a relatively small group of subjects (80). Gesheider et al.[17] have established the different effects of aging between low and high frequencies. Similarly Stevens et al.[18], when comparing the thresholds between two groups of subjects; up to 27 and over 65 years old, revealed the unquestionable effect of aging at 250 Hz, but an uncertain effect at 20 Hz. Great distortions of the results in both groups were observed on the scatter plot at the latter frequencies. In our study the age dependence was established for a broad range of frequencies by the ISO and standard methods.

The effect of height on the VPT's values has been observed in some studies[5,6,7]. However, height had a greater effect on toes; for hands height was not an important predictor. These differences were explained with the hypothesis that the length of the nerve increases susceptibility to peripheral neuropathy[5] or by the relation of receptor density in the skin to body size[19]. On the other hand several authors did not observe any relationships between height and VPTs [5, 20]. In our study the influence of the height was negligible, the positive correlation was found in sparse frequencies of the low frequency range.

Body weight actually was a much more important factor influencing VPTs. The effect of weight was significant over almost the entire frequency range, but at frequencies lower than 250 Hz correlation was positive, at higher frequencies – negative. The literature data deny the relationship between weight and VPTs[5,6].

It has been suggested that smoking and gender could be important factors that increase vibration perception threshold, although study results are sparse. Skov et al.[5] were not able to show any consistent association between vibrotactile thresholds and gender or race. According to Hilz[6] gender was a determinant only in individuals over 50 years (subgroup of men had slightly higher thresholds than women). Torgen et al.[7] found significant differences in VPTs on the left hand and foot. Smoking influenced VPTs neither in tour study nor in previous examinations[7], except in the study by Gerr and Letz[21], where the authors show that the smokers had slightly elevated vibration thresholds. Skin temperature is believed to be an important factor, particularly in the high frequency range[21,22]. In our study the temperature has not been analyzed, because it was maintained at a constant level during examination.

In conclusion, several factors including age, weight and to the lesser degree – height might affect the vibration perception thresholds. Therefore they should be incorporated in the models of calculation of VPTs normative values. ISO method seems to supply more information about skin receptors and different kinds of touch sense than the standard method currently used in Poland. In order to perform international cohort studies there is a need to establish a unified standardised method of vibrotactile sense assessment.

REFERENCES

1. Wenemark M. Lundstrom R. Hagberg M. Nilsson T. Vibrotactile perception thresholds as determined by two different devices in a working population. *Scand J Work Environ Health*. 1996, 22: pp. 204–10

2. Morioka M, Griffin M. Dependence of vibrotactile thresholds on the psychophysical measurement method. *Int Arch Occup Environ Health* 2002, 75: pp. 78–84

3. Whitehouse D, Griffin M. A comparison of vibrotactile thresholds obtained using different diagnostic equipment: the effect of contact conditions. *Int Arch Occup Environ Health* 2002, 75: pp. 85–89

4. Lundstrom R, Stromberg T, Lundborg G. Vibrotactile perception threshold measurements for diagnosis of sensory neuropathy: Description of a reference population. *Int Arch Occup Environ Health* 1992, 64: pp. 201–207

5. Skov T, Steenland K and Deddens J. Effect of age and height on vibrotactile threshold among 1,663 U.S. workers *Am J Ind Med* 1998, 34: pp. 438–444

6. Hilz MJ, Axelrod FB, Herman K, Haertl U, Deutsch M and Neundorfer B. Normative values of vibratory perception thresholds in 530 children, juveniles and adults aged 3–79 years. *J Neurol Sci* 1998, 159: pp. 219–225

7. Torgen M, Swerup Ch. Individual factors and physical work load In relation to sensory thresholds in a middle-aged general population sample". *Eur J Appl Physiol*, 2002, 86: pp. 418–427

8. Dillon W. R., Goldstein M. (1984): *Multivariate Analysis. Methods and Applications*. John Wiley & Sons, Inc., New York

9. Sen A., Srivastava N. (1991): *Regression Analysis. Theory, Methods, and Applications*. Springer-Verlag, New York

10. Jobson J. D. (1990): *Applied Multivariate Data Analysis. Volume I: Regression and Experimental Design*. Springer-Verlag, New York

11a. Zamyslowska Szmtyke E. Sliwinska-Kowalska M. Dudarewicz A. Gajda A. Standaryzacja nowej metodyki badania czucia wibracji. *Med. Pracy* 2001,: pp. 315–320 (Polish)

11b. Harada N. Griffin M. Factors Influencing vibration sense thresholds used to assess occupational exposures to hand transmitted vibration. *Br J Ind Med* 1991, 48: pp. 185–192

12. Grunert BK, Wertsch JJ, Matkoub HS, McCallum-Burke S. Reliability of sensory threshold measurement using a digital vibrogram. *J Occup Med* 1990, 32(2): pp. 100–2

13. Gerr F.E. Letz R. Reliability of a widely used test of peripheral cutaneous vibration sensitivity and a comparison of two testing protocols. *Br J Ind Med* 1988, 45, pp. 635–639

14. Lofvenberg J, Johansson RS. Regional differences and interindividual variability in sensitivity to vibration in the glabrous skin of the human hand. *Brain Res*, 1984, 28; 301(1): pp. 65–72

15. Lindsel JL, Griffin MJ. Normative data for vascular and neurological tests of the hand-arm vibration syndrome. *Int Arch Occup Environ Health*, 2002, 75: pp. 43–54

16. Gescheider GA, Bolanowski SJ, Hall KL, Hoffman KE, Verrillo RT. The effects of aging on information processing channels in the sense of touch: I. Absolute sensitivity. *Somatosensory and Motor Research*, 1994, 11: pp. 345–357

17. Stevens JC, Cruz A, Marks LE and Lakatos S. A multimodal assessment of sensory thresholds in aging. *J Gerontology: Psychological Sciences* 1998, 53B, 4: pp. 263–272.

18. Stevens JC, Choo K. Temperature sensitivity of the body surface over the life span. Somatosens Mot Res 1998; 15: pp. 13–28

19. Halonen P. Quantitative vibration perception thresholds in healthy subjects of working age. *Eur J Appl Physiol* 1986; 54: pp. 647–655

20. Gerr F, Letz R. Covariates of human peripheral nerve function II: Vibrotactile and thermal thresholds. *Neurotoxicol and Teratol* 1994, 16: pp. 105–112

21. Bartlett G, Stewart J, Tamblyn R, Abrahamowicz M. Normal distributions of thermal and vibration sensory thresholds. *Muscle Nerve* 1998; 21: pp. 367–374.

22. ISO 13091-1-2001 Mechanical vibration – Vibrotactile perception thresholds for the assessment of nerve dysfunction – Part 1: Methods of measurements on the fingertips.

Chapter 5. Physiological and health effects of low frequency vibration

In addition to annoyance and discomfort, low frequency vibration can cause physiological effects. These effects are discussed in the papers in this chapter.

This paper evaluated nine cases of workers who suffered vibration induced disease due to excessive hand arm vibration. The authors concluded that there is a relationship between the risk factors associated with the tool and its operation, and the clinical features in cases of vibration induced disease.

Low frequency vibration at a sufficiently high level can cause changes in the autonomic nervous system at frequencies as low as 0.01 Hz for exposure times of 15 minutes. Whole body accelerations over 0.2 m/s^2 have been shown to be generally unpleasant while those below 0.05 m/s^2 were found to be generally pleasant in the range 0.2 to 0.6 Hz..

Changes in heart rate, respiratory rate, salivation and subjective symptoms were measured on test subjects before and after exposure to low frequency vibration in the range 0.02 to 1.0 Hz for 21 minutes. The results were not really conclusive.

Occupational disease induced by hand transmitted vibration – The relationship between characteristics of case and kind of tool

Shin'ya Yamada and Hisataka Sakakibara
Dept. of Public Health, Nagoya University School of Medicine, Tsurumai-cho 65, Showa-ku, Nagoya 46600065, Japan

ABSTRACT
In statistics of workers' accidents under the Compensation Law in Japan, the number of cases of occupational disease due to hand-arm vibration in private enterprises was 2,120 from 1994 to 1997. These consisted of 712 cases (33.6%) by rock drill operation, 372 (17.5%) by chain saw, 323 (15.2%) by pick hammer, 241 (11.4%) by concrete vibrator, 80 (3.8%) by concrete breaker, 61 (2.9%) by bush cutter, 50 (2.4%) by chipping hammer, and 51 (2.4%) by portable grinder, among others. The clinical features of vibration disease differ greatly with the kind of tool. Each tool has its own characteristic of an engineering nature, which, combined with many factors in operation at the workplace, will induce characteristic clinical features. Nine cases from our laboratory cases are discussed in this connection.

NUMBER OF NEW CASES OF VIBRATION DISEASE BY KIND OF TOOL IN THE STATISTICS FROM WORKERS' ACCIDENT COMPENSATION (FROM 1994 TO 1997) IN JAPAN
The number of new cases recognized as occupational vibration disease in a year increased from 361 in 1965 to 2,595 in 1978, and then gradually decreased to 2,120 in the period 1994-1997 in private enterprises. From 1994, new cases due to hand-arm vibration exposure were reported together with the kind of vibrating tools used. Table I shows the number of cases by tool (tool No. 1 to 28) per year and the total number of cases from 1994 to 1997 (Ministry of Labour[1], Yamada[6]).

NINE CASES OF OCCUPATIONAL VIBRATION DISEASE TREATED IN OUR LABORATORY
From 1965 to the 1980s, many workers suffering from hand-transmitted vibration injury visited our laboratory for diagnosis and treatment. In the 1960s and 1970s, without any preventive medicine work regulation, many cases had severe symptoms, and their prognoses were not good in spite of taking long rest periods and receiving appropriate therapy.

Table I. Number of cases recognised as occupational vibration disease in private industry (from 1994 to 1997) (Workers Accident Compensation Statistics Ministry of Labour in Japan)

Tool No.	Predominant Tool	Main industry, in which cases occurred	Number of cases 94	95	96	97	Total number of cases from 1994 to 1997
1	Rock drill	Mine and construction	144	192	190	186	712 (32.1%)
2	Chain saw	Forestry	138	132	127	151	548 (24-7)
3	Coal pick hammer	Mine and construction	62	87	71	103	323 (14.5)
4	Concrete vibrator	Construction	36	64	68	73	241 (10.9)
5	Concrete breaker	Construction	11	20	24	25	80 (3.6)
6	Brush cutter	Forestry	14	16	16	15	61 (2.7)
7	Chipping hammer	Stone and metal	21	10	6	13	50 (2.3)
8	Portable grinder	Metal	10	17	13	11	51 (2.3)
9	Sander	Metal	12	5	7	10	34 (1.5)
10	Vibration drill	Metal	5	6	7	3	21 (0 9)
11	Impact wrench	Metal and automobile	5	4	6	3	18 (0.8)
12	Pedestal grinder	Metal	4	8	4	0	16 (0.7)
13	Portable tight tamper	Railway	2	1	2	6	11 (0.5)
14	Sand rammer	Metal	1	3	4	0	8 (0.4)
15	Electric hammer	Metal	2	1	2	0	6 (0.3)
16	Scaling hammer	Metal	3	0	1	0	4 (0.2)
17	Engine cutter	Metal	0	3	1	0	4 (01)
18	Jigsaw	Metal and wood	0	2	1	1	4 (0.2)
19	Riveter	Automobile, Ship Building and construction	0	0	1	2	3 (0.1)
20	Swing grinder	Metal	1	0	0	2	3 (0.1)
21	Hand hammer	Metal and wood	1	0	1	0	2 (0.1)
22	Chisel with multi needle	Metal	0	1	0	1	2 (0.1)
23	Chaulking hammer	Metal	0	0	0	1	1 (0.05)
24	Baby hammer	Metal	0	1	0	0	1 (0.05)
25	Portable barker	Wood	0	0	0	0	0 (0)
26	Floor grinder	Construction	0	0	0	0	0 (0)
27	Vibration shear	Metal	0	0	0	0	0 (0)
28	Others		3	5	4	5	17 (0.8)
Total			475	578	556	612	ml(100.0%)

From cases recognised as occupational disease under the Workers' Compensation Law, we selected nine cases (one case per tool) treated in our laboratory as shown in Tables IIa and IIb. Table IIa tabulates the tool with the risk factors from the tool and tool operation. Table IIb tabulates name, age, tool (severity of Raynaud's Phenomenon) R.P. dysfunction at first diagnosis. (severity with respect to vascular, neural, muscular, joint systems, finger dexterity and dysfunction assessment after stopping vibration exposure, i.e. severity of total features of the case). The symbols indicate the severity of symptoms: (+++) is severe (++) is moderate, (+) is light and (-) is slight or none (/) indicates not observed after stopping vibration exposure.

Table IIa. Risk factors in tool and tool operation (Tool No. corresponds to Tool No. in Table 1)

Ease No.	Tool (No. in Table 1)	Risk factors in tool and tool operation
1	Engine chain saw (2)	1) tree felling on mountain slope, 2) cold –5 to 5°C) 3) high level & long-term vibration exposure (5 to 7 hrs/day, 12 yrs), 4) bending posture, with heavy weight (15kg to 12kg) 5) gripping by both hands
2	Pneumatic rock drill (1)	1) digging out limestone in quarry 2) cold (–3 to 6°C), cold exhaust air 3) high level & long-term vibration exposure (5 to 7 hrs/day, 15 yrs), 4) bending posture with heavy weight (12 kg) 5) gripping by both hands
3	Pedestal grinder (11)	1) grinding edged tool, 2) cold (air: 0 to 5°C, and blower's cold wind), 3) high level & long-term vibration exposure (5 to 7 hrs/day, 10 yrs), 4) tightly gripping cutting tip, with muscle strain in whole body, with bending posture 5) gripping by both hands.
4	Rock drill (1) & Chipping Hammer (7)	1) mining stone in quarry 2) cold (0 to 5°C, cold exhaust air 3) high level vibration, long hrs exposure (pick 5, drill 3 hrs/day, 35 yrs) 4) tightly gripping chisel by L hand, supporting R hand, weight (5 kg) 5) gripping rock drill with both hands (12 kg)
5	Coal pick (3) (A) then Portable Grinder (8) (B)	A. 1) mining coal under ground 2) hot in mine (30 to 32°C) (safety factor) 3) moderate level & long term vibration exposure (5 to 6 hrs/day, 11 yrs) 4) bending posture forward with coal pick (12 kg), 5) both hand gripping B. 1) grinding in foundry 2) cold (0 to 5°C) in factory & home 3) moderate level & short term vibration exposure (4 to 5 hrs/day, 1.5 yrs) 4) bending posture with weight (5 kg) 5) gripping by both hands

Table IIa. continued

Ease No.	Tool (No. in Table 1)	Risk factors in tool and tool operation
6	Concrete immersion vibrator (4)	1) stirring non-hardening concrete in building site 2) cold (0 to 6°C) 3) moderate level & long-term vibration exposure (5 to 6 hrs/day, 9 yrs) 4) tightly gripping shaft by L hand, supporting by L hand, weight (3.5 kg) 5) advanced age
7	Vibration drill (10)	1) make hole and grinding hole in metal parts for electric wire making, 2) cool in factory (5-10°C), cold in town (0 to 8°C), 3) moderate level vibration exposure (4 to 5 hrs/day 10 yrs) 4) tight gripping, shaft by R hand & supporting by R elbow on table, weight (15 kg)
8	Sand rammer (14)	1) tamping sand in mould, 2) cool in factory, cold in town (–5 to 5°C), 3) moderate level & short-term vibration exposure (3 to 5 hn/day, 5 yrs), 4) tightly gripping shaft of tamper by L hand, weight 2.5 kg
9	Impact wrench (12)	1) closing & opening mould of concrete pile, 2) cold (3-6°C), exhaust air, 3) high level and long-term shock vibration exposure (6-8 hrs/day, 6 yrs), 4) supporting by left hand & gripping accel level by R hand, weight (4.6 kg), 5) tightly gripping by both hands

RELATIONSHIP BETWEEN RISK FACTORS IN TOOL AND TOOL OPERATION, AND CHARACTERISTICS OF DYSFUNCTION AND PROGNOSIS IN NINE CASES

In Case 1 (engine chain saw), Case 2 (pneumatic rock drill), and Case 3 (pedestal grinder), the risk factors in tool and tool operation include high vibration level, long-term exposure in a day and over years, cold environment (and cold by expansion of exhaust compressed air in Case 1), heavy weight of tool, tight gripping, bending posture and by both hands gripping. Cold in Cases 1 and 2 is from the atmosphere and expansion of exhausted compressed air, and in Case 3 from the atmosphere and ventilation. Their dysfunction appears in all items in Table IIb that is, R.P., vascular, neural, muscular and joint systems, and loss of finger dexterity. The severity ranged from grade l l l to + in both hands. The dysfunction 10 years after stopping vibration exposure did not show improvement.

In Case 3 muscular dysfunction was more severe that in Cases 1 and 2. It was caused by excessive strain from tightly gripping and supporting the cutter tip by both hands under high vibration exposure, while attempting to avoid damage to the cutter edge. This case is described in detail in the next section.

Table IIb. Dysfunction at first diagnosis and after stopping vibration exposure (+++: severe dysfunction ++: moderate +: light -: slight or none /: without observation)

Case No. Name Age	Tool (No. in Table 1)	R.P.	Dysfunction at first diagnosis: Vascular system	Neural system	Muscular system	Joint system	Finger dexterity	Dysfunction after stopping vibration exposure (years): 0-5	5-10	10-15	15-
1 T.N. 48	Engine chain saw (2)	R+++	+++	++	++	+	+++	+++	++	+	–
		L+++	+++	+++	++	++	+++				
2 T.K. 45	Pneumatic rock drill (1)	R++	++	++	++	+	+	+++	+	/	/
		L++	++	++	++	++	++				
3 N.N. 39	Pedestal grinder (11)	R++	++	+++	+++	++	++	++	+	/	/
		L++	++	+++	+++	++	++				
4 Y.N. 34	Rock drill (1) & chipping hammer (7)	R+	++	+	+	–	–	self-employed worker, unable to stop operation, dysfunction continued			
		L++	+++	+++	+	+	+				
5 M.N. 44	Coal pick (3) then Portable grinder (8)	R+	+	+++	++	+	+	++	+	/	/
		L+	+	+++	++	++	++				
6 H.H. 62	Concrete immersion vibrator (4)	R–	++	+	+	–	–	++	/	/	/
		L+	+++	+++	+++	+++	++				
7 T.O. 38	Vibration drill (10)	R+	++	+	++	+	+	/	/	/	/
		L–	–	–	–	–	–				
8 E.I. 40	Sand rammer	R–	–	–	–	–	–	+	/	/	/
		L+	+	+	+	+	+				
9	Impact wrench (12)	R–	+++	++	++ cramp	+	+				

Case 4 operated a rock drill and chipping hammer. The risk factors were cold environment and cold from exhaust compressed air, long hours of exposure (drill for 3 hours and pick for 5 hours), tight gripping of the pneumatic drill with both hands, and tight gripping of the chisel by the left hand while supporting the handle of the chipping hammer by the right hand. His dysfunction was severe from chisel gripping on the left side and moderate on the other right side.

Case 5 operated a coal pick hammer and portable grinder. The risk factors were different from the other cases. He experienced R.P. after exposure to vibration, firstly long term in a warm condition, and secondly short term in a cold condition. This case is described in detail in the next section.

Case 6 operated a concrete immersion vibrator. The risk factors were cold environment, moderate vibration levels with prolonged vibration exposure and tight gripping of the shaft usually by the left hand, sometimes by the right hand. His dysfunction appeared to be severe in R.P. and in many systems on the gripping side, and moderate on the other side without R.P.

Case 7 operated a vibration drill to make a hole in metal moulds for wire processing. Risk factors included cool environment in the factory and cold in the town, a moderate level of vibration exposure, and tight gripping of the shaft by the right hand while supporting the right elbow on a table. Thus, muscle strain in the hand, arm and shoulder on the right side was extreme. Dysfunction was apparent only on the right side.

Case 8 operated a sand rammer to tamp sand into moulds. The risk factors were cool environment in the factory and cold in the town, moderate level of vibration, long-term exposure, and gripping by the left hand. Dysfunction of light severity was apparent in R.P and other systems on the gripping side.

Case 9 operated an impact wrench to set or loosen screws in moulds for concrete piles. Risk factors included a cold environment and cold from exhaust compressed air, high vibration levels and shock, exposure for long hours per day and for many years, and tight gripping by both hands. Dysfunction was severe in many systems, but without R.P. Severe muscle cramps in fingers and arms occurred frequently. Shock vibration may be the cause of muscle cramps.

CASES 3 AND 5 WITH DIFFERENT RISKS IN OPERATION

Case 3. N.N. aged 39. (Tool No. 11)

Occupation

At 22 years old, he worked in a lathe factory. He had two activities, operation of a lathe (6 hrs per day) and grinding tips of lathe cutters by a pedestal grinder (2 hrs per day) for 3 years. The vibration level was very high.

The time of grinding cutter tips gradually increased from 4 to 5 hrs over a period of 3 years. The company changed from an alternative job system to a full job system for high efficiency and his griding hours became 8.5 hours per day. In the grinding room, the ventilation fan worked at high speed to exhaust the cutting dust, so he always felt cold during grinding.

History of Symptoms

The work load in lathe operating was light. However, grinding cutter tips put a heavy strain on his fingers, hand, arm and other parts of his whole body, because in order to keep the tips steady against the grinding wheel and to avoid damage to the cutter edge he had to strain forward in a bending posture. Also, in the grinding room, he always felt cold in his hands and legs while grinding.

With increasing grinding hours, he experienced weariness in his arms. shoulders, neck and back and coldness of hands as well as numbness, tingling and cyanosis in fingers, pain in shoulders and stiffness of muscle in neck and back. Feverishness and pain in his muscles at night disturbed his sleep. Cold hands and frequent attacks of white finger (Stockholm Vascular Stage R III, L III) appeared after 1.5 years from beginning the full-job system. He then stopped his work and visited our laboratory.

Physical examination

Vascular and neurological findings; 1) Cold induced attacks of white finger, 2) decrease of peripheral circulation, 3) decrease in sensibility in upper extremities,4) decrease of grip, pinch force and tapping ability in both hands, 5) stiffness and pain in muscles of arm, neck, shoulder and back, 6) radiation of pain from cervical nerve plexus by compression. Normal findings in blood and urine chemistry, circulatory function test, X-ray test and orthopaedic test for discernment. No other diseases were noted.

Diagnosis and compensation

The diagnosis was hand-arm vibration syndrome due to grinding cutting tips. He took a rest and hospital therapy which were covered by Workers' Accident Compensation.

Prognosis

After stopping exposure to vibration, he was treated by thermal and medical therapy in a hot spa hospital for six months. Hot spa therapy proved effective in reducing his symptoms. After leaving the hospital, he moved to a warm city and took a light job. In the fourth winter season after leaving the hospital, he experienced a mild white finger attack. After 5 years, the stiffness in his muscles disappeared. But the reduction in grip force, pinch force, and sensorineural function remained.

CASE 5.T.N. AGED 32 (TOOLS NO. 3 & 8)

Occupation

At 25 years old, he worked in an underground coal mine in Kyushu (in southwest Japan). He operated a coal pick hammer for 8 hours a day for 11 years. The coal pick hammer weighed 7.2 kg and delivered 1,230 strokes per minute. Atmospheric temperature in the workplace ranged from 30 to 32°C in all seasons. His living area was warm in winter. After the closing of the coal mine, he moved to central Japan in autumn. It was cold in winter. He began operating a portable grinder in a foundry, grinding castings for 8 hours a day.

History of symptoms

In the coal mine, he felt weariness and pain in fingers, arms and shoulder joints after coal pick operation, and slight hand chills in winter. Three months after moving, he felt hand chills in daily life and work. Soon, he felt pain and tingling in fingers and hands after working and during the night. In the second winter, he noticed three white fingers on his left hand (dominant side) and then two in the right. R.P. attacks were frequent. He lost dexterity in finger movement. The pain and tingling in his hands and arms disturbed his sleep. The hospital doctor doubted that his symptoms originated from exposure to vibration in grinding work, and the labour Standard Office also had its doubts: "Does such short-term exposure in grinding work cause severe vibration syndrome? Are there any other possible causes for his symptoms?"

Physical examination

Vascular and neurological findings were as follows; I) cold-induced attacks of white finger, 2) decrease of peripheral circulation, 3) remarkable decrease in sensibility in extremities, especially upper extremities,4) decrease of grip and pinch force and tapping ability in both hands, and 5) stiffness in muscles of arm, neck, shoulder and back. Normal findings in blood and urine chemistry, circulatory function test, X-ray test and orthopaedic test for discernment. No other diseases were noted.

Opinion

Judging from his complaints concerning operation of the coal pick and the hand grinder, it may be supposed that exposure to vibration and the heavy weight of the pick hammer affected his peripheral vascular, sensori-neural and musculoskeletal systems. But he was not aware of such dysfunction in daily life except for some impairment of fine touch and movement. It may be presumed that exposure to vibration created hypersensitivity of the blood vessels in his finger to cold, but warm conditions in daily life and work did not induce any vascular contraction due to the cold. After moving to central Japan and engaging in grinding, the combined effect of cold and more intense vibration will have enhanced the hypersensitivity of peripheral blood vessels to cold from the previous exposure in the coal mine. The short-term exposure to vibration in the foundry and the cold encountered in daily life will have precipitated the attacks of white finger.

Diagnosis and compensation

The diagnosis was hand-arm vibration syndrome due to pick hammer operation. He took a rest and hospital therapy under the Workers' Accident Compensation Law.

Prognosis

After stopping exposure to vibration, he was given thermal and medical therapy in a hospital setting. After three months, he changed jobs. His symptoms very slowly diminished over four winter seasons. In the fifth winter, he experienced a mild white finger attack, but the coldness in his hands, and loss of finger dexterity remained. His sensorineural dysfunction had failed to disappear.

DISCUSSION

Complex relationship among risk factors in tool and tool operation

The risk factors causing vibration disease in tool and tool operations in the cases described above were high vibration level and shock, long-exposure (hours per day, by years), heavy weight of tools, tight gripping of the handle or shaft or processing material, bending posture in operating, cold conditions (in working or residential environment, cold from exhaust compressed air, cold air current by ventilation). The effects of these factors always appear in such a complex manner and so unquantitatively that the diagnosis and recognition of occupational vibration disease from epidemiological research often faces difficulties. The relationship between the risk factors and the clinical characteristics in nine cases is as follows.

Relationship between risk factors in tool and the clinical characteristics of cases with vibration disease

Vibration is the most essential risk factor, and it is evaluated by the frequency weighted acceleration level as described in ISO guideline (1986)[2] for the prevention of vibration injury on the basis of the relationship between vibration exposure dose and the incidence of R.P. A relationship between vibration exposure dose and dysfunction of other vascular, neural, muscular, skeletal systems, dexterity system and subjective symptoms have been reported by a few researchers (1992 Lundstrom[3], 1994 Bovenzi[4], 1996 Yamada[5]).

Lengthy daily exposure to vibration over years intensifies the effects of vibration. Exposure hours were 5 to 6 hours/day in Cases 5 and 6, 5 to 7 hour/day in Cases 1, 2 and 3, and 6 to 8 hours/day in Case 9. Exposure periods were over 10 years in Cases 1, 2, 3, 5 and 7 hence symptoms were severe. In Case 4, exposure hours were very long (pneumatic chisel 5 hours, and rock drill 3 hours a day), and severe symptoms resulted in the short term (3 years).

Cold is an important factor, firstly because it intensifies vascular dysfunction and secondly causes low nutrition in tissue and stagnation of tissue fluid, resulting in tissue degeneration (in nervous and connected tissues).

Vibration exposure together with cold readily causes hypersensitivity of blood vessels to cold, and in temperatures lower than 10°C R.P. is precipitated. Eight cases, (all but Case 5(A)) worked and lived under temperature conditions lower than 10°C in winter and many lower than 5°C. The R.P. was severe in Cases 1, 2 and 3, moderate in 4 and 5, and light in 6,7 and 8. Case 5, who was exposed to vibration for a long time in warm conditions and then for a short time in cold conditions, is described fully in the earlier section.

Case 9, exposed to a high vibration level, has no R.P., but suffers from severe vascular dysfunction. From this result, it may be suspected that the mild temperature in his factory and living area delayed his first attack of R.P., but will appear sooner or later.

Bearing weight in a bending posture tends to intensify muscle and tendon strain in hand, arm, shoulder, neck, back and leg. Muscle and tendon strain intensified by weight decreases the blood flow in each tissue and causes dysfunction of muscles and the neuromuscular apparatus. Cases 1, 2, 4 and 5 continually handled tool weights of over 10 kg.

Grip force loss was significant in all cases. Case 3 gripped very tightly the cutter tips and suffered a form of stiffness and pain in the muscles throughout his body. Cases 6, 7 and 8 gripped the tool only by one hand, and were affected by R.P and dysfunction in many systems only on the gripping side.

The clinical features of occupational vibration disease in the nine cases showed great variation. These differences were caused by a combination of frequency-weighted vibration level and other risk factors. In the diagnosis and recognition of occupational vibration disease and in the analysis of epidemiological research among workers exposed to vibration, it is the relationship between the characteristics of the case and the risk factors in tool and tool operation that is very important.

CONCLUSION

There is a relationship between the risk factors in tool and tool operation, and the clinical features in cases of vibration induced disease. This variability needs to be recognised by epidemiological researchers.

REFERENCES

1. Ministry of Labor (Japan). (1999). *Statistics of Labor Accident Compensation.*

2. ISO: ISO 5349. (1986). *"Guideline for the Measurement and Assessment of Human Exposure to Hand Transmitted Vibration"*, Ceneva

3. Lundstrom R., Hagberg M. and Nilsson T. (1992). "Dose-Response Relationship for Hand-Arm Vibration Syndrome with Respect to Sensorineural Disturbances Among Platters and Assemblers", (*Proceedings of 6th International Conference on Hand-Arm Vibration*). Bonn. pp. 865-874.

4. M Bovenzi and Italian Study Group on Physical Hazards in Stone Industry. (1994). "Hand-arm vibration syndrome and dose-response relation for vibration induced white finger among quarry drillers and stone carvers", *Occup. Environ. Med.*, Vol. 51, pp. 603-611.

5. S Yamada, H Sakakibara, M Futatsuka. (1996). "Vibration exposure dose dependency of clinical stage, examination results and symptoms in vibration syndrome", *Centr eur J publ Hth,* Vol. 4, No. 2, pp. 133-136.

6. S Yamada (2002) "National regulations for diagnostics in health surveillance, therapy and compensation of hand transmitted injury", *Int Arch. Occup. Environ. Health.*, Vol. 75, No. 1-2, pp. 120-128.

The evaluation of horizontal whole-body vibration in the low frequency range

Masashi Uchikune
Dept. of Precision Machinery Engineering College of Science & Technology, Nihon University, 7-24-1 Narashinodai Funabashi-shi, Chiba 274-8501, Japan

ABSTRACT
International Standard ISO 2631 gives the reaction to horizontal vibration for the occupant of a building, a public transport system, the oscillations of the ocean and so on. These effects have not yet been evaluated by physiological methods as the criterion curves in the Standard are based on psychological evaluations of comfort and discomfort, rather than physiological investigations. Physiological and psychological effects of low frequency horizontal vibration on the whole-body were as follows: Changes in the autonomic nervous system were observed, in which the system tended to change from the state of predominance of the parasympathetic nervous system to that of the sympathetic nervous system, when the frequency exceeded the range 0.2 to 0.4Hz, typically around 0.3Hz.

INTRODUCTION
This study investigated the difference between the excitation by triangular and sinusoidal waves of physiological and psychological effects. Psycho-physiological effects have been shown objectively for exposures below 0.1 Hz.

The study has adopted a physiological method of measuring the reflex of the autonomic nervous system and has adopted a psychological measuring method of subjective evaluation in the five and seven grades over twenty items.

By examining the psycho-physiological effects base of the exposure to low frequencies, the data obtained illustrates the measurement technique. In addition, the new data on physiological effects could be applied to the Standard for the evaluation of a lower range. (Fig 3 shows that, for example, fatigue decreased proficiency boundaries in the Standard are only for 1 Hz and above.)

METHOD
The study was performed using multi-input vibration testing equipment in the large structure testing building at this Faculty. This vibrator operated under computer control at frequencies ranging from 0.01 to

50Hz with a maximum amplitude of 200mm. A legless seat was mounted on the vibration table, and a subject was exposed to the vibrations, sitting with legs hanging down. The amplitudes of the vibration table were 15, 25, 50,100,125, 150,175 and 200mm. Frequencies ranged upwards from 0.01Hz. Rms values of vibratory acceleration ranged up from approximately 0.000042 m/s^2. Exposure time was 15-minutes, and the vibration parameters were changed randomly. Physiological effects were examined by investigating the effects on the cardiovascular system, respiratory movement and salivation, in order to confirm the effects on the autonomic nervous system (sympathetic and parasympathetic nervous systems). Measurements were performed for 30-seconds at 3-minute intervals after starting the vibration in the cases of heart rate and respiratory rate. In the case of saliva secretion, the dental cotton roll was replaced at 3-minute intervals to measure the secreted quantity over this time. Figure 1 shows the timing of measurements during whole-body vibration exposure.

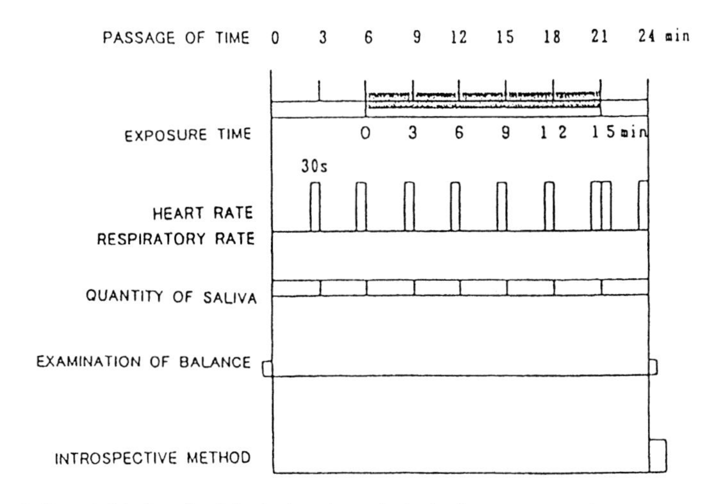

Figure 1. Time table for physiological and psychological measurements

RESULTS AND CONCLUSIONS

The effects on the autonomic nervous system showed as changes in heart rate, respiratory rate and quantity of saliva secretion. The heart rate ratios and respiratory ratios tended to decrease in the case of vibration with frequencies of less than 0.08Hz, but increase in the case of frequencies above 0.2Hz. The ratio of saliva secretion tended to increase in the cases of vibration with frequencies of 0.01, 0.015, 0.02, 0.03Hz. Thus the autonomic nervous system tended towards predominance of the parasympathetic system.

Respiratory rate ratios tended to decrease in the case of vibration with frequencies of less than 0.2Hz, but increase in the case of frequencies of more than 0.2Hz, when stimulus amplitudes were 125 and 175mm. The ratio of saliva secretion tended to increase in the case of vibration with a frequency of 0.1 Hz and amplitude of 125 or 175mm, and it tended to be close to the reference value, or decrease, in the case of vibrations at other frequencies. On the subjects, the small changes at 125mm-0.56 m/s^2,175mm-0.78 m/s^2, showed a restraining tendency. By using this data, numerical evaluations of saliva secretion from whole-body vibration have been developed to give calculations for the value of accumulated saliva over 15 minutes (Xp) and over three minute (Y) Intervals as shown in Table I. Figure 2 shows cumulative values and fluctuations of saliva production. Calculated values

Table I. The following set of equations was used

Example: 150mm, 0.2Hz, 0.17m/s2 (A quantity of saliva secretion ratio) The equation becomes

$$Y_p = y_p - x_p\, y_a/x_a \quad (1)$$

X_p: value of sccumulation for 15-minute, y_a: each time, X_a: every 15-minute, y_p: value of accumulation each time Y_p is function of a parameter X_p The value of fluctuation Y can be calculated using the equation.

min	ratio		y_p	Y_p
3	0.682	-0.318	-0.318	Y_3=0.318-3 X (-0.827)/15=-0.1526
6	0.774	-0.226	-0.544	Y_6=0.544-6 X (-0.827)/15=-0.2132
9	0.925	-0.075	-0.619	Y_9=0.619-9 X (-0.827)/15=-0.1228
12	0.884	-0.116	-0.735	Y_{12}=-0.735-12 X (-0.827)/15—0.0734
15	0.908	-0.092	-0.827	Y_{15}=-0.827-15 X (-0.827)/15=0

A cumulative value of quantity of saliva secretion ratio= -0.827:X_p

A fluctuation value of quantity of saliva secretion ratio= I -0.2132 I =0.21:Y

Example: 150mm, 0.015Hz, 0.00094m/s^2 (A quantity of saliva secretion ratio)

min	ratio		y_p	Y_p
3	1.249	+0.249	+0.249	Y3=0.249-3 X (0.767)/15=0.0956
6	1.447	+0.447	+0.696	Y6=0.696-6 X (0.767)/15=0.3892
9	1.226	+0.226	+0.922	Y9=0.922-9 X (0.767)/15=0.4618
12	0.929	-0.071	+0.851	Y12=0.851-12 X (0.767)/15=0.2374
15	0.916	-0.084	+0.767	Y15=0.767-15 X (0.767)/15=0

A cumulative value of quantity of saliva secretion ratio= +0.767:X_p

A fluctuation value of quantity of saliva secretion ratio= I +0.4618 I =0.46:Y

The results of these calculations are depicted in Fig. 2 for the following two type (Autonomic nervous system: a) the state of the sympathetic nervous system with evaluation of the physiological effects. b) the state of the parasympathetic nervous system with evaluation the physiological effects.

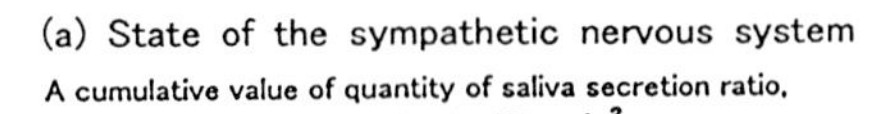

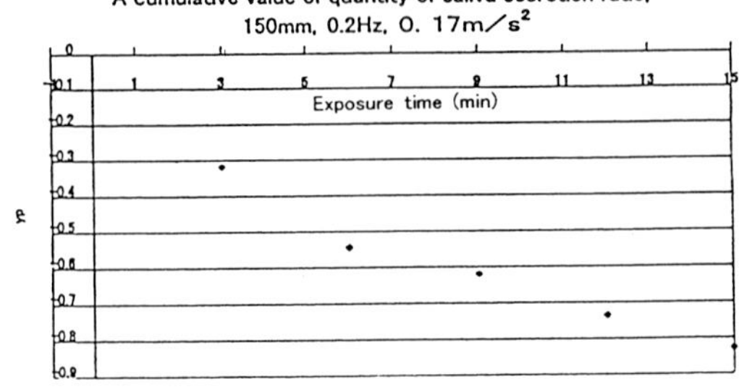

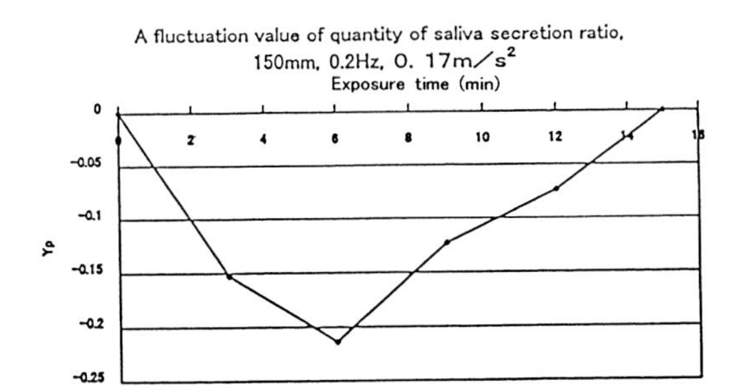

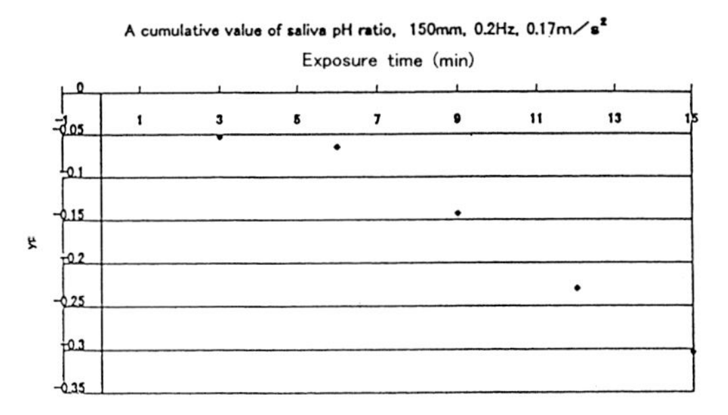

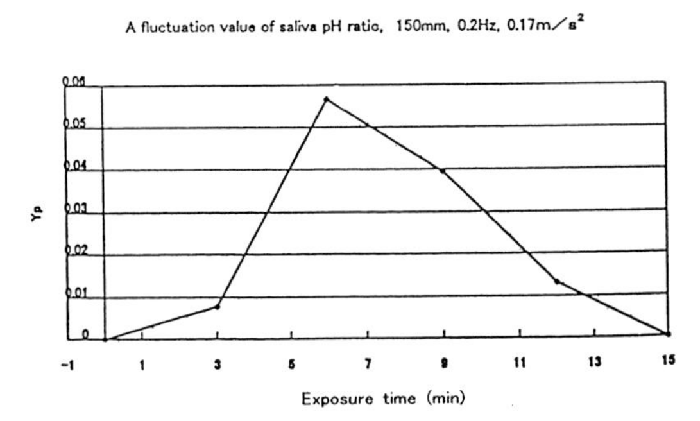

Figure 2. *continued opposite*

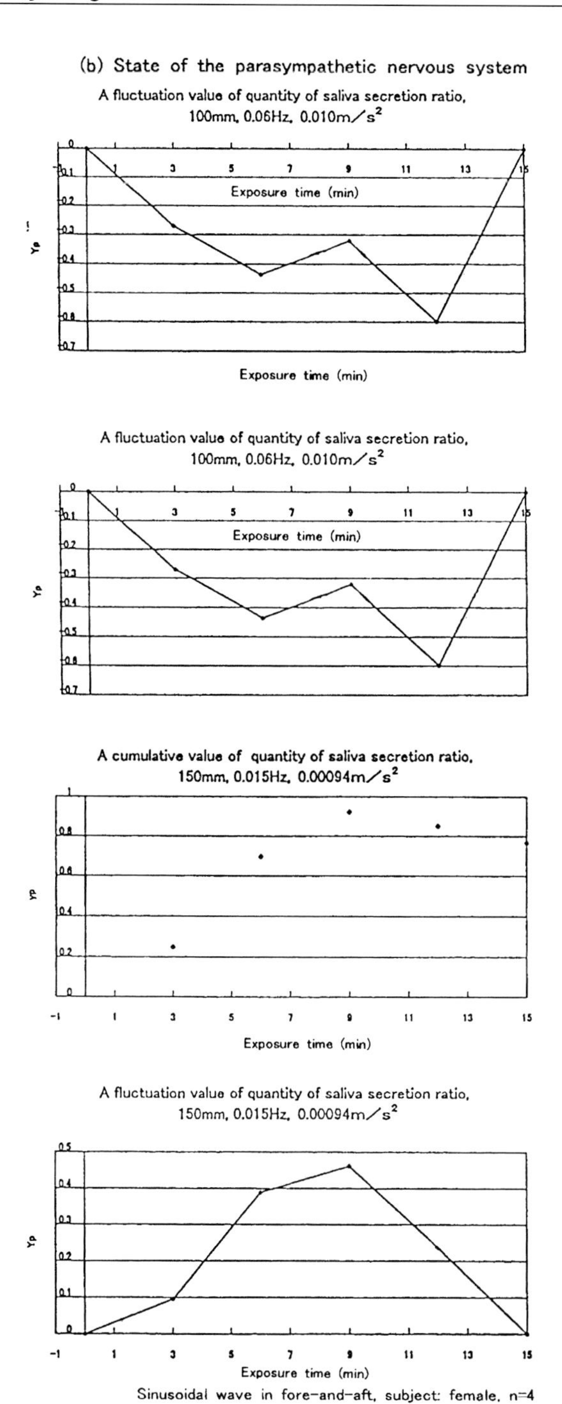

Figure 2. A cumulative value and a fluctuation value of the calculation for saliva

(Xp, Y) can also be derived for other physiological parameters as shown in Table II. The magnitude of Y is considered an effective indicator for the absolute value for the state of the autonomic nervous system for whole-body vibration in the low frequency range below 1 Hz.

Table II. The calculations of Xp and Y for the Heart-rate ratio and the Respiratory-rate ratio.

	male: lateral (y-axis)	X_p	**Y**
Heart-rate ratio	150mm, 0.04Hz, 0.0067m/s^2	0	0.02
	125mm, 0.3Hz, 0.31m/s^2	+0.27	0.03
Respiratory-rate ratio	150mm, 0.04Hz, 0.0067m/s^2	–0.10	0.10
	125mm, 0.4Hz, 0.56m/s^2	+0.60	0.13
	female: fore and aft (x-axis)		
	50mm, 0.4Hz, 0.22m/s^2	+0.11	0.10

Psychological evaluation by the subjective method of numerical values showed that five and seven grades were converted to a scale numbered (1-5, 7) in which "3" and "4" represented "normal". It was believed that when the acceleration exceeded 0.20 m/s^2 it was "unpleasant" and that the low frequency area which gave "quite pleasant' was for 0.000070-0.05 m/s^2. Data was obtained at a frequency range from 0.01 to 0. 1 Hz.

This study showed that subjects responded to physiological influences when rms values of vibratory acceleration reached 0.56 m/s^2.

For sinusoidal excitation, as the rms values of vibratory acceleration increased, the sensation of vibration tended to change from " pleasant " to " unpleasant ". For the subjects, 0.67 m/s^2 or 0.78 m/s^2 decreased the evaluation value to 2.6. 2.8. The "feeling" parts for evaluations shown in Table III should be considered to be representative of their physiological effects.

Table III. Sensation areas due to sinusoidal stimulation (lateral: y-axis) by the Subjective method.

male: n=8	**Areas Stimulated**
100mm, 0.04Hz, 0.0045m/s^2	abdomen 40%, upper arm 40%, ankle 20%
15mm, 0.01Hz, 0.000042m/s^2	head 34%, abdomen 33%, ankle 33%
25mm, 0.8Hz, 0.45m/s^2	head 27%, upper arm 22%, ankle 17% neck-shoulders 11%, chest 11%
125mm, 0.4Hz, 0.56m/s^2	abdomen 12%, upper arm 16%, ankle 16% head 16%, chest 16%, neck-shoulders 12%
150mm, 0.4Hz, 0.67m/s^2	abdomen 19%, upper arm 17%, ankle 15% neck-shoulders 15%, head 12%, back 12%

There are differences between subjects. For y-axis vibration, X_p was unchanged or negative for frequencies below 0.04Hz, whilst it was positive for frequencies of 0.3 and 0.4Hz. The state of the autonomic nervous system is indicated by decreases or increases in the measurement. For example, stimulation of the sympathetic system results in reduced saliva production and increased heart rate, whilst the opposite changes occur for stimulation of the parasympathetic system. These changes are effective physiological indicators.

Sinusoidal	**male: y-axis**	X_p	**Y**
Saliva secretion ra.	100mm, 0.06Hz, 0.010m/s^2	+2.67	0.50
	125mm, 0.4Hz, 0.56m/s^2	-1.30	0.10
	female: x-axis		
	100mm, 0.06Hz, 0.010m/s^2	+1.93	0.60
	50mm, 0.4Hz, 0.22m/s^2	-1.98	0.26

male, female: senior of univ.

The number of areas stimulated, increased when the acceleration was greater than 0.45m/s^2.

For triangular excitation, the following data were obtained for physiological measurements. The heart-rate ratios were more than one in the case of vibration with amplitudes of 25, 50,100,125,150mm and frequencies of 0.2,0.3, 0.6Hz, and acceleration 0.11, 0.13, 0.14, 0.17, 0.25 m/s^2. Respiratory-rate ratios tended to increase in the case of frequencies of more than 0.1 Hz. At 0.2, 0.4, 0.6Hz, amplitudes were 25, 50, 100, 125,175mm, with acceleration at 0.05, 0.11, 0.14, 0.22, 0.25 m/s^2. The ratio of saliva secretion tended to decrease in the case of vibration with amplitudes of 25, 50,100,125,150, 175mm. In the range from 0.3 Hz to 0.2 Hz there were experimental points at accelerations of 0.13, 0.14, 0.22, 0.25 m/s^2. Ratios of movement area tended to increase with amplitudes of 25mm, 50mm, 100mm, 125mm, 150mm, 75mm, frequencies of 0.2 ,0.6Hz, and acceleration of 0.11, 0.14, 0.17, 0.25 m/s^2. For the reasons mentioned above, the autonomic nervous system was controlled by the state of the sympathetic nervous system. During vibration, accelerations of 0.063 m/s^2 were "quite pleasant ", resulting in different responses to accelerations of 0.25 m/s^2 and less than 0.063 m/s^2. Fig. 4 illustrates the regions of predominance of the sympathetic and parasympathetic nervous systems.

The T-test result was at 0.31 m/s^2 ($p<0.1$, level: 1-5, $p<0.05$, level: 1-7)

Subjective descriptions of the physiological tendency were "hard", "great", "strong", "violent", "sharp", "aggressive" and "speedy". The difference in evaluation for different stimulations, showed contributions of the effects of acceleration.

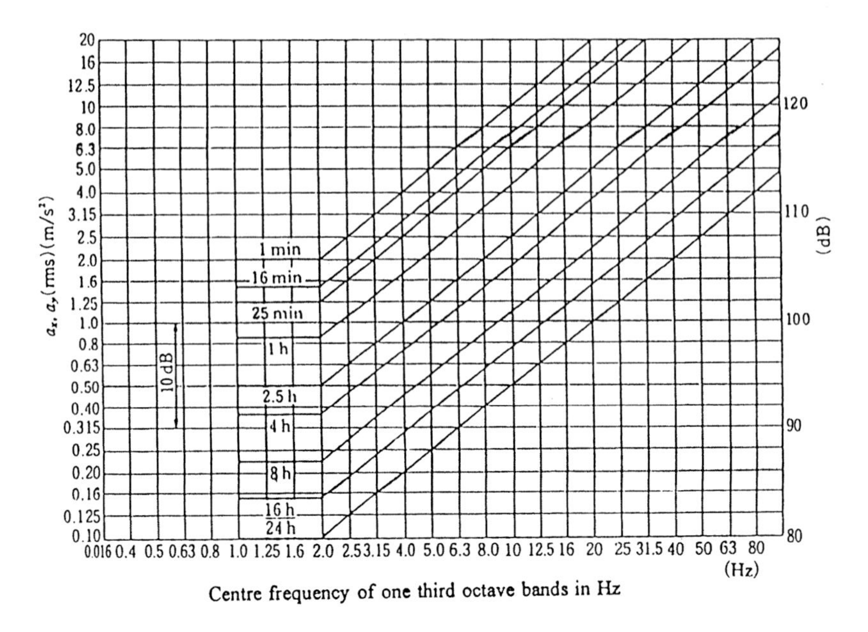

Figure 3. ISO for fatigue-decreased proficiency boundary in x, y-axis directions

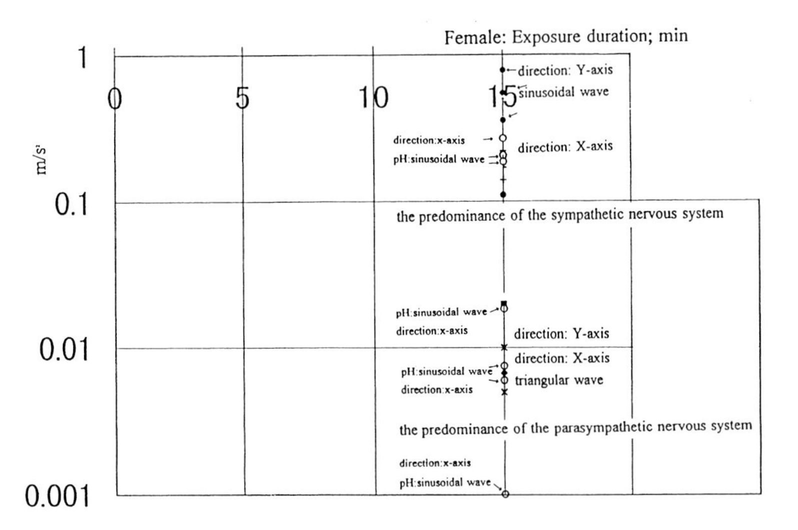

Figure 4. The predominance of the parasympathetic and sympathetic nervous system from a fluctuation of saliva secretion and pH ratio for a duration of 15-min. These data are an instance of the physiological effects. The data in the acceleration range 0.00094 m/s^2 to 0.78 m/s^2 are calculated from baseline value at 15-min. The state of the autonomic nervous system was changed below approximately 0.020 m/s^2 and at 0.10 m/s^2 or more. Reference ISO curve in Figure 3.

Evaluation of physiological changes enables us to understand the psycho-physiological effects and makes possible the application of these to the Standard.

REFERENCES

1. M. Uchikune, Y. Yoshida (1991). *IEA'91*, Paris, France, Proceedings.

2. M. Uchikune, Y. Yoshida and S. Shirakawa. (1994). *IEA'94*, Toronto, Canada, Proceedings.

3. M. Uchikune, Y. Yoshida and S. Shirakawa (1994), "Studies on the Effects of Low Frequency Horizontal Vibration to the Human Body", vol. 13, pp. 139-142 *J. Low Freq Noise Vib & Active Control* (Multi-Science Publishing Co. Ltd.)

4. M. Uchikune, S. Shirakawa and Y. Yoshida., (1996). *25th ICOH*, Stockholm, Sweden, Book part 1.

5. M. Uchikune, S. Shirakawa and Y. Yoshida., (1997). "Studies in physiological effect and psychological evaluation on a human body with the low frequency vibration", *Proceedings of IEA'97*, Vol. 2, pp. 602-604.

6. M. Uchikune, Y. Yoshida., (1997). "The effects on exposure of the whole-body to low frequency vibration in the range 0.01-0.6Hz", *Proceedings of the 8th International Meeting on Low Freq Noise Vib.*, pp. 182-186. (A MULTI-SCIENCE PUBLICATION).

7. M. Uchikune, S. Shirakawa., (1998). "Psychophysiological effects of the vibrating whole-body on low frequency vibration", *Proceedings of PIE'98*, pp. 62-63.

Study of the effects of whole-body vibration in the low frequency range

Masashi Uchikune
Dept. of Precision Machinery Engineering, College of Science & Technology, Nihon University, 7-24-1 Narashinodai, Funabashi-shi, Chiba, 274-8501 Japan

ABSTRACT
The current ISO criterion curve (ISO 2631) evaluated a psychological method on comfortable and uncomfortable feelings. To this end, it was necessary to have data for both the physiological reaction and the psychological evaluation of humans in the low frequency range. The purpose of this study is to clarify the physiological and psychological effects of low frequency horizontal whole body vibration. Heart rate, respiratory rate, salivation and subjective symptoms were measured before starting and 21-mins after starting at frequencies ranging from 0.01 to 0.8 Hz. The results were as follows: Changes in the autonomic nervous system were observed, and the system tended to change from the state of predominance of the parasympathetic nervous system to that of the sympathetic nervous system when the frequency values outside the range 0.4 to 0.8 Hz, during 21-min. Throughout all the process of the measurements, it is hypothescied that these variations have been caused by the physiological and psychological changes due to the frequency, delay time and acceleration based on the whole-body vibration.

1. INTRODUCTION
In this study, sinusoidal vibrations were used to find physiological and psychological effects. For exposure below 0.1 Hz, effects have objectively shown.

The study has adopted a physiological measuring method of the reflex of the autonomic nervous system and has adopted a physiological measuring method of evaluation using a serum point scale for the subjective assessment of twenty items.

The figures these new data on physiological effects have made it possible to show the applicability of the standard to a lower range of frequency.

This paper describes the development of equipment in the ultra low frequency range for human-body vibration studies.

2. METHOD
The study was performed using multi-input vibration testing equipment in the large structure testing building at this faculty. This vibrator was

able to vibrate at exact frequencies ranging from 0.01 to 50 Hz with a maximum amplitude of 200 mm, by computer control. A legless-chair was mounted on the vibration table, and the subject was exposed to the vibrations, sitting with his legs down. The amplitudes of the vibration table were 15, 25, 50, 100, 125, 150 and 175 mm: r.m.s; Values of vibratory acceleration ranged from approximately 0.000042 m/s^2 upwards. Exposure time was a 21-minutes, and the frequency and amplitude of vibration were changed randomly. Physiological effects were examined by investigating the effects on the cardiovascular system, respiratory rate and salivation, to confirm the effects on the autonomic nervous system (sympathetic and parasympathetic nervous systems). Measurements were performed for 30-seconds at 3-minute intervals after starting vibration in the cases of heart rate and respiratory rate. In the case of saliva secretion, the dental cotton roll was replaced at 3-minute intervals to measure the secreted quantity over 3-minutes. Table I shows a timetable used when measuring whole-body vibration responses.

Table I. Timetable of measurement sequence

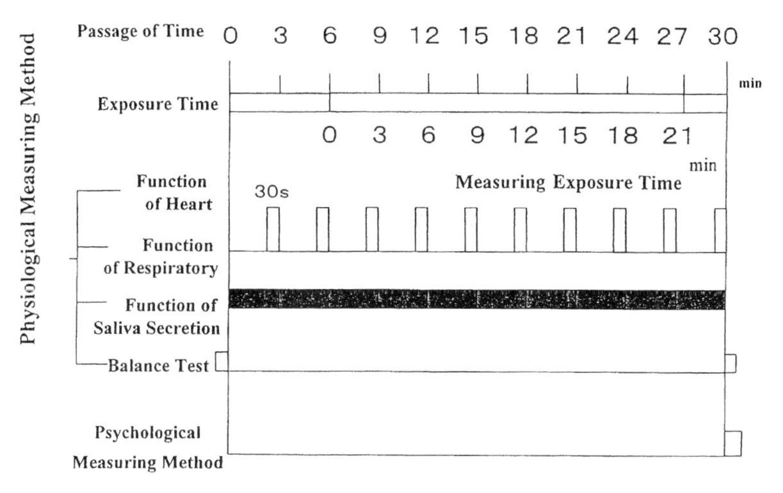

Motions of the body are detected by three sensory systems are caused by head movements or movements of the environment. The acceleration and frequency of the head were measured. The vibration and acceleration meter was designed, produced, and calibrated for this study. Figure 1 shows the development of a sensor, a circuit, and device for a vibration acceleration meter. It shows the system for the measurement of vibration acceleration on human-bodies over an ultra low frequency range. A vibration acceleration meter with a particularly high sensitivity is required to study the influence of the ultra low frequency range on human-bodies. The equipment constructed comprised an acceleration detection device, an active-type low pass filter, a power source regulator, and an operational amplifier. The meter is characterized by high sensitivity owing to the use of a semiconductor strain gauge, which has an output 91.5 times greater than that of a metallic resistance wire strain gauge (see Table II). An operational amplifier (LF356N) provided a gain of 2, 5 on 10 times. The active-type low pass filter (UF-4FL) had a cut-off frequency of 14 Hz because of noise.

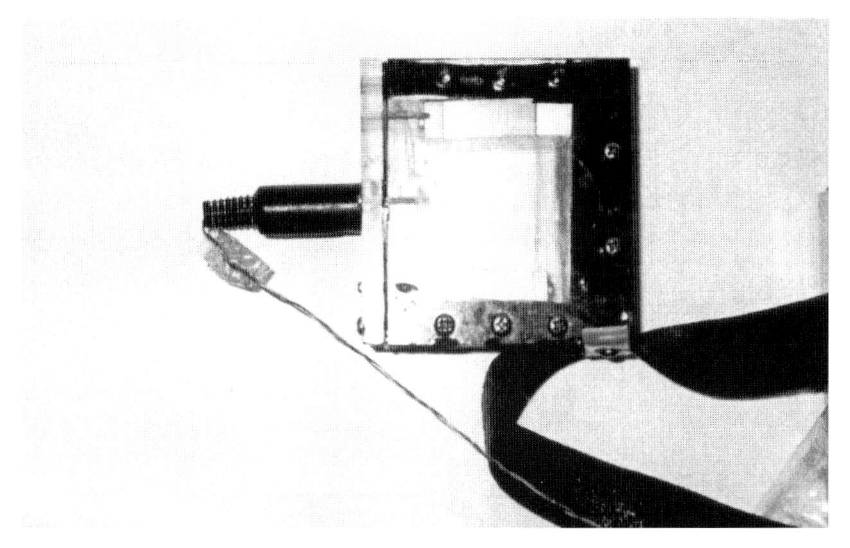

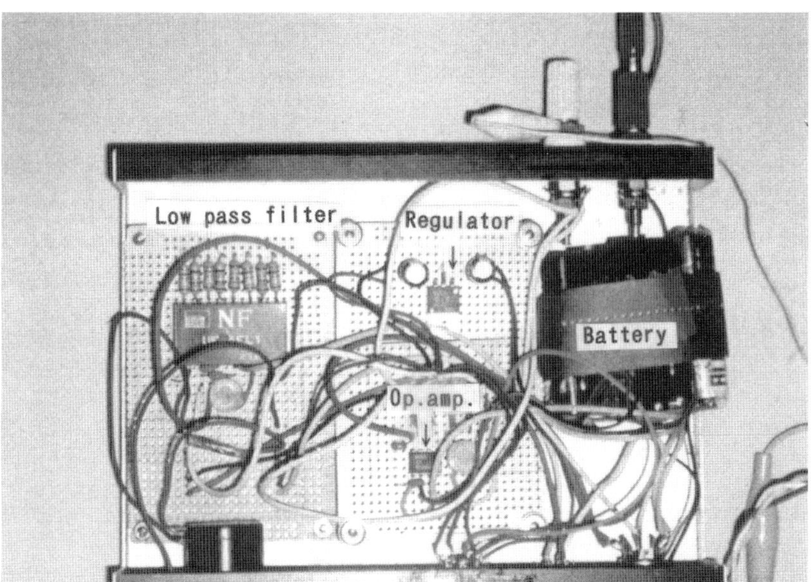

Figure 1. Appearance of the device for ultra low frequency mesurement.

Table II. The dimensions of a semiconductor strain gauge

Type	KSP-2-1K-E4
Gauge resistance	9153 Ω ± 2 %
Gauge factor	183 ± 3% at 294 K
Gauge length	length, width: 9 mm × 4 mm
Allowable current	10 mA

3. RESULTS AND CONCLUSIONS

The equipment was successfully calibrated on the range 0.01 to 0.1 Hz the range. The calibration used a sinusoidal vibration generation and found the calibration-value. This confirmed the validity of the system. Successful measurements in a frequency range from 0.02 Hz to 1.0 Hz were obtained, with acceleration levels from 0.0020 m/s^2 (r.m.s.: unweighted). Figure 2 shows the FFT analysis. The delay time from the vibration table to the subject's head was shown to be approximately 0.2 s below 0.010 m/s^2 and approximately 0.3 s below 0.018 m/s^2 (r.m.s.). Figure 3 shows the acceleration wave form of the head. The acceleration and the vibration of the head were found for the ultra low frequencies, in the lateral direction. The results showed that they increase the time of head-transmitted vibration with increasing frequency. In our previous experiment on seated subjects, it was found that the physiological change with the autonomic nervous system by whole-body vibration occurred first, so the function of the autonomic nervous system worked the sympathetic and the para-sympathetic

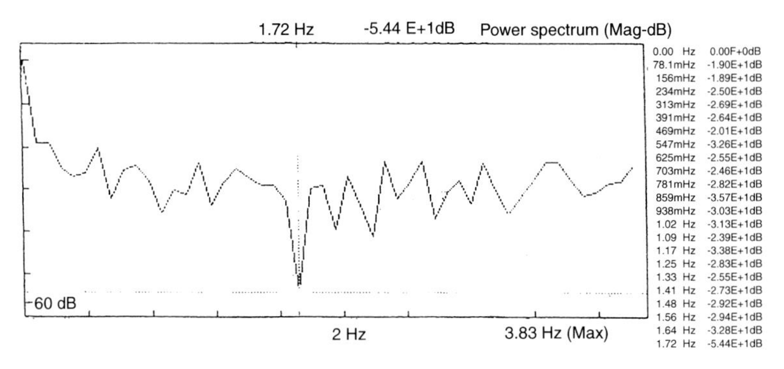

Figure 2. FFT analysis of an acceleration wave, 175 m amplitude, frequency of 0.02 Hz, acceleration 0.0020 m/s^2, with a sinosoidal waveform.

nervous system. Effects on the autonomic nervous system appeared as changes in heart-rate ratio, respiratory rate and quantity of saliva secretion. The heart-rate ratios and respiratory ratios tended to decrease in the case of vibration with frequencies of less than 0.06 Hz but increase in the case of frequencies of more 0.2 Hz, compared with the baseline value. The ratios of saliva secretion tended to increase in the case of vibration with a frequency of 0.015, 0.02, 0.04 Hz and the sympathetic nervous system at 0.11 (100 mm, 0.2 Hz), 0.22 (50 mm, 0.4 Hz), 0.45 (25 mm, 0.8 Hz), 0.56 m/s^2 (r.m.s.), including 0.4 Hz centered frequency of octave bands in the range 0.2 to 0.8 Hz for whole-body lateral vibration.

Table III shows the physiological results found at 18 and 21-min.

Figure 4 shows the respiratory and saliva ratios of the physiological responses from 3 to 21 mins.

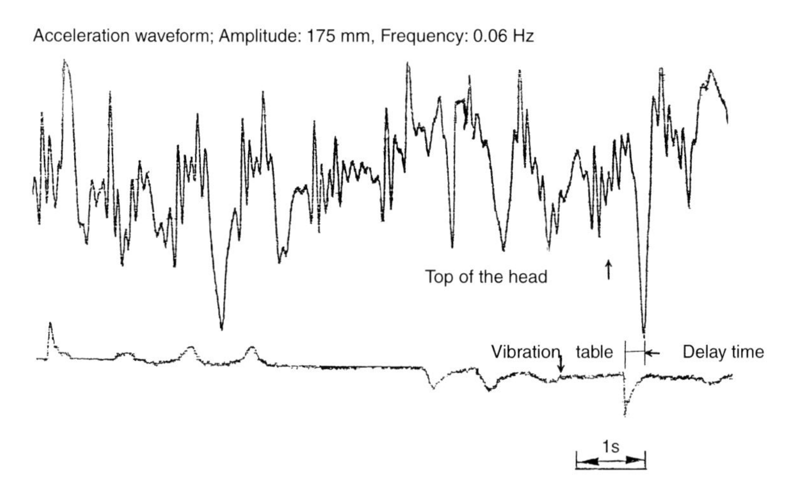

Figure 3. Acceleration waveform at the head, 175 mm amplitude, frequency of 0.06 Hz, acceleration 0.18 m/s^2, with a triangular waveform.

Table III. Example of physiological responses to vibration over 18 and 21-min.

Amp, Frequency, Acceleration:	**175 mm, 0.02 Hz, 0.0020 m/s^2** **18-min 21-min**	**25 mm, 0.8 Hz, 0.45 m/s^2** **18-min 21-min**
Heart-rate ratio	1.00 ± 0.06, 0.99 ± 0.06	1.03 ± 0.07, 1.02 ± 0.0
Respiratory-rate ratio	0.98 ± 0.10, 1.00 ± 0.10	1.02 ± 0.15, 1.01 ± 0.14
Saliva secretion ratio	1.00 ± 0.40, 1.25 ± 0.20*	0.56 ± 0.27*, 0.81 ± 0.3
	male: lateral (aged 22–26)	

Ratio = after stimulation/before stimulation Mean ± S.D. (n = 8), p: probability (*:p < .05)

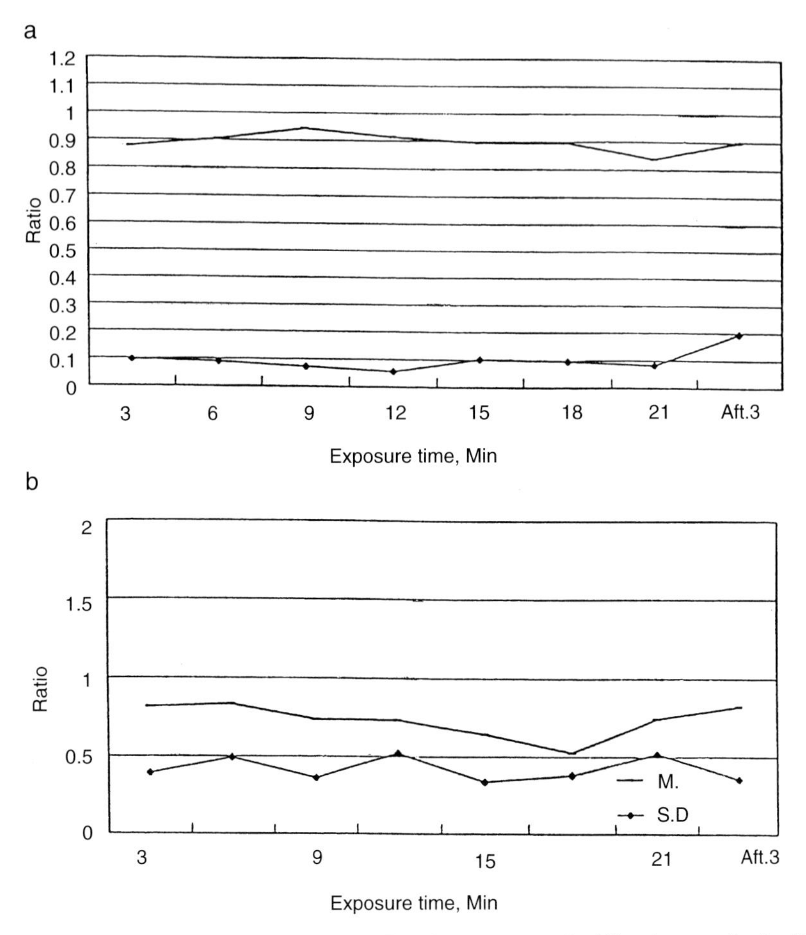

Figure 4. Physiological responses. (a) Respiratory-rate ratio. Vibration amplitude 150 mm. Frequency 0.06 Hz, Acceleration 0.015 m/s^2. Full line - mean, Dotted line - S.D. (b) Saliva secretion ratio: male. Vibration amplitude 125 mm , Frequency 0.4 Hz, Acceleration 0.56 m/s^2.

Figure 5 (b) shows the psychological evaluation by the subjective method. In this method level 1 corresponds to extremely uncomfortable. "Uncomfortable" was recorded at 0.56 m/s^2. With fore-and-aft vibration the effects appeared to occur at 0.4 Hz.

REFERENCES

1. Olle Backeteman. (1997. Gothenburg, Sweden). "Human comfort related to vibrations in buildings", *Proceedings of the 8th International Meeting on Low Frequency Noise and Vibration*, pp. 4–6.

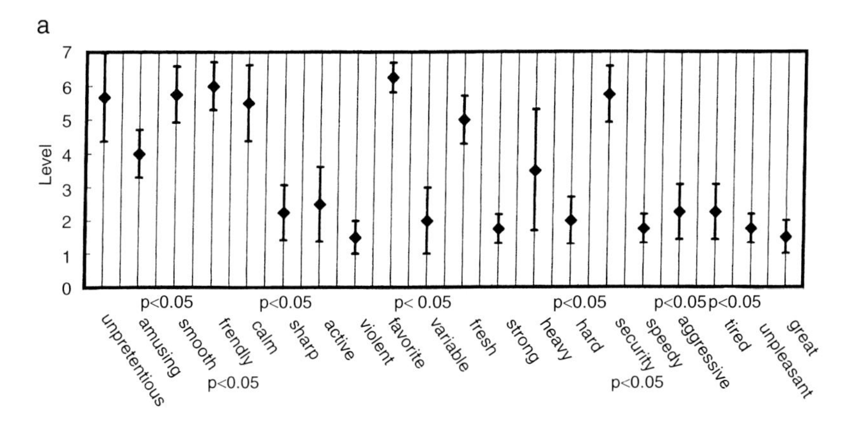

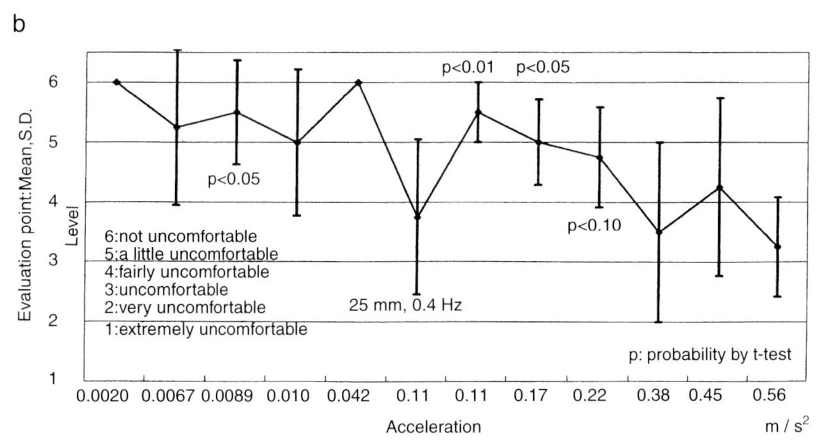

Figure 5. Evaluation of the Subjective Method. (a) Ratings of vibration (Amplitude 125 mm, Frequency 0.4 Hz, Acceleration 0.56 m/s^2 rms, 8 male subjects) Scale, 1; very, 2; fairly, 3; a little, 4; normal, 5, 6, and 7 have the opposite meanings. (b) Subjective rating of vibration comfort vs acceleration level at 0.4 Hz. Sinosiodal fore and aft vibration, 6 male subjects.

2. Michael J Griffin. (1997, Inuyama, Japan) "Human responses to vibration: current and proposed standards and the process of producing standards", *Proceedings of 5th Japan group meeting on human response to vibration*, pp. 1–26. (Nagoya University School of Medicine).

3. International Organization for Standardization, ISO 2631 Part 1, 1985.

4. International Organization for Standardization, ISO 2631 Part 1, 1997.

5. Dupuis H and Zerlett G. “Responses of whole-body vibration”. Translation: Matsumoto T, Okada, A, Ariizumi M, Nohara S and Inaba R; Nagoya Univ.-Press, 1989: pp. 62–109.

6. P. L. Pelmear and D. K. N. Leong (2002, York, UK). “EU Directive on Physical Agents – Vibration”, *Proceedings of the 10th International Meeting on Low Frequency Noise and Vibration and its Control*, pp. 149–160.